Microplastics

This book focuses on the detection, extraction, remediation techniques, and future perspectives of microplastics. It includes characteristics, fluctuations, distribution, and water remediation of microplastics using various functionalized nanomaterials. This book also covers the impact of microplastics discharged from domestic and various industrial fields such as pharmaceutical, clothing, polymer industries, etc., for the quantification of poisonous substances in water. Different techniques in water remediation and environment as well as in the determination of hazard, toxicity, and monitoring standards towards microplastics are also covered.

Features:

- Discusses the presence of microplastics in matrices prone to human consumption.
- Includes general information on microplastics, their origin, types, shapes, size, and nomenclature.
- Reviews microplastics and numerous types of detection and extraction methods.
- Illustrates the fundamental methods and techniques used for the remediation of microplastics.
- Explores the overall impacts of microplastics and their future perspectives.

This book is aimed at researchers, graduate students, and faculty members who can apply their knowledge in the emerging field of research on microplastic remediation. Despite our best efforts, mistakes and misconceptions may have occurred, for which we apologize. We welcome constructive criticism and suggestions to improve the presentation.

Microplastics
Detection, Extraction, Remediation Techniques, and Future Perspectives

Edited by
Nirmala Kumari Jangid and Rekha Sharma

CRC Press is an imprint of the
Taylor & Francis Group, an **informa** business

Designed cover image: © Shutterstock, Stock Photo ID: 2146152787, Photo Contributor Sansoen Saengsakaorat

First edition published 2025
by CRC Press
2385 NW Executive Center Drive, Suite 320, Boca Raton FL 33431

and by CRC Press
4 Park Square, Milton Park, Abingdon, Oxon, OX14 4RN

CRC Press is an imprint of Taylor & Francis Group, LLC

© 2025 selection and editorial matter, Nirmala Kumari Jangid and Rekha Sharma; individual chapters, the contributors

Reasonable efforts have been made to publish reliable data and information, but the author and publisher cannot assume responsibility for the validity of all materials or the consequences of their use. The authors and publishers have attempted to trace the copyright holders of all material reproduced in this publication and apologize to copyright holders if permission to publish in this form has not been obtained. If any copyright material has not been acknowledged please write and let us know so we may rectify in any future reprint.

Except as permitted under U.S. Copyright Law, no part of this book may be reprinted, reproduced, transmitted, or utilized in any form by any electronic, mechanical, or other means, now known or hereafter invented, including photocopying, microfilming, and recording, or in any information storage or retrieval system, without written permission from the publishers.

For permission to photocopy or use material electronically from this work, access www.copyright.com or contact the Copyright Clearance Center, Inc. (CCC), 222 Rosewood Drive, Danvers, MA 01923, 978-750-8400. For works that are not available on CCC please contact mpkbookspermissions@tandf.co.uk

Trademark notice: Product or corporate names may be trademarks or registered trademarks and are used only for identification and explanation without intent to infringe.

ISBN: 9781032782362 (hbk)
ISBN: 9781032782386 (pbk)
ISBN: 9781003486947 (ebk)

DOI: 10.1201/9781003486947

Typeset in Times
by Newgen Publishing UK

Contents

Preface..vii

About the Authors/Editors ix

Contributors xi

List of Abbreviations...xiii

Chapter 1 Microplastics – An Overview.. 1

*Anusha Srinivas, Ramesh Chandran K, Sreeja P C,
and Sapna Nehra*

Chapter 2 Detection and Extraction Techniques for Microplastics 21

Bhawana Jangir, Sarita Mochi, and Anjali Yadav

Chapter 3 Adsorptive Techniques for the Remediation of Microplastics 42

*Manisha Bhardwaj, Anjana Bhardwaj, Sukhdev Prajapati,
Rajendra Vishwakarma, Jaya Dwivedi, and Swapnil Sharma*

Chapter 4 Thermal Techniques for the Degradation and Remediation of
Microplastics 66

*Deepa Sharma, Diwakar Chauhan, Swapnil L. Sonawane,
and Purnima Jain*

Chapter 5 Membrane Filtration Technique for Remediation of
Microplastics 86

Anshul Yadav and Kavita Poonia

Chapter 6 Rapid Sand Filtration Technique for Remediation of
Microplastics 107

Aditi Pandey, Asha Bhausaheb Kadam, and Achal Mukhija

Chapter 7 Bioremediation of Microplastics...125

*Anjali Yadav, Shubhvardhan Singh, Bhawana Jangir,
Jaya Dwivedi, and Manish Srivastava*

vi

Contents

Chapter 8 Electrocoagulation for Remediation of Microplastics......................147

Anita Choudhary, Anshul Yadav, Priyanka Ghanghas, and Kavita Poonia

Chapter 9 Photocatalytic Degradation and Remediation of Microplastics 168

Manjinder Kour and Sapana Jadoun

Chapter 10 Challenges and Fate of Microplastics in Wastewater Treatment Processes 201

Chandrakanta Mall and Prem Prakash Solanki

Chapter 11 Future Perspectives of Microplastic towards Environmental Assessment...212

Hari Murthy

Index...231

Preface

"A Poison Like No Other: How the Microplastics Corrupted Our Planet and Our Bodies"

-Matt Simon

The "War on Plastic Waste" has steadily gained momentum as the depth of knowledge has increased on the far-reaching implications of plastic pollution on both environmental health and human health. A spotlight has been shone on microplastics (MPs) in recent years as reports of their presence in the remotest regions of Earth have emerged. Understanding the sources, fate, and impact of MPs is a continuous process, as contributions from various sectors of society and industry to MP pollution are investigated and understood. Microplastic pollution is an issue of global concern that requires urgent attention and action. Strategies to reduce further release of MPs into the environment are being generated and endorsed by policymakers. Researchers are actively trying to identify novel methods to treat plastic waste in the environment through either removal or degradation.

About the Authors/Editors

Dr. Nirmala Kumari Jangid works as an Assistant Professor of Chemistry at Banasthali Vidyapith, Banasthali (India). Her research focuses on materials chemistry, inorganic chemistry, and polymer chemistry (conducting polymers and their functionalization for applications in photocatalysis, optoelectronics, sensing, etc.). She likes to work in interdisciplinary fields. She completed her Ph.D. in 2015 from the Department of Chemistry, Mohanlal Sukhadia University, Rajasthan, India. She received her master's in Organic Chemistry from the Mohanlal Sukhadia University, Rajasthan, India, and her bachelor's degree from the University of Rajasthan, India. She has qualified for CSIR (Council for Scientific & Industrial Research) NET – JRF (National Eligibility Test-Junior Research Fellowship) in Chemical Sciences with All India Rank 73rd (June 2011). She has published over 3 patents, 15 peer-reviewed papers, 25 review articles, 30 book chapters, 10 chapters in an encyclopedia, an authored book, and an edited book. She has presented research work at 15 international and national conferences. She is an active reviewer in more than 25 journals of international repute. At present, five students are doing their research work under her supervision.

Dr. Rekha Sharma received her B.Sc. from the University of Rajasthan, Jaipur, in 2007. In 2012, she completed her M.Sc. in Chemistry from Banasthali Vidyapith. She was awarded a Ph.D. in 2019 by the same university under the supervision of Prof. Dinesh Kumar. Presently, she is working as an assistant professor in the Department of Chemistry, Banasthali Vidyapith, and has entered a specialized research career focused on developing water purification technology. With 4 years of teaching experience, she has published 16 articles in journals of international repute, an authored book with CRC Press, and over 60 book chapters in the field of nanotechnology. She has presented her work at more than 15 national and international conferences. Dr. Sharma has reviewed many renowned journals including *Science Direct*, *Trends in Carbohydrate Research*, and *Springer Nature*. She has been recognized as a Young Women Scientist by the Department of Science and Technology (DST), Government of Rajasthan. Her research interests include developing water purification technology by developing biomaterial-reduced NPs and polymers and biopolymers, incorporating metal oxide-based nanoadsorbents and nanosensors to remove and sense health-hazardous inorganic toxicants like heavy metal ions from aqueous media for water and wastewater treatment.

Contributors

Anjana Bhardwaj
Department of Chemistry, Banasthali
 Vidyapith
Rajasthan, India
Sukhdev Prajapati
Department of Chemistry, Banasthali
 Vidyapith
Rajasthan, India

Manisha Bhardwaj
Department of Chemistry, Banasthali
 Vidyapith
Rajasthan, India

Asha Bhausaheb Kadam
Research Center and Post Graduate
 Department of Botany, Dada Patil
 Mahavidyalaya Karjat, Ahmednagar,
 Affiliated to Savitribai Phule
 University
Pune, India

Ramesh Chandran K
School of Basic and Applied Sciences,
 Nirwan University
Jaipur, India,
and Innovation Centre, Mane Kancor
 Ingredients Private Limited
Cochin, India

Diwakar Chauhan
Department of Chemistry, Netaji Subhas
 University of Technology
Dwarka, India

Anita Choudhary
Department of Chemistry, Kishan Lal
 Public College
Rewari, India

Jaya Dwivedi
Department of Chemistry, Banasthali
 Vidyapith
Tonk, India

Priyanka Ghanghas
Department of Chemistry, Banasthali
 Vidyapith
Rajasthan, India

Sapana Jadoun
Sol-ARIS Chem Lab, Departamento
 de Química, Facultad de Ciencias,
 Universidad de Tarapacá, Avda
Arica, Chile

Purnima Jain
Department of Chemistry, Netaji Subhas
 University of Technology
Dwarka, India

Bhawana Jangir
Department of Chemistry, JECRC
 University
Jaipur, India

Manjinder Kour
Department of Microbiology and Cell
 Biology, Montana State University
Department of Chemistry and
 Biochemistry, Montana State
 University
Bozeman, Montana, USA

Chandrakanta Mall
Department of Chemistry, Institute of
 Science, Banaras Hindu University
Varanasi, India

xi

Sarita Mochi
Department of Chemistry, JECRC University
Jaipur, India

Achal Mukhija
Department of Chemistry, Banasthali Vidyapith
Rajasthan, India

Hari Murthy
Department of Electronics and Communication Engineering, School of Engineering and Technology, CHRIST Deemed to be University
Kumbalgodu, India

Sapna Nehra
School of Basic and Applied Sciences, Nirwan University
Jaipur, India

Aditi Pandey
Department of Chemistry, Banasthali Vidyapith
Rajasthan, India

Kavita Poonia
Department of Chemistry, Banasthali Vidyapith
Rajasthan, India

Deepa Sharma
Department of Chemistry, Netaji Subhas University of Technology
Dwarka, India

Swapnil Sharma
Department of Pharmacy, Banasthali Vidyapith
Rajasthan, India

Shubhvardhan Singh
Department of Chemistry, JECRC University
Jaipur, India

Sreeja P C
School of Basic and Applied Sciences, Nirwan University
Jaipur, India
and Innovation Centre, Mane Kancor Ingredients Private Limited
Cochin, India

Anusha Srinivas
School of Basic and Applied Sciences, Nirwan University
Jaipur, India
and Innovation Centre, Mane Kancor Ingredients Private Limited
Cochin, India

Manish Srivastava
Department of Chemistry, University of Allahabad
Prayagraj, India

Prem Prakash Solanki
Department of Chemistry, Institute of Science, Banaras Hindu University
Varanasi, India

Swapnil L. Sonawaneand
Department of Chemistry, Netaji Subhas University of Technology
Dwarka, India

Rajendra Vishwakarma
Department of Chemistry, Banasthali Vidyapith
Rajasthan, India

Anjali Yadav
Department of Chemistry, JECRC University
Jaipur, India

Anshul Yadav
Department of Chemistry, Banasthali Vidyapith
Rajasthan, India

Abbreviations

ABS	Acrylonitrile Butadiene Styrene
AC	Alternating Current
AEM	Anion Exchange Membranes
AFM	Atomic Force Microscopy
AMX	Amoxicillin
ATP	Adenosine Triphosphate
ATR-FTIR	Attenuated total reflectance-Fourier transform infrared spectroscopy
BAF	Biological Activated Carbon Filtration
BDD	Boron-doped Diamond
BOD	Biological Oxygen Demand
BPA	Bisphenol A
BPE	Bipolar Electrodes
CA	Cellulose Acetate
CB	conduction band (CB)
CE	Cell entrapment
CEMs	Cation Exchange Membranes
COD	Chemical Oxygen Demand
DC	Direct Current
DC-EC	Direct Current-Electrocoagulation
DDT	Dichloro-diphenyl-trichloro ethane
DEP	Di-electrophoresis
DW	Drinking Water
DWTP	Drinking Water Treatment Plant
DWTPs	Drinking Water Treatment Plants
EC	Electrocoagulation
EDS	Energy Dispersive X-ray Spectroscopy
EF	Electro-Fenton
EFG	Electric Field Gradients
EO	Electrochemical Oxidation
EP	Electrophoresis
EPS	Extracellular Polymeric Substances
ES	Effective Size
FICP	Faradaic Ion Concentration Polarization
FO	Forward Osmosis (FO),
FO-MBR	Forward Osmosis Membrane Bioreactor
FTIR	Fourier Transform Infrared Spectroscopy
GC	Gas Chromatography (GC)
GPC/SEC	Gel Permeation Chromatography/Size Exclusion Chromatography
GS	Gas Separation
HCH	Hexachlorocyclohexane
HDPE	High-Density Polyethylene
HPLC	High-Performance Liquid Chromatography

HRT	Hydraulic Retention Time
IEMs	Ion Exchange Membranes
LDPE	Low-Density Polyethylene
LM	Liquid Membranes
MACZ	Magnetic-Activated Biochar-Zeolite Composite
MBR	Membrane Bioreactor
MCM-41	Mobil Composition of Matter No. 41
MCPs	Microplastic Compound Pollutants
MEA	Membrane-Electrode Assembly
MEC	Microbial Electrolysis Cell
MF	Microfiltration
MFC	Microbial Fuel Cell
MLSS	Mixed Liquid Suspended Solids
MMPW	Manufacture of Mismanaged Plastic Waste
MNP	Micronanoplastic
MOFs	Metal-Organic Frameworks
MP	Microplastic
MPs	Microplastics
MS	Mass Spectrometry
MST	Membrane separation technology
MWP	Mixed Waste Plastics
NF	Nanofiltration
NIR	Near-infrared spectroscopy
NMR	Nuclear Magnetic Resonance
NOM	Nominal Organic Matter
NPs	Nanoplastics
NTU	Nephelometric Turbidity Units
PA	Polyamide
PAC	Powdered Activated Carbon
PAN	Polyacrylonitrile
PBS	Polybutyl succinate
PC	hotocatalysis
PCs	Polycarbonates
PCB	Polychlorinated Biphenyls
PDMS	Polydimethylsiloxane
PE	Polyethylene
PEG	Polyethylene Glycol
PES	Polyethersulfone
PET	polypropylene Terephthalate
PHA	Polyhydroxyalkanoate
PHB	Polyhydroxybutyrate
PLA	Polylactic Acid
PMMA	Poly-methyl Methacrylate
PODH	Polyoxadiazole-cohydrazide
POP	Persistent Organic Pollutant
PP	Polypropylene

Abbreviations

PPEs	Personal Protection Equipment
PS	Polystyrene
PTFE	Polytetrafluoroethylene
PU	Polyurethane
PVA	Polyvinyl Alcohol
PVC	Polypropylene Chloride
PVDF	Polyvinylidene fluoride
RHE	Reversible Hydrogen Electrode
RO	Reverse Osmosis
RO-MBR	Reverse Osmosis Membrane Bioreactor
ROS	Reactive Oxygen Species
RSF	Rapid Sand Filters
SEM	Scanning Electron Microscopy
SRT	Solids Retention Time
SS	Suspended Solids
SSF	Slow Sand Filtration
TEM	Transmission Electron Microscopy
TOPSIS	Technique for Order of Preference by Similarity to Ideal Solution
TPI	Textile Processing Industry
TSS	Total Suspended Solids
UC	Uniformity Coefficient
UF	Ultrafiltration
UV	Ultra-violet
VB	Valence Band
VOCs	Volatile Organic Compounds
WHO	World Health Organization
WW	Wastewater
WWTPs	Wastewater Treatment Plants
XPS	X-ray Photoelectron Spectroscopy
ZSM-5	Zeolite Socony Mobil–5

1 Microplastics
An Overview

Anusha Srinivas, Ramesh Chandran K,
Sreeja P C, and Sapna Nehra

1.1 INTRODUCTION

In todays interconnected world, plastic plays a vital role in almost every facet of our lives. Its adaptability has brought it to the forefront of our daily lives, from easy packaging to strong construction materials. However, while we welcome plastics convenience, we unintentionally release a silent threat: microplastics.

Microplastics are tiny bits of plastic that are sometimes invisible to the naked eye and measure smaller than 5mm in length. They can originate from several sources, including the breakdown of larger plastic items like bags and bottles or the release of smaller plastic particles, such as microbeads found in personal care products, into the environment. Invisible to the naked eye, these microscopic particles have integrated into our environment to the point where they may be found in the food we consume, the water we drink, and the air we breathe. Perversely, our reliance on plastic has led to the creation of microplastics, the origins of which are as diverse as the products they formerly comprised [1, 2].

There are several different types of microplastics, each with its unique characteristics and origins. These include fibers, pieces, pellets, and microbeads. Particles like these can have two possible origins: primary and secondary. Microbeads and other produced microplastics are examples of primary microplastics, while weathering, mechanical abrasion, and degradation processes produce secondary microplastics from the breakdown of bigger plastic items. Primary sources of microplastics include a wide range of consumer products, such as cosmetics, cleaning agents, and industrial abrasives, which contain intentionally added microplastic particles. Secondary sources encompass a broader spectrum of plastic items, including plastic bags, bottles, packaging materials, and synthetic textiles, which degrade over time into smaller particles. The breakdown of bigger pieces of plastic like bags, bottles, and fishing gear into smaller pieces can also release microplastics into the environment. This breakdown can occur by photodegradation, mechanical abrasion, or microbiological degradation.

DOI: 10.1201/9781003486947-1

Once unleashed into the environment, microplastics embark on a journey fraught with peril, as they navigate through ecosystems, traversing oceans, rivers, soil, and even the atmosphere. Along the way, they encounter a myriad of organisms, from microscopic plankton to majestic whales, each susceptible to the insidious allure of these synthetic particles [3]. The impact of microplastics reverberates throughout the natural world, disrupting delicate ecosystems and threatening the very foundation of life itself. Marine organisms, in particular, fall victim to the deceptive allure of microplastics, mistaking them for food and ingesting them with dire consequences. As these plastics accumulate within their bodies, they pose a myriad of threats, from physical blockages and internal injuries to the transfer of toxic substances [4-6].

Microplastics originate from a myriad of sources, both human-made and natural. The most common of these is the gradual degradation of bigger plastics into smaller ones, as happens naturally over time in environments with plenty of water and sunshine. This includes things like bottles, packing materials, and fishing gear. When consumers flush goods containing microbeads, such as exfoliating scrubs and toothpaste, down the toilet, they add a substantial amount of microplastic pollution to the environment. Synthetic fibers from clothing and textiles shed during washing and wearing, while industrial processes like manufacturing and recycling release microplastics into the environment through air and water pollution. Additionally, tire wear particles from vehicle tires, plastic coatings, agricultural practices utilizing plastic mulches and fertilizers, and marine paints also contribute to the proliferation of microplastics in various ecosystems. Understanding the diverse array of sources is vital for developing comprehensive strategies to mitigate microplastic pollution and safeguard environmental and human health [7, 8].

The pervasive presence of microplastics in our environment raises significant concerns as it poses adverse effects on both animal and human well-being. The microscopic plastic particles pose a threat to marine life's reproductive and nutrition systems, as well as important organs, due to their sharp edges and diminutive size [9, 10]. Over time, prolonged exposure to microplastics can result in detrimental health consequences for humans due to the presence of organic and inorganic contaminants [11]. For example, pigments like chromophores and metals utilized in the production of colored plastics can adversely affect the nerve and reproductive systems of both humans and animals [12]. Microplastics measuring less than 130 μm can accumulate within human tissues through various transport mechanisms, thereby releasing toxins, additives, and monomers upon entry into the body, which have been associated with carcinogenic behavior [13]. In addition, microplastics can cause changes in liver function and harm the lungs when consumed. Consequently, it becomes imperative to comprehend and regulate microplastic pollution to safeguard the health of humans, animals, and aquatic ecosystems alike. This chapter explores the various aspects of microplastics, including their origins, movement, and ultimate destination, effects on ecosystems and humans, and methods for controlling them. Figure 1.1 demonstrates the various application fields, detection techniques, environmental fate, types, and impact on the ecosystem and humans.

Microplastics – An Overview

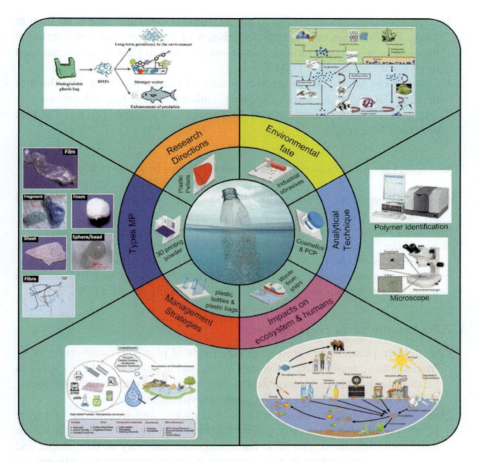

FIGURE 1.1 Schematic representation of various aspects of microplastics.

1.2 TYPES AND FORMS OF MICROPLASTICS

Microplastics, encompassing a wide range of plastic particles smaller than 5 mm, exhibit diverse characteristics and forms, each with unique environmental implications [14]. For a better understanding of microplastics' distribution, fate, and effects on ecosystems and human health, it is essential to know how they are classified, what shapes they take, and what chemicals they are composed of.

1.2.1 Classification based on Size

Macroplastics: Macroplastics are plastic debris larger than 5mm in size, often visible to the naked eye [4]. Waste products like plastic bottles, bags, and fishing equipment can decompose into smaller pieces as a result of natural processes including weathering and mechanical wear and tear. [15, 16].

Mesoplastics: Mesoplastics refers to plastic particles ranging in size from 5mm to 1mm. The breakage of bigger plastic objects or their purposeful production at this size range can both produce these particles, such as plastic microbeads used in personal care products [17-19].

Microplastics: Microplastics are tiny pieces of plastic less than 1 millimeter in size. They can be either "primary" microplastics, like microbeads, or "secondary" microplastics, which are formed when bigger pieces of plastic degrade [20, 21].

Microplastics are further categorized into two subgroups based on size:

Primary Microplastics: Particles smaller than 1mm, intentionally manufactured for various applications.

Secondary Microplastics: Particles between 5mm and 1mm in size [22, 23].

Nanoplastics: With a dimension of fewer than 100 nanometers (0.1 micrometers), nanoplastics are the tiniest kind of plastic particles. The subsequent disintegration of microplastics by means such as photodegradation and fragmentation might lead to the formation of these incredibly minute particles [24, 25].

1.2.2 CLASSIFICATION BASED ON FORMS

Fibers: Microplastic fibers are elongated strands of plastic, typically originating from synthetic textiles like polyester, nylon, and acrylic [26]. These fibers can be shed during the washing of clothes, leading to their release into wastewater systems and subsequently entering aquatic environments.

Fragments: The disintegration, mechanical abrasion, and weathering of bigger plastic products produce microplastic fragments, which are particles of plastic trash with irregular shapes [27]. These fragments can vary in size, shape, and color, depending on the polymer composition and environmental conditions.

Pellets: The production of plastic goods begins with microplastic pellets, which are little spherical or cylindrical beads of plastic resin. Other names for these pellets are nurdles and pre-production pellets [28]. These pellets can escape into the environment during production, transportation, and handling processes, posing risks to marine and terrestrial ecosystems.

Microbeads: Microplastic microbeads are little spherical bits of plastic, usually smaller than 1 millimeter in diameter, that are utilized as exfoliants in cosmetic items including face washes, toothpaste, and facial scrubs [29]. These microbeads can enter aquatic environments directly through wastewater discharge, where they pose risks to marine life through ingestion and bioaccumulation.

1.2.3 CLASSIFICATION BASED ON CHEMICAL COMPOSITIONS AND VARIABILITY

The fact that microplastics could include a wide range of different synthetic polymers emphasizes how difficult it will be to solve this persistent environmental problem. Understanding the chemical makeup of microplastics is essential for evaluating their ecological effects, enabling focused mitigation plans, and promoting environmentally friendly methods of reducing plastic pollution.

Polyethylene: A large amount of microplastic contamination is caused by polyethylene (PE), which is widely used in consumer items and packaging [30]. Tiny fragments produced by its deterioration find their way into both terrestrial and marine ecosystems.

Polypropylene: Polypropylene (PP), which is widely used in textiles and food containers, breaks down into microplastics that are frequently discharged via normal wear and tear or the disintegration of bigger objects [31].

Polyethylene terephthalate: Due to product fragmentation or fiber shedding, polyethylene terephthalate (PET), which is commonly found in beverage bottles and synthetic fibers, adds to the microplastic pollution problem [32].

Polyvinyl chloride: The building and packaging material polyvinyl chloride (PVC) degrades and releases microplastics into the environment [33].

Polystyrene: Microplastics are created when polystyrene (PS), which is used in packaging materials and disposable containers, slowly breaks down and remains in the environment [34].

Polyurethane: The material polyurethane (PU), which is widely used in foam goods and shoes, breaks down or wears out and releases microplastics, which makes environmental cleanup difficult [35].

Other synthetic polymers such as nylon (PA) [36], acrylic (PMMA) [37], and polycarbonate (PC) [38] can also be included in microplastics, and they all add to the intricate web of environmental contamination.

1.3 DETECTION AND ANALYTICAL TECHNIQUES FOR MICROPLASTICS

Microplastics, being ubiquitous in the environment, require robust and reliable detection and analytical techniques for their identification, quantification, and characterization [39]. This section provides an overview of sampling methods and analytical techniques commonly employed in microplastic research, along with the challenges and advancements in microplastic detection and analysis.

1.3.1 SAMPLING METHODS

1. *Sediment Sampling:* Sediment sampling involves collecting sediment samples from aquatic environments, such as rivers, lakes, and oceans, using various techniques like coring, grab sampling, and box coring [40]. Sediment samples are processed in the laboratory to extract and isolate microplastic particles using density separation methods, such as sedimentation or centrifugation.

2. *Fractionated Filtration:* To remove microplastics from water, it is necessary to filter them using filters with specific pore diameters. Filtration methods vary based on the water sample volume, flow rate, and desired size fraction.

After filtration, microplastic particles are visually identified and quantified under a microscope [41].

3. *Air Sampling:* Air sampling involves collecting airborne particles using air pumps and filters placed in indoor and outdoor environments. Filters are subsequently analyzed in the laboratory to identify and quantify microplastic particles using microscopy or spectroscopy techniques [42].

1.3.2 ANALYTICAL TECHNIQUES FOR MICROPLASTIC IDENTIFICATION AND QUANTIFICATION

1. *Microscopy:* It is usual practice to visually identify and characterize microplastic particles according to their size, shape, color, and surface characteristics using optical microscopy, which encompasses stereomicroscopy and polarized light microscopy. It is usual practice to visually identify and characterize microplastic particles according to their size, shape, color, and surface characteristics using optical microscopy, which encompasses stereomicroscopy and polarized light microscopy [43]. Microscopy allows for rapid screening and qualitative analysis of microplastics but may have limitations in detecting smaller particles or differentiating between microplastics and natural particles.

2. *Spectroscopy:* As a result of their molecular makeup, microplastics can be chemically identified and characterized using spectroscopic techniques such as Fourier-transform infrared spectroscopy (FTIR) and Raman spectroscopy. These techniques provide valuable information about the polymer type and additives present in microplastic particles, aiding in source attribution and contaminant analysis [44].

3. *Chromatography:* Gas chromatography-mass spectrometry (GC-MS) and liquid chromatography-mass spectrometry (LC-MS) are analytical techniques used for the quantitative analysis of organic contaminants adsorbed on microplastic surfaces. Chromatography allows for the separation and detection of target compounds, such as persistent organic pollutants (POPs) and plasticizers, providing insights into the potential risks associated with microplastic ingestion [45, 46].

1.3.3 CHALLENGES AND ADVANCEMENTS IN MICROPLASTIC DETECTION AND ANALYSIS

1. *Standardization:* Lack of standardized protocols for microplastic sampling, extraction, and analysis hinders comparability and reproducibility across studies, highlighting the need for harmonized methodologies and quality assurance measures [47].

2. *Small Particle Detection:* Detection and quantification of nano plastics (particles < 100 nm) pose challenges due to their small size and low abundance in environmental samples. Advanced analytical techniques, such as nanoparticle tracking analysis (NTA) and single-particle inductively coupled plasma mass spectrometry (SP-ICP-MS), are emerging for the detection of nanoplastics [48, 49].

Microplastics – An Overview

Advances in chromatography and spectroscopy techniques enable comprehensive analysis of microplastic-associated contaminants, enhancing our understanding of their environmental impacts. Microplastics in environmental samples are detected and analyzed using a mix of sampling procedures and analytical techniques. Overcoming challenges such as standardization, small particle detection, and contaminant analysis is essential for advancing microplastic research and informing evidence-based policies and management strategies aimed at mitigating microplastic pollution. Our knowledge of microplastics' origins, destinations, and effects on ecosystems and human health will continue to expand as detection and analytical methods improve.

1.4 DISTRIBUTION AND ENVIRONMENTAL FATE OF MICROPLASTICS

Microplastics, ubiquitous in various environmental compartments, exhibit complex distribution patterns and fate pathways influenced by a multitude of factors. To determine the effects of microplastics on ecosystems and human health and to create efficient methods for managing and reducing their presence, it is essential to comprehend their dispersal and environmental destiny.

1.4.1 DISTRIBUTIONS

Oceans: Oceans serve as major reservoirs for microplastics, with high concentrations found in surface waters, sediments, and even deep-sea environments. Ocean currents can carry microplastics great distances, where they can then build up in convergence zones like gyres and coastal regions [50].

Freshwater Bodies: Microplastics are also prevalent in freshwater bodies, including rivers, lakes, and streams, where they originate from urban runoff, industrial discharges, and wastewater effluents [51]. Hydrodynamics, closeness to pollution sources, and sedimentation rates are some of the variables that affect the regional and temporal variation in microplastic concentrations.

Soil: Microplastics have been detected in soils worldwide, with sources including agricultural practices, landfills, and atmospheric deposition. Microplastics can interact with soil particles, affecting soil structure, water retention, and nutrient cycling [52]. Their distribution in soil profiles depends on factors such as soil texture, organic matter content, and land use practices.

Atmospheric deposition: Even in densely populated places, microplastics have the potential to travel great distances in the air and land on land masses. Atmospheric deposition of microplastics is influenced by factors such as wind patterns, precipitation, and proximity to emission sources [53].

Marine Biota: Microplastics can enter marine organisms through ingestion, inhalation, and dermal contact, leading to bioaccumulation and biomagnification along marine food webs. Microplastics have been detected in a wide range of marine species, from zooplankton to apex predators, with potential impacts on their health, behavior, and reproductive success [54].

Terrestrial Biota: Microplastics can also accumulate in terrestrial organisms, including soil invertebrates, plants, and terrestrial vertebrates, through ingestion, soil contact, and bioaccumulation [55]. However, research on the impacts of microplastics on terrestrial biota is still in its infancy compared to marine environments.

1.4.2 ENVIRONMENTAL FATE PATHWAYS

Sedimentation: Microplastics in aquatic environments can undergo sedimentation and accumulate in bottom sediments over time, where they may remain buried or resuspend back into the water column under certain conditions [50].

1.5 IMPACTS ON ECOSYSTEMS AND HUMANS

Microplastics, pervasive in the environment, pose significant threats to ecosystems and human health through a multitude of direct and indirect pathways [26]. Understanding the impacts of microplastics and biofilm on ecosystems and humans is crucial for developing effective mitigation and management strategies to address this growing environmental challenge [56-58]. This section explores the diverse impacts of microplastics on both ecosystems and human populations.

1.5.1 IMPACTS ON ECOSYSTEM

Biodiversity Loss: Microplastics can have detrimental effects on biodiversity by disrupting ecosystems and food webs. Microplastics have the potential to impact population dynamics and species interactions through physically harming, nutritionally deficient, and behaviorally altering a broad variety of organisms, from zooplankton to apex predators [59–61].

Habitat Alteration: Microplastics can accumulate in aquatic and terrestrial habitats, altering sediment composition, soil structure, and biogeochemical processes [62]. Microplastic pollution may also impair ecosystem services, such as nutrient cycling, water filtration, and carbon sequestration, with cascading effects on ecosystem functioning and resilience [63].

Chemical Contamination: Microplastics can absorb and store harmful substances from their surroundings, such as heavy metals, persistent organic pollutants (POPs), and plastic additives [64]. These contaminants can desorb into organisms upon ingestion of microplastics, leading to toxicological effects, reproductive impairment, and developmental abnormalities in wildlife.

1.5.2 IMPACTS ON HUMAN

Food Chain Contamination: Meat, water, and agricultural goods tainted by microplastics can make their way into people's diets. Ingestion of microplastics by humans has been documented, with potential health risks associated with exposure to microplastic-associated contaminants, such as endocrine disruptors, carcinogens, and neurotoxins [65].

Microplastics – An Overview 9

Respiratory Health: Inhalation of airborne microplastics, particularly microplastic fibers shed from synthetic textiles and atmospheric deposition, poses potential risks to respiratory health. Inhalation of microplastics may cause pulmonary inflammation, oxidative stress, and respiratory tract damage, with implications for lung function and susceptibility to respiratory diseases [66, 67].

Dermal Exposure: Dermal contact with products containing microplastics, such as personal care items and synthetic textiles, can lead to the transfer of microplastics onto the skin and the absorption of associated chemicals [68]. Chronic dermal exposure to microplastics and plastic additives may contribute to skin irritation, allergic reactions, and long-term dermatological conditions.

Economic Costs: Environmental remediation, public health care, and the loss of ecosystem services are only a few areas where microplastic contamination has a major impact on society's wallet. The economic burden of microplastic pollution extends to industries reliant on clean environments, such as fisheries, tourism, and agriculture, which may suffer from reduced productivity and market demand [69–71].

Social Equity: The environmental degradation and health inequities linked to microplastic exposure may fall heaviest on marginalized and low-income communities, which are already struggling to cope with the effects of microplastic pollution [72]. Addressing social equity considerations in microplastic management is essential for promoting environmental justice and ensuring equitable access to clean and healthy environments for all [69, 73].

It is critical that we immediately implement all-encompassing plans to reduce microplastic contamination and safeguard ecosystems, human health, and society from the far-reaching effects of microplastics [74, 75]. To tackle the complex issues caused by microplastics, the public, researchers, and legislators must work together to establish rules that are both effective and sustainable, encourage responsible behavior, and raise awareness about the dangers of microplastic pollution. By addressing the impacts of microplastics holistically, we can work towards a more resilient and sustainable future for ecosystems and human societies alike [69, 76, 77].

1.6 MITIGATION AND MANAGEMENT STRATEGIES OF MP POLLUTION

Regulations, new technologies, public education campaigns, and cross-sector collaboration are all part of the puzzle when it comes to controlling and reducing microplastic (MP) contamination [78]. To govern the manufacture, use, and disposal of plastics including microplastics regulatory frameworks at the municipal, national, and international levels are necessary [79]. Limits or prohibitions on single-use plastics and microbeads in cosmetics and toiletries, along with extended producer responsibility (EPR) programs, can all contribute to a decrease in the amount of microplastics released into the environment [80]. Also, before discharging their effluents into water bodies, wastewater treatment plants can use cutting-edge filtering and treatment technology to catch and eliminate microplastics. Technological innovations for microplastic removal and remediation are also advancing rapidly, offering promising solutions for mitigating microplastic pollution [81]. These include the development of efficient filtration systems, such as mesh screens and sediment traps, as well as innovative technologies like electrocoagulation, ultrasonic treatment, and advanced oxidation processes for microplastic removal from water and wastewater [82]. Furthermore, research efforts are underway to explore the potential of biodegradable plastics and alternative materials as substitutes for conventional plastics, thereby reducing the environmental burden of microplastic pollution. To educate the public, businesses, and lawmakers about the dangers of microplastic contamination and encourage them to alter their habits, public awareness campaigns are crucial [83]. Encouraging practices such as plastic reduction, proper waste management, and recycling can help minimize the generation of microplastics and mitigate their environmental impacts [84]. In addition, encouraging community involvement in microplastic monitoring and cleanup projects and citizen science programs can enable individuals to fight microplastic pollution on their own and as part of a larger collective. Collaborative partnerships among governments, academia, industry, non-governmental organizations (NGOs), and other stakeholders are critical for addressing the complex challenges of microplastic pollution effectively [85]. By fostering interdisciplinary collaborations, sharing knowledge and resources, and mobilizing collective action, stakeholders can work together to develop and implement evidence-based strategies for mitigating microplastic pollution at local, regional, and global scales [86, 87]. Ultimately, achieving meaningful progress in mitigating microplastic pollution requires a concerted and coordinated effort across multiple fronts, guided by the principles of sustainability, innovation, and environmental stewardship.

1.7 EMERGING RESEARCH DIRECTIONS IN MICROPLASTICS

As our understanding of microplastics continues to evolve, new research directions are emerging to address emerging challenges and knowledge gaps in this field. These research directions encompass a wide range of interdisciplinary topics, spanning from the sources and fate of microplastics to their ecological and human health impacts. This section highlights some of the key emerging research directions in microplastics:

Nanoplastics: Nanoplastics, particles smaller than 100 nanometers, represent a frontier area of research in microplastics science. Emerging studies are focusing on the detection, characterization, and environmental fate of nanoplastics, as well as their potential biological and ecological impacts [88–90]. Advanced analytical techniques, such as nanoparticle tracking analysis (NTA) and single-particle inductively coupled plasma mass spectrometry (SP-ICP-MS), are being developed to address the challenges of detecting and quantifying nanoplastics in environmental samples [91].

Atmospheric Microplastics: Atmospheric microplastics, transported through the air over long distances, are receiving increased attention as a potential pathway for microplastic pollution.

A growing body of literature is exploring the origins, pathways, and final resting places of microplastics in the air, both on land and at sea, and how these pollutants affect ecosystems and human health [92]. Atmospheric modeling studies are also being conducted to understand the transport pathways and fate of microplastics in the atmosphere.

Microplastics in Food Chains: Understanding the pathways and dynamics of microplastic uptake and transfer in food chains is a critical research area with implications for ecosystem health and human exposure [93, 94]. New studies are looking into how microplastics can accumulate and become more toxic in aquatic and terrestrial food chains, and how they can be passed on from animals that eat them to those that eat them [10, 95]. Innovative techniques, such as stable isotope analysis and biomarker studies, are being employed to trace the uptake and fate of microplastics in organisms and food webs [96].

Microplastics and Microbial Interactions: The creation of biofilms and the colonization of microplastics by microbes alter their environmental fate, behavior, and ecological consequences [97, 98]. Emerging research is investigating the interactions between microplastics and microbial communities, including the composition, diversity, and activity of microbial biofilms on microplastic surfaces [99]. To understand where microplastics go and what effects they have in both marine and terrestrial ecosystems, it is essential to comprehend the activities that microbes play, including breakdown, aggregation, and sorption.

Microplastics and Human Health: Emerging research is focusing on the potential health effects of microplastic exposure on human populations, including ingestion, inhalation, and dermal contact pathways [100]. Studies are investigating the distribution and accumulation of microplastics in human tissues and organs, as well as the release of microplastic-associated contaminants into the body [101–103]. To determine the potential dangers to human health from long-term contact with microplastics and compounds linked to plastic, researchers are performing toxicological evaluations and epidemiological investigations [104].

Our knowledge of microplastics is growing as a result of new lines of inquiry into their production, movement, eventual demise, and effects on ecosystems and human populations. By filling in these gaps in our understanding and tackling these new problems, researchers can help reduce the negative effects of microplastic pollution on our environment and our health. Collaborative efforts across disciplines and sectors are essential for advancing research in microplastics and developing effective strategies for sustainable microplastic management.

1.8 CONCLUSION

To sum up, ecosystems, human health, and society face significant issues due to the ubiquitous presence of microplastics in the environment. From their origins in plastic production and usage to their widespread distribution in aquatic and terrestrial environments, microplastics represent a pressing environmental issue that requires urgent attention and concerted action.

Through this comprehensive overview, we have explored the diverse facets of microplastics, including their sources, forms, distribution, fate, impacts, detection methods, and mitigation strategies. We have examined how microplastics affect ecosystems by disrupting food webs, altering habitats, and introducing chemical contaminants. We have also delved into microplastic exposure's potential human health implications, including ingestion, inhalation, and dermal contact pathways.

Despite the challenges posed by microplastic pollution, there is cause for optimism as researchers, policymakers, industry stakeholders, and the public increasingly recognize the urgency of addressing this issue. By fostering interdisciplinary collaboration, advancing scientific research, implementing regulatory measures, and promoting public awareness, we can work towards sustainable solutions for mitigating microplastic pollution and safeguarding the health of ecosystems and communities worldwide.

In the face of emerging challenges and uncertainties, continued research and innovation are essential for deepening our understanding of microplastics and developing effective strategies for their management and remediation. Our shared commitment to sustainability, environmental stewardship, and collaborative action will guide us towards a world free of microplastics and thriving ecosystems that support human health.

Ultimately, the journey towards a cleaner, healthier planet free from the scourge of microplastic pollution requires commitment, collaboration, and determination from all sectors of society. By working together, we can overcome this obstacle and ensure that future generations live in a more stable and environmentally friendly world.

REFERENCES

1. Sreelakshmi, T., and K. C. Chitra. "Microplastics contamination in the environment: An ecotoxicological concern." International Journal of Zoological Investigations 7, (2021): 230–258. https://doi.org/10.33745/ijzi.2021.v07i01.019
2. Carpenter, Edward J., Susan J. Anderson, George R. Harvey, Helen P. Miklas, and Bradford B. Peck. "Polystyrene spherules in coastal waters." *Science* 178, (1972): 749–750. https://doi.org/10.1126/science.178.4062.749
3. Surendran, U., M. Jayakumar, P. Raja, Girish Gopinath, and Padmanaban Velayudhaperumal Chellam. "Microplastics in terrestrial ecosystem: Sources and migration in soil environment." *Chemosphere* 318, (2023): 137946. https://doi.org/10.1016/j.chemosphere.2023.137946
4. Faruqi, M. Humam Zaim, and Faisal Zia Siddiqui. "Microplastics and Nanoplastics in the Environment." In *Analysis of Nanoplastics and Microplastics in Food*, (2020): 47–60, Editors: Leo M.L. Nollet, Khwaja Salahuddin Siddiqi, CRC Press. https://doi.org/10.1201/9780429469596-3
5. Blackburn, Kirsty, and Dannielle Green. "The potential effects of microplastics on human health: What is known and what is unknown." *Ambio* 51, (2022): 518–530. https://doi.org/10.1007/s13280-021-01589-9
6. Pironti, Concetta, Maria Ricciardi, Oriana Motta, Ylenia Miele, Antonio Proto, and Luigi Montano. "Microplastics in the environment: Intake through the food web, human exposure and toxicological effects." *Toxics* 9, (2021): 1–29. https://doi.org/10.3390/toxics9090224
7. An, Lihui, Qing Liu, Yixiang Deng, Wennan Wu, Yiyao Gao, and Wei Ling. "Sources of microplastic in the environment." In *Microplastics in Terrestrial*

Environments: Emerging Contaminants and Major Challenges, (2020): 143–159, Editors: Defu He, Yongming Luo, Springer Nature Switzerland. https://doi.org/10.1007/698_2020_449

8. Browne, Mark A. "Sources and pathways of microplastics to habitats." In *Marine Anthropogenic Litter*, (2015): 229-244, Editors: Melanie Bergmann Lars Gutow Michael Klages, Springer. https://doi.org/10.1007/978-3-319-16510-3_9

9. Barboza, Luís Gabriel Antão, and Barbara Carolina Garcia Gimenez. "Microplastics in the marine environment: Current trends and future perspectives." *Marine Pollution Bulletin* 97, (2015): 5–12. https://doi.org/10.1016/j.marpolbul.2015.06.008

10. Carbery, Maddison, Wayne O'Connor, and Thavamani Palanisami. "Trophic transfer of microplastics and mixed contaminants in the marine food web and implications for human health." *Environment International* 115, (2018): 400–409. https://doi.org/10.1016/j.envint.2018.03.007

11. Franzellitti, Silvia, Laura Canesi, Manon Auguste, Rajapaksha HGR Wathsala, and Elena Fabbri. "Microplastic exposure and effects in aquatic organisms: A physiological perspective." *Environmental Toxicology and Pharmacology* 68, (2019): 37–51. https://doi.org/10.1016/j.etap.2019.03.009

12. Campanale, Claudia, Carmine Massarelli, Ilaria Savino, Vito Locaputo, and Vito Felice Uricchio. "A detailed review study on potential effects of microplastics and additives of concern on human health." *International Journal of Environmental Research and Public Health* 17, (2020): 1–26. https://doi.org/10.3390/ijerph17041212

13. Wright, Stephanie L., and Frank J. Kelly. "Plastic and human health: A micro issue?" *Environmental Science & Technology* 51, (2017): 6634–6647. https://doi.org/ 10.1021/acs.est.7b00423

14. Koelmans, Albert A., Paula E. Redondo-Hasselerharm, Nur Hazimah Mohamed Nor, Vera N. de Ruijter, Svenja M. Mintenig, and Merel Kooi. "Risk assessment of microplastic particles." *Nature Reviews Materials* 7, (2022): 138–152. https://doi.org/10.1038/s41578-021-00411-y

15. Andrady, Anthony L. "Persistence of plastic litter in the oceans." In *Marine Anthropogenic Litter*, (2015): 57–72, Editors: Melanie Bergmann Lars Gutow Michael Klages, Springer. https://doi.org/10.1007/978-3-319-16510-3

16. Gong, Wenwen, Yu Xing, Lihua Han, Anxiang Lu, Han Qu, and Li Xu. "Occurrence and distribution of micro-and mesoplastics in the high-latitude nature reserve, northern China." *Frontiers of Environmental Science & Engineering* 16, (2022): 113. https://doi.org/10.1007/s11783-022-1534-7

17. Andrady, Anthony L. "The plastic in microplastics: A review." *Marine Pollution Bulletin* 119, (2017): 12–22. https://doi.org/10.1016/j.marpolbul.2017.01.082

18. Kefer, Simone, Oliver Miesbauer, and Horst-Christian Langowski. "Environmental microplastic particles vs. engineered plastic microparticles-a comparative review." *Polymers* 13, (2021): 1–26. https://doi.org/10.3390/polym13172881

19. Loganathan, Yukeswaran, and Moni Philip Jacob Kizhakedathil. "A review on microplastics-an indelible ubiquitous pollutant." *Biointerface Research Applied Chemistry* 13, (2023): 126. https://doi.org/10.33263/BRIAC132.126

20. Díaz-Mendoza, Claudia, Javier Mouthon-Bello, Natalia Lucia Pérez-Herrera, and Stephanie María Escobar-Díaz. "Plastics and microplastics, effects on marine coastal areas: A review." *Environmental Science and Pollution Research* 27, (2020): 39913–39922. https://doi.org/10.1007/s11356-020-10394-y

21. Rodríguez-Seijo, Andrés, and Ruth Pereira. "Morphological and physical characterization of microplastics." In *Comprehensive Analytical Chemistry*, (2017): 49–66,

Editors: Teresa A. P. Rocha-Santos, Armando C. Durate, Elsevier. https://doi.org/10.1016/bs.coac.2016.10.007

22. Bajt, Oliver. "From plastics to microplastics and organisms." *FEBS Open Bio* 11, (2021): 954–966. https://doi.org/10.1029/2018JC014719

23. Pinlova, Barbora, and Bernd Nowack. "From cracks to secondary microplastics-surface characterization of polyethylene terephthalate (PET) during weathering." *Chemosphere*, (2024): 141305. https://doi.org/10.1016/j.chemosphere.2024.141305

24. Fu, Wanyi, Jiacheng Min, Weiyu Jiang, Yang Li, and Wen Zhang. "Separation, characterization and identification of microplastics and nanoplastics in the environment." *Science of the Total Environment* 721, (2020): 137561. https://doi.org/10.1016/j.scitotenv.2020.137561

25. Thakur, Babita, Jaswinder Singh, Joginder Singh, Deachen Angmo, and Adarsh Pal Vig. "Biodegradation of different types of microplastics: Molecular mechanism and degradation efficiency." *Science of the Total Environment* 877, (2023): 162912. https://doi.org/10.1016/j.scitotenv.2023.162912

26. Huang, Wei, Biao Song, Jie Liang, Qiuya Niu, Guangming Zeng, Maocai Shen, Jiaqin Deng, Yuan Luo, Xiaofeng Wen, and Yafei Zhang. "Microplastics and associated contaminants in the aquatic environment: A review on their ecotoxicological effects, trophic transfer, and potential impacts to human health." *Journal of Hazardous Materials* 405, (2021): 124187. https://doi.org/10.1016/j.jhazmat.2020.124187

27. Deocaris, Chester C., Jayson O. Allosada, Lorraine T. Ardiente, Louie Glenn G. Bitang, Christine L. Dulohan, John Kenneth I. Lapuz, Lyra M. Padilla, Vincent Paulo Ramos, and Jan Bernel P. Padolina. "Occurrence of microplastic fragments in the Pasig River." *H2Open Journal* 2, (2019): 92–100. https://doi.org/10.2166/h2oj.2019.001

28. Turner, Andrew, and Luke A. Holmes. "Adsorption of trace metals by microplastic pellets in fresh water." *Environmental Chemistry* 12, (2015): 600–610. https://doi.org/10.1071/EN14143

29. Miraj, Shaima S., Naima Parveen, and Haya S. Zedan. "Plastic microbeads: Small yet mighty concerning." *International Journal of Environmental Health Research* 31, (2021): 788–804. https://doi.org/10.1080/09603123.2019.1689233

30. Dümichen, Erik, Anne-Kathrin Barthel, Ulrike Braun, Claus G. Bannick, Kathrin Brand, Martin Jekel, and Rainer Senz. "Analysis of polyethylene microplastics in environmental samples, using a thermal decomposition method." *Water Research* 85, (2015): 451–457. https://doi.org/10.1016/j.watres.2015.09.002.

31. Li, Dunzhu, Yunhong Shi, Luming Yang, Liwen Xiao, Daniel K. Kehoe, Yurii K. Gun'ko, John J. Boland, and Jing Wang. "Microplastic release from the degradation of polypropylene feeding bottles during infant formula preparation." *Nature Food* 1, (2020): 746–754. https://doi.org/10.1038/s43016-020-00171-y

32. Zhang, Junjie, Lei Wang, Leonardo Trasande, and Kurunthachalam Kannan. "Occurrence of polyethylene terephthalate and polycarbonate microplastics in infant and adult feces." *Environmental Science & Technology Letters* 8, (2021): 989–994. https://doi.org/10.1021/acs.estlett.1c00559

33. Yan, Yuanyuan, Zhanghao Chen, Fengxiao Zhu, Changyin Zhu, Chao Wang, and Cheng Gu. "Effect of polyvinyl chloride microplastics on bacterial community and nutrient status in two agricultural soils." *Bulletin of Environmental Contamination and Toxicology* 107, (2021): 602–609. https://doi.org/10.1007/s00128-020-02900-2

34. Hwang, Jangsun, Daheui Choi, Seora Han, Se Yong Jung, Jonghoon Choi, and Jinkee Hong. "Potential toxicity of polystyrene microplastic particles." *Scientific Reports* 10, (2020): 7391. https://doi.org/10.1038/s41598-020-64464-9

35. Coralli, Irene, Isabel Goßmann, Daniele Fabbri, and Barbara M. Scholz-Böttcher. "Determination of polyurethanes within microplastics in complex environmental samples by analytical pyrolysis." *Analytical and Bioanalytical Chemistry* 415, (2023): 2891–2905. https://doi.org/10.1007/s00216-023-04580-3

36. Peng, Chu, Xuejiao Tang, Xinying Gong, Yuanyuan Dai, Hongwen Sun, and Lei Wang. "Development and application of a mass spectrometry method for quantifying nylon microplastics in environment." *Analytical Chemistry* 92, (2020): 13930–13935. https://doi.org/10.1021/acs.analchem.0c02801

37. Nikpay, Mitra. "Polystyrene and Polymethylmethacrylate microplastics embedded in Fat, Oil, and Grease (FOG) deposits of sewers." *Pollution* 8, (2022): 1338–1347. https://doi.org/10.22059/POLL.2022.342517.1464

38. Shi, Yanqi, Peng Liu, Xiaowei Wu, Huanhuan Shi, Hexinyue Huang, Hanyu Wang, and Shixiang Gao. "Insight into chain scission and release profiles from photodegradation of polycarbonate microplastics." *Water Research* 195, (2021): 116980. https://doi.org/10.1016/j.watres.2021.116980

39. Lee, Jieun, and Kyu-Jung Chae. "A systematic protocol of microplastics analysis from their identification to quantification in water environment: A comprehensive review." *Journal of Hazardous Materials* 403, (2021): 124049. https://doi.org/10.1016/j.jhazmat.2020.124049

40. Tuit, C. B., and A. D. Wait. "A review of marine sediment sampling methods." *Environmental Forensics* 21, (2020): 291–309. https://doi.org/10.1080/15275922.2020.1771630

41. Pittroff, Marco, Yanina K. Müller, Cordula S. Witzig, Marco Scheurer, Florian R. Storck, and Nicole Zumbülte. "Microplastic analysis in drinking water based on fractionated filtration sampling and Raman microspectroscopy." *Environmental Science and Pollution Research* 28, (2021): 59439–59451. https://doi.org/10.1007/s11356-021-12467-y

42. Enyoh, Christian Ebere, Andrew Wirnkor Verla, Evelyn Ngozi Verla, Francis Chizoruo Ibe, and Collins Emeka Amaobi. "Airborne microplastics: A review study on method for analysis, occurrence, movement and risks." *Environmental Monitoring and Assessment* 191, (2019): 1–17. https://doi.org/10.20944/preprints201908.0316.v1

43. Kalaronis, Dimitrios, Nina Maria Ainali, Eleni Evgenidou, George Z. Kyzas, Xin Yang, Dimitrios N. Bikiaris, and Dimitra A. Lambropoulou. "Microscopic techniques as means for the determination of microplastics and nanoplastics in the aquatic environment: A concise review." *Green Analytical Chemistry* 3, (2022): 100036. https://doi.org/10.1016/j.greeac.2022.100036

44. Guo, Xin, Helen Lin, Shuping Xu, and Lili He. "Recent advances in spectroscopic techniques for the analysis of microplastics in food." *Journal of Agricultural and Food Chemistry* 70, (2022): 1410–1422. https://doi.org/10.1021/acs.jafc.1c06085

45. Müller, Axel, Caroline Goedecke, Paul Eisentraut, Christian Piechotta, and Ulrike Braun. "Microplastic analysis using chemical extraction followed by LC-UV analysis: A straightforward approach to determine PET content in environmental samples." *Environmental Sciences Europe* 32, (2020): 1–10. https://doi.org/10.1186/s12302-020-00358-x

46. Hermabessiere, Ludovic, Charlotte Himber, Béatrice Boricaud, Maria Kazour, Rachid Amara, Anne-Laure Cassone, Michel Laurentie, Paul-Pont, Soudant, Dehaut, and Duflos. "Optimization, performance, and application of a pyrolysis-GC/MS method for the identification of microplastics." *Analytical and Bioanalytical Chemistry* 410, (2018): 6663–6676. https://doi.org/10.1007/s00216-018-1279-0

47. Hermsen, Enya, Svenja M. Mintenig, Ellen Besseling, and Albert A. Koelmans. "Quality criteria for the analysis of microplastic in biota samples: A critical review." *Environmental Science & Technology* 52, (2018): 10230–10240. https://doi.org/10.1021/acs.est.8b01611

48. Filipe, Vasco, Andrea Hawe, and Wim Jiskoot. "Critical evaluation of Nanoparticle Tracking Analysis (NTA) by NanoSight for the measurement of nanoparticles and protein aggregates." *Pharmaceutical Research* 27, (2010): 796–810. https://doi.org/10.1007/s11095-010-0073-2

49. Jiménez-Lamana, Javier, Lucile Marigliano, Joachim Allouche, Bruno Grassl, Joanna Szpunar, and Stephanie Reynaud. "A novel strategy for the detection and quantification of nanoplastics by single particle inductively coupled plasma mass spectrometry (ICP-MS)." *Analytical Chemistry* 92, (2020): 11664–11672. https://doi.org/10.1021/acs.analchem.0c01536

50. Kane, Ian A., and Michael A. Clare. "Dispersion, accumulation, and the ultimate fate of microplastics in deep-marine environments: A review and future directions." *Frontiers in Earth Science* 7, (2019): 80. https://doi.org/10.3389/feart.2019.00080

51. Forrest, Shaun A., Madelaine P. T. Bourdages, and Jesse C. Vermaire. "Microplastics in freshwater ecosystems." In *Handbook of Microplastics in the Environment*, (2020): 1–19. https://doi.org/10.1016/B978-0-12-822850-0.00017-X

52. de Souza Machado, Anderson Abel, Chung W. Lau, Werner Kloas, Joana Bergmann, Julien B. Bachelier, Erik Faltin, Roland Becker, Anna S. Görlich, and Matthias C. Rillig. "Microplastics can change soil properties and affect plant performance." *Environmental Science & Technology* 53, (2019): 6044–6052. https://doi.org/10.1021/acs.est.9b01339

53. Huang, Yumei, Tao He, Muting Yan, Lian Yang, Han Gong, Wenjing Wang, Xian Qing, and Jun Wang. "Atmospheric transport and deposition of microplastics in a subtropical urban environment." *Journal of Hazardous Materials* 416, (2021): 126168. https://doi.org/10.1016/j.jhazmat.2021.126168

54. Botterell, Zara L. R., Nicola Beaumont, Tarquin Dorrington, Michael Steinke, Richard C. Thompson, and Penelope K. Lindeque. "Bioavailability and effects of microplastics on marine zooplankton: A review." *Environmental Pollution* 245, (2019): 98–110. https://doi.org/10.1016/j.envpol.2018.10.065

55. Dissanayake, Pavani Dulanja, Soobin Kim, Binoy Sarkar, Patryk Oleszczuk, Mee Kyung Sang, Md Niamul Haque, Jea Hyung Ahn, Michael S. Bank, and Yong Sik Ok. "Effects of microplastics on the terrestrial environment: A critical review." *Environmental Research* 209, (2022): 112734. https://doi.org/10.1016/j.envres.2022.112734

56. Wang, Jianlong, Xuan Guo, and Jianming Xue. "Biofilm-developed microplastics as vectors of pollutants in aquatic environments." *Environmental Science & Technology* 55, (2021): 12780–12790. https://doi.org/10.1021/acs.est.1c04466

57. Ge, Jianhua, Mingjun Wang, Peng Liu, Zixuan Zhang, Jianbiao Peng, and Xuetao Guo. "A systematic review on the aging of microplastics and the effects of typical factors in various environmental media." TrAC Trends in Analytical Chemistry 162, (2023): 117025. https://doi.org/10.1016/j.trac.2023.117025

58. Chaukura, Nhamo, Kebede K. Kefeni, Innocent Chikurunhe, Isaac Nyambiya, Willis Gwenzi, Welldone Moyo, Thabo TI. Nkambule, Bhekie B. Mamba, and Francis O. Abulude. "Microplastics in the aquatic environment—the occurrence, sources, ecological impacts, fate, and remediation challenges." *Pollutants* 1, (2021): 95–118. https://doi.org/10.3390/pollutants1020009

59. Enyoh, Christian Ebere, Leila Shafea, Andrew Wirnkor Verla, Evelyn Ngozi Verla, Wang Qingyue, Tanzin Chowdhury, and Marcel Paredes. "Microplastics exposure

routes and toxicity studies to ecosystems: An overview." *Environmental Analysis, Health and Toxicology* 35, (2020): 1–10. https://doi.org/10.5620/eaht.e2020004

60. Galloway, Tamara S., Matthew Cole, and Ceri Lewis. "Interactions of microplastic debris throughout the marine ecosystem." *Nature Ecology & Evolution* 1, (2017): 0116. https://doi.org/10.1038/s41559-017-0116

61. Ma, Hui, Shengyan Pu, Shibin Liu, Yingchen Bai, Sandip Mandal, and Baoshan Xing. "Microplastics in aquatic environments: Toxicity to trigger ecological consequences." *Environmental Pollution* 261, (2020): 114089. https://doi.org/10.1016/j.envpol.2020.114089

62. Huang, Qi, Yuyang Lin, Qiyin Zhong, Fei Ma, and Yixin Zhang. "The impact of microplastic particles on population dynamics of predator and prey: Implication of the lotka-Volterra model." *Scientific Reports* 10, (2020): 4500. https://doi.org/10.1038/s41598-020-61414-3

63. Horton, Alice A., and David K. A. Barnes. "Microplastic pollution in a rapidly changing world: Implications for remote and vulnerable marine ecosystems." *Science of The Total Environment* 738, (2020): 140349. https://doi.org/10.1016/j.scitotenv.2020.140349

64. Okoye, Charles Obinwanne, Charles Izuma Addey, Olayinka Oderinde, Joseph Onyekwere Okoro, Jean Yves Uwamungu, Chukwudozie Kingsley Ikechukwu, Emmanuel Sunday Okeke, Onome Ejeromedoghene, and Elijah Chibueze Odii. "Toxic chemicals and persistent organic pollutants associated with micro-and nanoplastics pollution." *Chemical Engineering Journal Advances* 11, (2022): 100310. https://doi.org/10.1016/j.ceja.2022.100310

65. Al Mamun, Abdullah, Tofan Agung Eka Prasetya, Indiah Ratna Dewi, and Monsur Ahmad. "Microplastics in human food chains: Food becoming a threat to health safety." *Science of the Total Environment* 858, (2023): 159834. https://doi.org/10.1016/j.scitotenv.2022.159834

66. Ahmad, Mushtaq, Jing Chen, Muhammad Tariq Khan, Qing Yu, Worradorn Phairuang, Masami Furuuchi, Syed Weqas Ali, Asim Nawab, and Sirima Panyametheekul. "Sources, analysis, and health implications of atmospheric microplastics." Emerging Contaminants 9, (2023): 100233. https://doi.org/10.1016/j.emcon.2023.100233

67. Yao, Xuewen, Xiao-San Luo, Jiayi Fan, Tingting Zhang, Hanhan Li, and Yaqian Wei. "Ecological and human health risks of atmospheric microplastics (MPs): A review." *Environmental Science: Atmospheres* 2, (2022): 921–942. https://doi.org/10.1039/D2EA00041E

68. Ageel, Hassan Khalid, Stuart Harrad, and Mohamed Abou-Elwafa Abdallah. "Occurrence, human exposure, and risk of microplastics in the indoor environment." *Environmental Science: Processes & Impacts* 24, (2022): 17–31. https://doi.org/10.1039/D1EM00301A

69. Kumar, Rakesh, Anurag Verma, Arkajyoti Shome, Rama Sinha, Srishti Sinha, Prakash Kumar Jha, Ritesh Kumar et al. "Impacts of plastic pollution on ecosystem services, sustainable development goals, and need to focus on circular economy and policy interventions." *Sustainability* 13, (2021): 9963. https://doi.org/10.3390/su13179963

70. Roy, Poritosh, Amar K. Mohanty, and Manjusri Misra. "Microplastics in ecosystems: Their implications and mitigation pathways." *Environmental Science: Advances* 1, (2022): 9–29. https://doi.org/ 10.1039/D1VA00012H

71. Prata, Joana C., Ana L. Patrício Silva, João P. Da Costa, Catherine Mouneyrac, Tony R. Walker, Armando C. Duarte, and Teresa Rocha-Santos. "Solutions and integrated strategies for the control and mitigation of plastic and microplastic pollution." *International*

Journal of Environmental Research and Public Health 16, (2019): 2411. https://doi.org/10.3390/ijerph16132411

72. Bennett, Nathan J., Juan José Alava, Caroline E. Ferguson, Jessica Blythe, Elisa Morgera, David Boyd, and Isabelle M. Côté. "Environmental (in) justice in the Anthropocene ocean." *Marine policy* 147, (2023): 105383. https://doi.org/10.1016/j.marpol.2022.105383

73. Conlon, Katie. "Responsible materials Stewardship: Rethinking waste management globally in consideration of social and ecological externalities and increasing waste generation." *Advances in Environmental and Engineering Research* 5, (2024): 1–21. https://doi.org/10.1016/j.marpol.2023.105856

74. Thacharodi, Aswin, Ramu Meenatchi, Saqib Hassan, Naseer Hussain, Mansoor Ahmad Bhat, Jesu Arockiaraj, Huu Hao Ngo, Quynh Hoang Le, and Arivalagan Pugazhendhi. "Microplastics in the environment: A critical overview on its fate, toxicity, implications, management, and bioremediation strategies." *Journal of Environmental Management* 349, (2024): 119433. https://doi.org/10.1016/j.jenvman.2023.119433

75. McKinney, Amanda. "A planetary health approach to study links between pollution and human health." *Current Pollution Reports* 5, (2019): 394–406. https://doi.org/10.1007/s40726-019-00131-6

76. Ghosh, Shampa, Jitendra Kumar Sinha, Soumya Ghosh, Kshitij Vashisth, Sungsoo Han, and Rakesh Bhaskar. "Microplastics as an emerging threat to the global environment and human health." *Sustainability* 15, (2023): 10821. https://doi.org/10.3390/su151410821

77. Senathirajah, Kala, and Thava Palanisami. "Strategies to reduce risk and mitigate impacts of disaster: Increasing water quality resilience from microplastics in the water supply system." *ACS ES&T Water* 3, (2023): 2816–2834. https://doi.org/10.1021/acsestwater.3c00206

78. Kurniawan, Tonni Agustiono, Ahtisham Haider, Ayesha Mohyuddin, Rida Fatima, Muhammad Salman, Anila Shaheen, Hafiz Muhammad Ahmad et al. "Tackling microplastics pollution in global environment through integration of applied technology, policy instruments, and legislation." *Journal of Environmental Management* 346, (2023): 118971. https://doi.org/10.1016/j.jenvman.2023.118971

79. Deme, Gideon Gywa, David Ewusi-Mensah, Oluwatosin Atinuke Olagbaju, Emmanuel Sunday Okeke, Charles Obinwanne Okoye, Elijah Chibueze Odii, Onome Ejeromedoghene et al. "Macro problems from microplastics: Toward a sustainable policy framework for managing microplastic waste in Africa." *Science of the Total Environment* 804, (2022): 150170. https://doi.org/10.1016/j.scitotenv.2021.150170

80. Gionfra, Susanna, Clémentine Richer, and Emma Watkins. "The role of policy in tackling plastic waste in the aquatic environment." In *Plastics in the Aquatic Environment-Part II: Stakeholders' Role Against Pollution*, Editors: Friederike Stock, Georg Reifferscheid, and Evgeniia Kostianaia Cham: Springer International Publishing, (2020): 119–138. https://doi.org/10.1007/698_2020_484

81. Krishnan, Radhakrishnan Yedhu, Sivasubramanian Manikandan, Ramasamy Subbaiya, Natchimuthu Karmegam, Woong Kim, and Muthusamy Govarthanan. "Recent approaches and advanced wastewater treatment technologies for mitigating emerging microplastics contamination–A critical review." *Science of the Total Environment* 858, (2023): 159681. https://doi.org/10.1016/j.scitotenv.2022.159681

82. Ahmed, Shams Forruque, Nafisa Islam, Nuzaba Tasannum, Aanushka Mehjabin, Adiba Momtahin, Ashfaque Ahmed Chowdhury, Fares Almomani, and M. Mofijur. "Microplastic removal and management strategies for wastewater treatment

Microplastics – An Overview

plants." *Chemosphere* 347, (2024): 140648. https://doi.org/10.1016/j.chemosphere.2023.140648

83. Sandu, Cristina, Emoke Takacs, Giuseppe Suaria, Franco Borgogno, Christian Laforsch, Martin M. G. J. Löder, Gijsbert Tweehuysen, and Letitia Florea. "Society role in the reduction of plastic pollution." In *Plastics in the Aquatic Environment-Part II: Stakeholders' Role Against Pollution*, Editors: Friederike Stock, Georg Reifferscheid, and Evgeniia Kostianaia Cham: Springer International Publishing, (2020): 39–65. https://doi.org/10.1007/698_2020_483

84. Calero, Mónica, Verónica Godoy, Lucía Quesada, and María Ángeles Martín-Lara. "Green strategies for microplastics reduction." *Current Opinion in Green and Sustainable Chemistry* 28, (2021): 100442. https://doi.org/10.1016/j.cogsc.2020.100442

85. Hung, Ling-Ya, Shun-Mei Wang, and Ting-Kuang Yeh. "Collaboration between the government and environmental non-governmental organisations for marine debris policy development: The Taiwan experience." *Marine Policy* 135, (2022): 104849. https://doi.org/10.1016/j.marpol.2021.104849

86. Lusher, Amy L., Rachel Hurley, Hans Peter H. Arp, Andy M. Booth, Inger Lise N. Bråte, Geir W. Gabrielsen, Alessio Gomiero et al. "Moving forward in microplastic research: A Norwegian perspective." *Environment International* 157, (2021): 106794. https://doi.org/10.1016/j.envint.2021.106794

87. Donzelli, Gabriele, and Nunzia Linzalone. "Use of scientific evidence to inform environmental health policies and governance strategies at the local level." *Environmental Science & Policy* 146, (2023): 171–184. https://doi.org/10.1016/j.envsci.2023.05.009

88. Wang, Junyu, Xiaoli Zhao, Fengchang Wu, Lin Niu, Zhi Tang, Weigang Liang, Tianhui Zhao, Mengyuan Fang, Hongzhan Wang, and Xiaolei Wang. "Characterization, occurrence, environmental behaviors, and risks of nanoplastics in the aquatic environment: Current status and future perspectives." *Fundamental Research* 1, (2021): 317–328. https://doi.org/10.1016/j.fmre.2021.05.001

89. Ali, Imran, Qianhui Cheng, Tengda Ding, Qian Yiguang, Zhang Yuechao, Huibin Sun, Changsheng Peng, Iffat Naz, Juying Li, and Jingfu Liu. "Micro-and nanoplastics in the environment: Occurrence, detection, characterization and toxicity–A critical review." *Journal of Cleaner Production* 313, (2021): 127863. https://doi.org/10.1016/j.jclepro.2021.127863

90. Shen, Maocai, Yaxin Zhang, Yuan Zhu, Biao Song, Guangming Zeng, Duofei Hu, Xiaofeng Wen, and Xiaoya Ren. "Recent advances in toxicological research of nanoplastics in the environment: A review." *Environmental Pollution* 252, (2019): 511–521. https://doi.org/10.1016/j.envpol.2019.05.102

91. Marigliano, Lucile, Bruno Grassl, Joanna Szpunar, Stéphanie Reynaud, and Javier Jiménez-Lamana. "Nanoplastic labelling with metal probes: Analytical strategies for their sensitive detection and quantification by icp mass spectrometry." *Molecules* 26, (2021): 7093. https://doi.org/10.3390/molecules26237093.

92. Sridharan, Srinidhi, Manish Kumar, Lal Singh, Nanthi S. Bolan, and Mahua Saha. "Microplastics as an emerging source of particulate air pollution: A critical review." *Journal of Hazardous Materials* 418, (2021): 126245. https://doi.org/10.1016/j.jhazmat.2021.126245

93. Rose, Pawan Kumar, Sangita Yadav, Navish Kataria, and Kuan Shiong Khoo. "Microplastics and nanoplastics in the terrestrial food chain: Uptake, translocation, trophic transfer, ecotoxicology, and human health risk." *TrAC Trends in Analytical Chemistry* 167, (2023): 117249. https://doi.org/10.1016/j.trac.2023.117249

94. Krause, Stefan, Viktor Baranov, Holly A. Nel, Jennifer D. Drummond, Anna Kukkola, Timothy Hoellein, Gregory H. Sambrook Smith et al. "Gathering at the top?

Environmental controls of microplastic uptake and biomagnification in freshwater food webs." *Environmental Pollution* 268, (2021): 115750. https://doi.org/10.1016/j.envpol.2020.115750

95. Bhatt, Vaishali, and Jaspal Singh Chauhan. "Microplastic in freshwater ecosystem: Bioaccumulation, trophic transfer, and biomagnification." *Environmental Science and Pollution Research* 30, (2023): 9389–9400. https://doi.org/10.1007/s11356-022-24529-w

96. Benson, Nsikak U., Omowumi D. Agboola, Omowunmi H. Fred-Ahmadu, Gabriel Enrique De-la-Torre, Ayodeji Oluwalana, and Akan B. Williams. "Micro (nano) plastics prevalence, food web interactions, and toxicity assessment in aquatic organisms: A review." *Frontiers in Marine Science* 9, (2022): 851281. https://doi.org/10.3389/fmars.2022.851281

97. Rummel, Christoph D., Annika Jahnke, Elena Gorokhova, Dana Kühnel, and Mechthild Schmitt-Jansen. "Impacts of biofilm formation on the fate and potential effects of microplastic in the aquatic environment." *Environmental Science & Technology letters* 4, (2017): 258–267. https://doi.org/10.1021/acs.estlett.7b00164

98. He, Siying, Meiying Jia, Yinping Xiang, Biao Song, Weiping Xiong, Jiao Cao, Haihao Peng et al. "Biofilm on microplastics in aqueous environment: Physicochemical properties and environmental implications." *Journal of Hazardous Materials* 424, (2022): 127286. https://doi.org/10.1016/j.jhazmat.2021.127286

99. Yang, Yuyi, Wenzhi Liu, Zulin Zhang, Hans-Peter Grossart, and Geoffrey Michael Gadd. "Microplastics provide new microbial niches in aquatic environments." *Applied Microbiology and Biotechnology* 104, (2020): 6501–6511. https://doi.org/10.1007/s00253-020-10704-x

100. Rahman, Arifur, Atanu Sarkar, Om Prakash Yadav, Gopal Achari, and Jaroslav Slobodnik. "Potential human health risks due to environmental exposure to nano- and microplastics and knowledge gaps: A scoping review." *Science of the Total Environment* 757, (2021): 143872. https://doi.org/10.1016/j.scitotenv.2020.143872

101. Kutralam-Muniasamy, Gurusamy, V. C. Shruti, Fermín Pérez-Guevara, and Priyadarsi D. Roy. "Microplastic diagnostics in humans: "The 3Ps" progress, problems, and prospects." *Science of the Total Environment* 856, (2023): 159164. https://doi.org/10.1016/j.scitotenv.2022.159164

102. Pironti, Concetta, Maria Ricciardi, Oriana Motta, Ylenia Miele, Antonio Proto, and Luigi Montano. "Microplastics in the environment: Intake through the food web, human exposure and toxicological effects." *Toxics* 9, (2021): 224. https://doi.org/10.3390/toxics9090224

103. Liu, Zhengguo, and Xue-yi You. "Recent progress of microplastic toxicity on human exposure base on in vitro and in vivo studies." Science of the total Environment 903, (2023): 166766. https://doi.org/10.1016/j.scitotenv.2023.166766

104. Sun, Anqi, and Wen-Xiong Wang. "Human exposure to microplastics and its associated health risks." *Environment & Health* 1, (2023): 139–149. https://doi.org/10.1021/envhealth.3c00053

2 Detection and Extraction Techniques for Microplastics

Bhawana Jangir, Sarita Mochi, and Anjali Yadav

2.1 INTRODUCTION

Plastics have now become an integral component of human life [1]. Plastics are derived from various methods of polymerization of a variety of monomers. The unique properties of plastics such as flexibility, lightweight, portability, and easy transportation make them potential candidates for various consumer goods and packaging industries at the commercial scale [2]. The durability of plastic makes it an attractive property for extensive applications in various fields, but its properties such as resistance to degradation make its management a significant challenge [3]. Despite the increased international awareness and efforts to solve the problem of plastic pollution, the aggregation of plastic materials in our surroundings remains a significant issue. The continuous disposal of plastic waste has resulted in increasing levels of microplastic (MP) pollution, which is affecting the environment as a whole.

Polymers possessing particle sizes less than 5 mm are known as MPs [4]. Upon moving into the environment, MPs undergo several processes that cause mechanical breakdown, resulting in diverse characteristics depending on size, color, shape, source, particle density, and specific gravity [5–7]. These nonbiodegradable MPs in the environment can severely damage ecological health and biodiversity. Qi et al. reported an experiment that evaluated the effect of mixing 1% MPs with soil on wheat plant growth [8]. It was observed that the presence of MPs like polyethylene terephthalate (PET) significantly harmed the quality and growth of wheat. Similarly, using the extraction method, the impact of MP concentration in the soil on tomato plant growth and maturation was investigated. While it was observed that the presence of MPs in soil enhanced the growth of plants, the maturation process was delayed.

The effect of inadvertently consumed MPs by humans through fruits, vegetables, and preserved foods has also been studied extensively [9]. Reports on fruits and vegetables in Italy highlighted significant concerns regarding MP adulteration in the food system [10]. The daily intake of MPs through food systems was estimated by

DOI: 10.1201/9781003486947-2

21

the study to be between 87,600 and 124,900 particles per gram. This alarming level of contamination is attributed to a potential translocation chain from plant and soil pathways to the broader ecosystem. The studies on packaged foods and disposable cups highlight the pervasive nature of MP contamination in everyday items [11]. The average MP content found in these items and the substantial annual release of MPs underscores the urgent need for effective measures to reduce plastic use and mitigate its impact. Microplastics, due to their small size, pose a significant risk of ingestion by various organisms throughout the food chain. This ingestion can cause the inclusion of damaging chemicals and additives into the ecosystem, causing environmental and health concerns [12].

The presence of MPs in the marine environment presents multiple challenges and has wide-ranging implications. These impacts span aesthetic, economic, industrial, and environmental dimensions, highlighting the need for comprehensive management strategies to address this issue. The discovery of small plastic fragments in the open ocean in the 1970s marked the initial recognition of plastic pollution in marine [13]. However, it was not until the past decade that a renewed scientific interest in MPs emerged, revealing their widespread and ubiquitous presence within the marine environment. This increased focus on MPs has underscored their potential to cause harm to marine biota and has prompted global research efforts to address this pressing environmental issue [14, 15]. While research on the toxic effect of MPs on human health is still in its early stages, there is growing evidence that chronic exposure to MPs can have detrimental effects on marine species [16, 17]. As marine organisms ingest MPs, these particles can accumulate in their tissues. When humans consume seafood and other marine products, there is a high likelihood that MPs can enter the human diet [18, 19]. Thus, it is imperative to understand the techniques for extraction, separation, and detection of MPs to address these issues.

2.2 EXTRACTION TECHNIQUES

In MP research, the processes of extraction and separation are crucial for analyzing and understanding the presence and impact of MPs in the ecosystem. Extraction refers to the process of isolating and purifying MP fragments from the sample medium, which could be water, sediment, biota, or other environmental samples.

Due to the vivid and wide aggregation of plastic debris in the ocean, most attention and research have focused on the marine ecosystem [5]. Three hundred beaches in New Zealand were found to be contaminated with virgin polyolefin plastics, however, MPs were extracted and detected two decades later for the first time in beach sediments [20]. Terrestrial ecosystems, including sediments, soil, compost, and sand, are especially prone to plastic contamination and play a crucial role in the production and spread of MPs. Despite this, these ecosystems are relatively unexplored compared to marine environments. Sediments and soils, which are deposited in a variety of sites by wind and runoff, are considered to be long-term sinks of MPs. Developing a reliable method for extracting and identifying MPs from sediments and soils is crucial for understanding their distribution, mass, and ecological effects in marine, freshwater, and terrestrial ecosystems. To date, a variety of methods have been developed for the extraction of MPs from sediments and soils. These methods range from simple,

Detection and Extraction Techniques 23

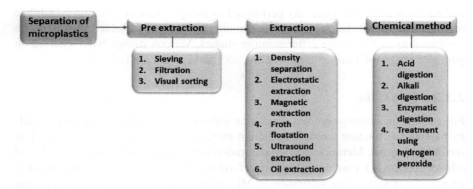

FIGURE 2.1 Methods for extraction of microplastics.

one-step extractions to more complex techniques that exploit the unique properties of MPs. The extraction of MPs from soils is indeed more complex than from sediments and sand due to the heterogeneous nature of soils, which include a mixture of organic matter, minerals, and various particle sizes. As a result, the extraction process often requires additional steps, particularly methods for digesting organic matter after the initial extraction. This chapter further covers the various methods used and developed for the extraction of MPs shown in Figure 2.1.

2.3 PRE-EXTRACTION: SIEVING AND FILTRATION

2.3.1 Sieving

Sieving is indeed a common pretreatment procedure used to separate MPs from soil, sand, and sediments based on granulometric fractions. In this method, MPs are separated from water samples using sieves of different variety and mesh sizes. This technique is usually preferred for the extraction of particles that are visible to human eyes and are larger than 5 mm [21]. Soil samples should be dried before analysis to improve recovery of MPs. It is crucial to highlight the importance of precisely reporting the results mentioning if the work is done in dry or wet samples to ensure clarity, consistency, comparability, and standardization. It is always recommended to report quantities of MPs as mass/volume of either wet or dry sample rather than describing the number of MPs in a weight of sample. The samples of sediments carrying MPs and other particles are collected from the environment by using appropriate sampling technique. In general, the filtration setup comprises a sieve with a defined pore size chosen considering the desired size range for particle separation. The sample is passed through the sieves by applying gravity and vacuum filtration. The particles having a size smaller than the pore size of the sieve pass through it while larger size particles are retained on the top surface of the sieve. The particles obtained on the top surface of the sieve consisting of MPs and other particles are collected meticulously and shifted to a container for further processing. The filtrate-containing particles may be considered waste or subjected to additional steps of processing. In

MP research, sieving is generally performed by using a sieve or filtration membrane of single mesh size, but in a few cases, multiple sieves of variable mesh sizes are combined to facilitate size differentiation studies. Various studies have reported on pore size ranges of 1 mm, 5 mm, 330 μm, 335 μm, and 0.2 μm [22–25].

2.3.2 Filtration

Filtration is an extensively used separation technique to remove solids from liquids by passing the mixture through a medium that allows only the liquid to pass while retaining the solid. Membrane-based separation is indeed one of the most suitable and efficient methods for isolating MPs from food matrices. This method relies on semi-permeable membrane for separation of MPs from complex food samples depending on difference in particle size, shape, and chemical properties. This is an easy-to-use technique that does not require a complicated setup. It completely depends on a pressure difference to drive liquid through a membrane, effectively separating MPs from the rest of the sample matrix. This technique has attracted significant attention for separating MPs from food matrices due to its potential to eventually replace more energy consuming, complex, and maintenance intensive conventional techniques. This methodology provides a better and more effective alternative to fulfilling the growing demand for sustainable and cost-effective methods in environmental and food safety applications [26, 27]. Additionally, the process of separation can be classified on the basis of particles that are filtered, mainly ultrafiltration, microfiltration, nanofiltration, and reverse osmosis.

There are several reports on MPs found in food samples. A study was conducted on the presence of MPs in a few branded milk packets [28]. The milk samples were heated and treated at 0.5 bar pressure before filtration to avoid blocking of filters by fat and casein. Microfilters of variable sizes such as 0.22 μm, 0.45 μm, 11 μm, and 5 μm were selected where 11 μm was the most suitable pore size for separation. The result highlighted 97.5% fibers and 2.5% granules containing mostly polyethersulfone and polysulfone polymers. It was observed that many commonly used polymers such as polyethylene, polyester, polypropylene, and PET were not detected in any of the milk samples. It was concluded that sulfone polymers could be effectively isolated by membrane filters from milk samples.

Extensive efforts have also been made to find suitable methods to identify MPs from agricultural soil due to their adverse effects on soil organisms, plant growth, and nutrient cycle. Various approaches have also been investigated to get efficient separation and precise quantification of MPs in fishpond soil, bamboo soil, vegetable, and tea garden soil [29]. These studies focused on separation of MPs depending on the efficiency of various membrane materials and pore sizes such as quartz (pore size: 2.2 μm), glass fibre (pore size: 1.6 μm), nylon (pore size: 20 μm), and polytetrafluoroethylene (PTFE) (pore size: 2 μm). The reports showed that quartz and glass fiber-based filters were found to be more efficient in removing fibers, debris, and nonpolymer materials.

2.3.3 Visual Sorting

Separation through visual identification is an important step for the analysis of MPs to remove nonmicroplastic debris from the sample. Such debris generally includes biological waste products like shell fragments and seaweed, and anthropogenic contaminants such as oil residues, metals, paints, and wood. Large-size MPs can sometimes be identified and sorted by tweezers especially when they are brightly colored or have unique shapes. This technique is less effective for smaller MPs (<1mm), and there is a risk of missing less conspicuous particles, leading to underestimation of MPs present in the sample. For more precise sorting, an optical microscope is employed. Using a microscope, researchers can identify and pick out MPs of variety of sizes and shapes with the help of tweezers. Distinguishing MPs from other materials is difficult, making the process time-consuming and prone to misidentification [23, 24, 30]. Indeed, the similarity in appearance between MPs and other components highlights the limitations of visual sorting. Sometimes fragments of shells are mistaken as MPs due to their white color; similarly brittle polyethylene (PE) fragments can be mistaken for shell fragments. This shows that visual sorting through optical microscopy may lead to false results. Identification and separation of MPs is a challenging task due to their small size and variable composition. However, certain physical attributes such as hardness, plasticity, and surface properties can help in distinguishing them from other components [6]. An alternative method including the hot needle test has been developed to avoid the misidentification of MPs. In this method the hot needle is applied to the surface of plastic particles leading to melting of plastic with a characteristic smell [31]. Nile red has been extensively used for fluorescent staining of MPs for identification and separation through fluorescent microscope [32, 33]. However, this method has limited applications, as fibers are difficult to stain and low hydrophobic MPs like polycarbonate, polyvinyl chloride, polyurethane, and polyethylene terephthalate are weakly fluoresced.

2.4 EXTRACTION TECHNIQUES

2.4.1 Density Separation using Hypersaline Solutions

Density separation is a widely used method for extraction of MPs from large sediments and sand samples, primarily in marine environments, due to its efficiency and effectiveness. This technique is used to separate MPs from sediments or other inorganic substances that remain undamaged during chemical or enzymatic treatment. This technique, widespread in research for the last two decades, includes mixing materials of varying densities into an intermediate density liquid, leading the dense material to float and separate from denser components.

Density of MP is significantly impacted by the type of polymer, the concentration of additives, and even chemicals and organisms that have been adsorbed [34]. In this technique, a sample is thoroughly mixed with saturated salt solution or high concentration salt in the proper volumetric ratio and allowed for density-based separation for a predetermined amount of time (Figure 2.2). Sodium chloride has been widely used as an extraction salt solution for MPs due to its environmentally friendliness and

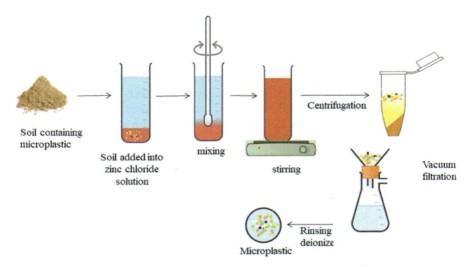

FIGURE 2.2 Density separation technique for removal of microplastic from soil.

cost effectiveness, despite its limitation in isolating denser polymers. Lower density polymers like polypropylene or polyamide can be easily isolated but polymers such as polyvinyl chloride and polyethylene terephthalate of higher densities cannot be isolated by using this method. Various studies have been reported showing the application of various salts for extraction of MPs from soils, considering the uncertainty of MP density. For example, Nuelle et al. developed a two-step density separation method [35]. First, the material was reduced to 80% by fluidization in a NaCl solution. This was followed by flotation with a small amount of NaI solution. Many studies have reported on the use of a saturated solution of sodium iodide (NaI), Sodium Bromide (NaBr), Zinc Bromide ($ZnBr_2$), and Zinc Chloride ($ZnCl_2$) for the suspension medium [36, 37].

When the density of MPs is uncertain, using higher-density salt solutions can enhance the efficiency of MP extraction and flotation. Recent studies have indeed proposed Calcium Chloride ($CaCl_2$) as a safer alternative for the extraction and flotation of MPs. The rate of recycling of high-density separation solution is significantly higher. The selection of suspension medium depends on the rate of recovery, cost, and its environmental impact. The recycling rate of the high-density separation solution is significantly higher than that of NaCl. The choice of suspension medium should depend on the recovery rate and price and its environmental impact. A novel method based on wet peroxide oxidation has been developed in which 96% ethanol is used instead of salts as a suspension medium [38]. The sample is stirred in ethanol for 3 min at 600 rpm and then allowed to settle. This results in flotation of 97% biogenic matter separated from MPs that have settled at the bottom. It is clear that all of the density separation protocols referred to in the literature have potential drawbacks, and in order to select the most appropriate technique, the efficacy, the environmental effects, and the cost effectiveness must be considered.

2.4.2 Electrostatic Extraction

Soil minerals and other particles often exhibit electrical conductivity, while plastics are nonconductive. This difference in electrical properties can be leveraged to separate the MPs from samples by applying an external electric field. Electrostatic separation is indeed a dry processing technique particularly useful to separate primary and secondary raw materials in various industrial applications. Felsing et al. performed a study on the electrostatic behavior of plastic particles and subsequently modified a small electrostatic separation device to improve the efficiency of separation of MP from sediments [39]. The device was found to have the efficiency to remove up to 99% of the original sample mass without losing MPs. The method also eliminated biological components without altering particle characteristics. However, recent reports highlight the limitations of electrostatic separation of MPs in soil samples specifically related to the sample preparation procedure. Key challenges include the tedious sample-drying process and the difficulty in managing humidity before freeze-drying. Despite its environmentally friendly nature and no use of chemicals, this technique may require additional processes such as density separation and digestion for more effective separation of MPs.

2.4.3 Magnetic Extraction

Exploring magnetic separation for MPs in environmental and drinking water samples involves utilizing iron nanoparticles (NPs) that bind to plastic, creating a hydrophobic tail. Iron NPs processed with hydrophobic hydrocarbons using cetyltrimethoxysilane (HDTMS) offer an efficient approach for extraction of MPs in the presence of external magnetic field [40]. This technology, magnetizing plastics with bound iron particles, attracts MPs resulting in a 92% recovery rate. Iron NPs have been selected due to their low cost, ferromagnetic properties, and high surface area. While this method maintains sample structural integrity, gaps in separation efficacy prompt researchers to view it as a pretreatment step rather than a standalone technique. Notably, studies on solvent or temperature variations in food samples are lacking. Further research is essential to comprehensively evaluate and optimize MP separation techniques.

2.4.4 Froth Flotation

The froth flotation method for MP separation exploits the density and hydrophobicity of the material. It is an effective separation process for hydrophilic and hydrophobic materials using simple air bubbles [41]. The hydrophobic particles attach to air bubbles and move upward, retaining hydrophilic particles submerged in the liquid phase. Various factors affect the flotation efficiency of materials including particle size, shape, roughness, surface chemistry, stability, and reactivity. Generally surfactants are used further to stabilize air bubbles and to enhance the froth formation. Several types of surfactants have been used to decrease the surface energy of the solid sample and thereby improve their hydrophobicity through chemisorption and physisorption mechanisms.

There are three methods of froth flotation: gamma flotation, reagent adsorption, and surface modification [41]. Gamma flotation exploits the surface tension of the flotation medium and is generally used to separate low surface energy solids. The complicated surface modification procedure takes time and suffers from the unpredictable floatability of plastics, leaving a gap between the theory of surface modification and its practical application [42]. Huang et al. used pinacol (97.71% pure) as a foaming agent and potassium permanganate as a surface modifier to achieve a 95% recovery rate for PVC and PMMA (polymethyl methacrylate) [43].

2.4.5 Ultrasound Extraction

Ultrasound extraction is a novel approach for MP extraction [27]. This method efficiently separates MP of polyethylene and styrene when operated at 39-41 kHz for 15 min. This method is proposed for marine environments due to cost-effectiveness, minimal sample damage, and improved solvent penetration compared to chemical and enzymatic alternatives. Similar ultrasound treatments were applied to beverages (honey, beer, milk, and soft drinks), revealing MP presence (32-67 MP/L). Furthermore, ultrasound extraction was explored for soil, vegetable processing, and industrial wastewater treatment, showcasing its versatility in addressing MP contamination with efficient results.

2.4.6 Oil-Extraction

The oil extraction method exploits the oleophilic properties of plastic polymers. The hydrophobic nature of oil assists in the separation of plastics from environmental samples and increases the rate of recovery. MPs can be isolated through the water-soil interface since the bulk of soil or sediment particles are hydrophilic in nature. Oil-polymer clusters always have lower total density than water. This method involves shaking of dry sediments with water, adding oil and then keeping the mix in a shaker for about 30 seconds to allow the sediment and MP to stick to the oil. Water layer is decanted before vacuum filtration. The oil layer is filtered and the filters are washed with alcohol to remove any oil residue that may interfere with further examination.

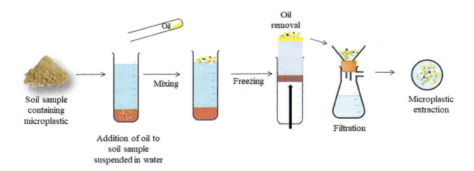

FIGURE 2.3 Oil based extraction of microplastics in soil sample.

Detection and Extraction Techniques

For samples rich in biomass, it has been suggested to perform digestion before the oil-extraction protocol shown in Figure 2.3.

The viscosity, density, and surface tension of the oil determine the oil's attraction for MPs. The higher the viscosity, the better the interaction between the oil and the micro-polymer. Canola oil has been utilized to isolate MPs due to the lipophilic nature of plastics [44]. It showed 92-97% recovery rate for various samples like seawater and beach sediments. But traces of canola oil on MPs can indeed complicate the identification process. The presence of oils and other organic contaminants can obscure the physical characteristics of the MPs and interfere with analytical techniques used to identify and quantify them. This problem can be resolved by using olive oil to isolate MPs from soil. Scopteani et al. used olive oil in which the sample was mixed vigorously in a PTFE cylinder, allowed it to settle for 2 h and frozen at -40 °C [45]. From the ice column, the oil layer can be easily removed. This methodology reported a recovery rate of 90–92%, 95–97%, and 94–95% for low-, medium-, and high-density MPs, respectively.

2.5 DIGESTION METHODS

Removal of other interfering organic matter is a crucial step prior to precise identification and characterization of MPs visually or spectroscopically. It is possible that MPs are embedded in the biological tissues of the sample, making it complicated for physical methods to separate and detect them. Since biotic materials are commonly mistaken for plastics, overestimation is another significant problem in the detection of MPs. Therefore, a digestion method is required that can break down biotic materials without changing the structural and chemical integrity of MPs [46]. Nevertheless, the need of digestion depends on the quantity of organic impurities present in the sample. It is strongly advised to perform the digestion step when the identification of sample is carried out by visual inspection.

2.5.1 ACID DIGESTION

Acid digestion can be employed for the breakdown of organic matter present in the sample. Before separation, a variety of acids, such as HCl and HNO_3, are used to dissolve organic matter present in the sample. However, a few polymers such as nylon and PET are less resistant to acids and can degrade when treated at high concentration and temperatures [47]. Yet, in order to remove biological matter efficiently in a reasonable period of time, an ideal temperature and concentration need to be optimized. HNO_3 is considered to be the most effective reagent for acid digestion of organic matter present in the sample. Yet, nitric acid may leave behind oily residues or tissue fragments. Melted PS (polystyrene), LDPE (low-density polyethylene), and HDPE (high-density polyethylene) cause yellowing of polymers such as PP (polypropylene), PVC (polyvinyl chloride), and PET [48–51]. According to Naidoo et al., PVC, polyester, HDPE, PS, and PE remain stable in HNO_3 (55%) for a month at room temperature, with degradation of PVC and whitening of nylon [52]. On the other hand, HCl is not preferred for acid digestion of samples as it distorts the surface of polyethylene terephthalate and polyvinyl chloride [53]. Moreover, the outcomes

30   Microplastics

of HCl-mediated acid digestion are inconsistent and less effective at degradation of organic matter [54, 55]. Desforges demonstrated in his study that treating a sample in HNO_3 for one hour completely dissolved the zooplankton tissues, but a mixture of HNO_3 and HCl broke down zooplankton into smaller fragments [56]. Karami et al. reported a digestion efficiency > 95% by HCl at 25°C but with melting of PET [53].

2.5.2 ALKALI DIGESTION

However, several studies report that powerful acids have the ability to degrade specific polymers such as polystyrene and polyamide. Alkaline digestion is a good alternative to acid digestion with greater potential. In this technique, organic matter is broken down in the presence of strong bases such as KOH or NaOH. The effect of strong alkali has been studied to find the usefulness for separation and identification of MP residues in biological samples. Alkali digestion, however, may also leave greasy residues or damage plastics [47] or re-depositing residues on plastic surfaces, making it more complicated to characterize the material using vibrational spectroscopy [23]. In order to separate MP, some fish were digested by adding 10 M KOH solution to the samples [57]. The process of digestion took 2–3 weeks. It was highlighted that most of the plastic residues released after digestion process are resistant to KOH except cellulose acetate plastic base material. Karami et al. reported a highly efficient digestion of MPs in the presence of 10% KOH incubated for 48–72 h at 40 °C [53]. NaOH has also been used for the alkali digestion of MPs. However, NaOH was found to have limited applications due to its degradation efficiency for plastic polymers such as polyethylene terephthalate, polyvinyl chloride, polycarbonate, and cellulose acetate along with the organic matter. Thus, KOH is the preferred choice for alkali digestion of MPs samples as it is very efficient in the digestion of biological matter without altering the chemical composition of MPs.

2.5.3 ENZYMATIC DIGESTION

Since enzymes do not break down plastic polymers like chemical digestion does, they have been extensively utilized to break down biological entities. However, enzymatic digestion is a slow process, and each enzyme must be maintained at specific pH and temperature to degrade biological matter efficiently [30, 58]. The enzymes used in the digestion process are cellulose, chitinase, proteinase-K, lipase, and protease. Cole et al. reported the use of proteinase K enzyme to separate MPs from the biogenic tissues [54]. The sample was treated with enzyme at 50 °C for 2 h in the presence of sodium perchlorate ($NaClO_4$). The method showed a digestion efficiency of >97%. The digestive ability of the proteinase K enzyme was tested in the presence of $CaCl_2$ followed by further treatment with H_2O_2 to remove organic matter [59]. The rate of recovery was found to be 97% but the plastic surface was contaminated with calcium deposition. In the initial stage of digestion, a single enzyme is used, and in the later stages, a combination of enzymes is chosen to break down the organic matter. Courtene-Jones et al. reported 72--88% digestion efficiencies in the presence of trypsin, collagenase, and papain without affecting the composition of MPs [60].

Detection and Extraction Techniques

A basic enzymatic purification protocol was developed that included a series of treatment steps such as the use of detergent followed by three enzymatic treatments (protease, cellulose, and chitinase) and the use of H_2O_2. This methodology was found to have efficiency in reducing organic matter up to 98% with a recovery rate of 83%. Enzymatic digestion is preferred as it is a mild and selective treatment process, but the selection of a suitable enzyme for soil samples is a crucial step [61].

2.5.4 Hydrogen Peroxide Treatment

H_2O_2 has been widely used as a reagent for digestion of organic matter present in samples of sediment, seawater, and soil without altering the chemical composition of MPs [62]. It was reported that a concentration of 30% H_2O_2 oxidized the organic matter very efficiently within 7 days without affecting the plastic polymers. H_2O_2 makes slight changes to the texture of polymers converting them into smaller and thinner components. In a study on MPs in marine snow, Zhao et al. noted that treating organic-rich samples with 30% hydrogen peroxide frequently produces a dense foam that may suspend a sizable amount of the sample above the reagent [63]. Thus, they suggested using 15% H_2O_2 for 24 hours at 75 °C, which is said to be equally as effective as 30% H_2O_2; however, temperature-sensitive polymers may undergo some changes at such high temperatures. Tagg et al. suggested lowering the exposure period rather than the concentration to prevent the MP samples from being altered [64]. Fenton's reagent catalyzes the oxidation of organic components with H_2O_2 using ferrous cations. It was effective in isolating MP particles from wastewater and does not modify the surface of PE, PP, or PVC MPs. On the other hand, variations in the nylon fragment stock are thought to be the cause of the differences in PA particles rather than the impact of Fenton's reagent. However, the impact on weathered plastics was not tested here and may produce different outcomes than when using pristine particles. Hurley et al. carried out a series of tests using 30% H_2O_2 at 60 and 70 °C, Fenton's reagent at room temperature, 1 M NaOH, 10 M NaOH at 60 °C, and 10% KOH at 60 °C to remove organic matter from complex, organic-rich environmental matrices [65]. They discovered that oxidizing soils and sewage sludge at room temperature using Fenton's reagent has the highest rate of organic removal while not affecting any of the MP granules that were examined. Additionally, they investigated the impact of the digestion procedure in conjunction with density separation by water and NaI. The recovery rate was shown to be unaffected significantly by the sequence in which organic matter removal and density separation were performed. To exclude any effects of the highly reactive H_2O_2 on weathered MPs on environmental samples, it may be advisable to test weathered MPs instead of pristine granules. Furthermore, because of the intense exothermal reaction, caution must be taken to ensure that organic-rich samples treated with Fenton's reagent do not surpass temperatures beyond 60 °C in order to prevent thermal deterioration. Hurley et al., recommend not to let reaction temperatures exceed 40 °C. It was reported that Fenton's reagent can effectively reduce organic matter from soil samples, as this method is relatively cost and time efficient.

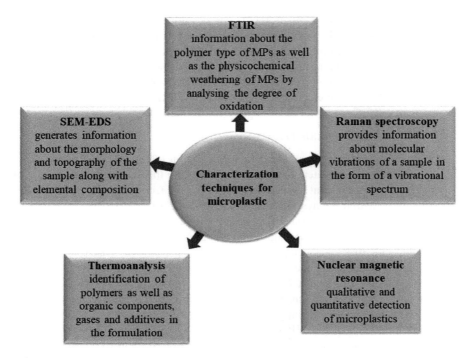

FIGURE 2.4 Characterization techniques used to detect microplastics.

2.6 ANALYTICAL TECHNIQUES FOR DETECTION OF MICROPLASTICS

After purification and digestion of samples, MPs need to be characterized and quantified. In this step, the identity of the separated particle as a true or fake MP is verified. MPs are identified based on size, form, composition, and quality. There are different techniques used for identification and characterization such as SEM-EDS, FTIR, Raman, NIR, and NMR spectroscopy (Figure 2.4). Details about the surface of MPs and additives present on it can be identified by a powerful microscope, SEM-EDS. FTIR and Raman spectroscopy are the best techniques for the chemical characterization of MPs. However, these techniques are time-consuming and have a limitation of sample size. These limitations are overcome by gas chromatography but it affects other characteristics of MPs. Hence, each characterization technique only recognizes a subset of the necessary MP characteristics.

2.6.1 SEM-EDS

Scanning Electron Microscopy (SEM) irradiates the sample with high-intensity electron beam for imaging. The signal generated by the interaction of sample with the electron beam emits secondary electrons, which are utilized to generate information about the morphology and topography of the sample. High-resolution images are

Detection and Extraction Techniques

generated by this technique. Energy Dispersive X-ray Spectroscopy (EDS) makes the SEM not only useful for imaging but also for elemental analysis [38]. This interaction between electron and sample generates characteristic X-ray photons used for elemental-specific characterization. It is well reported in the literature that EDS analysis of polystyrene, polypropylene, and polyethylene-based MPs are mostly associated with peaks corresponding to nitrogen elements which could be due to the presence of biomass. The presence of intense nitrogen in MPs signifies bioaccumulation indicating a strong interaction between biota and MPs [66]. SEM-EDS is a fast-screening technique for nonplastic vs. plastic samples and can easily identify small particles that could not be detected visually. MPs show strong carbon peaks associated with some other elemental peaks of weak intensity, while nonplastic samples do not show the presence of strong peaks for carbon elements. Cooper et al. revealed that SEM images of plastic particles confirm the embrittlement of MPs by physical and chemical weathering [67]. Weathering causes grooves, fractures, horizontal notches, pits, and vermiculate mixtures on plastic surfaces making them vulnerable to further breaking down into smaller particles. The presence of MPs in milk samples has also been identified by using SEM analysis [28]. Most of the MPs were blue in color with microholes and cracks on the surfaces.

2.6.2 FTIR

The most widely used method for the detection of MPs is Fourier Transform Infrared (FTIR) spectroscopy because it is a straightforward, dependable, nondestructive technique and generates unique band patterns using distinct infrared spectra for various types of plastic samples [66]. Depending on the molecular structure of the MPs, an infrared light source (wave number range 400–4000 cm^{-1} for Mid-IR) will either absorb and quantify radiation in the reflection or transmission mode [68]. FTIR can provide an accurate information about the polymer type of MPs as well as the physicochemical weathering of MPs by analyzing the degree of oxidation [67]. FTIR can only detect MPs upto 20 µm, but this limitation can be overcome by using micro-FTIR, which has the ability to analyze samples below 10 µm. Micro-FTIR mapping includes the sequential measurement of infrared spectra at various points on the sample surface to produce a chemical image of the sample [69]. Various studies have been conducted to analyze the MPs in the food chain. Mintenig et al. conducted a study to evaluate MPs in drinking water samples and found an average concentration of particles/m^3 of water with sizes 5–150 µm [70]. The presence of MPs in food and beverages such as honey, beer, and soft drinks has also been analyzed by FTIR. A report highlighted the presence of MPs in bottled drinking water by integrating FTIR and Nile Red stain [71]. Nile Red produced fluorescence at specific wavelengths after absorption on the MP surface. ATR-FTIR is another reliable and fast technique for the characterization of MPs isolated from water and biological samples. Jung et al. reported the presence of MPs of size <50 µm in beach sand samples by using ATR-FTIR [72]. It was concluded that when the size of MPs decreases, the concentration of MPs in the sample increases, increasing the effectiveness of MP detection. ATR-FTIR was able to distinguish between low-density polyethylene and high-density polyethylene, which is difficult to distinguish among other types of MPs consumed by marine organisms. In addition, the ATR mode generates stable spectra from irregular surfaces of MPs. However, ATR has some limitations. It can damage the highly weathered and fragile MPs due to high pressure produced by the probe and MP particles could adhere to the probe tip due to electrostatic force.

2.6.3 RAMAN SPECTROSCOPY

Raman spectroscopy is another widely used technique for identification of MPs. It is a vibration spectroscopy technique that relies on inelastic scattering of light and provides information about molecular vibrations of a sample in the form of a vibrational spectrum. Similar to a fingerprint of chemical structure, the Raman spectrum enables the identification of the components contained in the sample [73]. High sensitivity to nonpolar functional groups, narrow spectral bands, and low interference due to water make this technique more beneficial than FTIR spectroscopy [74]. Raman spectroscopy has the efficiency to detect MPs as small as 1 μm and provides chemical and structural characteristics that can not be found by other spectroscopic techniques [75]. It is observed by Raman microspectroscopy that the materials used in packaging release MPs into the water [77]. MPs of size larger than 5 μm were automatically detected by a newly designed single-particle explorer software. The analysis provides a spectral map corresponding to each type of MP present in the sample. The spectral map obtained from the sample was compared with the available library for improved point mapping. There are two major drawbacks of classical Raman scattering for MP analysis: (1) inherent weakness of the signal and (2) fluorescence interference due to the presence of impurities such as biological components, degradation products, and coloring agents. It is necessary either to extend the integration time or to increase the duration of sample measurement to enhance the weak signal, which results in laser-induced degradation of the sample. Fluorescence causes a rise in baseline, which overshadows the Raman signal. To resolve the issue of fluorescence interference, it is suggested to use an appropriate cleaning procedure to remove the contaminants. There are several methods proposed for the pretreatment of MPs such as employing acids, bases, and oxidative and enzymatic agents [30, 76]. An automated algorithm was utilized by Ghosal et al. to remove the fluorescence background from Raman spectroscopy [77]. Polymer peaks obtained after subtracting the background fluorescence can be matched with the help of library matching software, which can provide information regarding chemical composition and morphology of MPs.

2.6.4 NUCLEAR MAGNETIC RESONANCE

^{1}H NMR spectroscopy can be employed for qualitative and quantitative detection of MPs. There are only a few reports to date that have used NMR for quantitative analysis of MPs. Quantitative analysis by NMR is based on the principle that the integral value of a signal is proportional to the number of protons that give rise to a particular resonance. Dissolution of MPs in a deuterated solvent is required for analysis in NMR spectroscopy. There are various determination methods available for MP analysis by NMR spectroscopy: relative determination, absolute determination by using internal or external standards, and calibration curve method [78]. The calibration curve method is most suitable for MP analysis in environmental samples because the exact composition

Detection and Extraction Techniques

of the sample is not required. Peez et al. reported the use of the calibration curve method for quantitative analysis of polyethylene, poly terephthalate, and polystyrene-based MPs in analytes [79]. For validation of the method, several parameters been determined linearity, limit of quantification, stability, precision, and accuracy.

2.6.5 Thermo-analytical Methods

The thermal degradation of MPs is an extensively used technique for the identification of polymers as well as organic components, gases, and additives in the formulation. Thermo analysis of the sample is carried out in a single run without the use of any solvent. The technique generally works on pyrolysis products generated at specific temperatures in the absence of oxygen. These products are quantitatively analyzed by gas chromatography (GC) and mass spectrometry (MS). This technique is widely used for identification of MP contamination in the food industry due to its reliability, accuracy, and cost-effectiveness. Watteau et. al. reported the presence of MPs in soil planted with wheat by using GC-MS and Transmission Electron Microscopy (TEM) [80]. The results showed MPs of sizes <200 µm along with organic components after pyrolysis treatment at 650 °C with 30 s holding time. Apart from MPs, barium, titanium, and other organic minerals were also detected in the sample. This technique is highly sensitive and can identify traces of MPs in the sample. The shape, size, and color of MPs are considered to be a huge barrier in their quantitative estimation. Additionally, this technique detects MPs in the millimeter range, which makes it a low-resolution and less-efficient technique.

2.7 CONCLUSION

The inclusion of MPs in various food commodities has become a huge global concern. Even though MPs are one of the well-recognized pollutants, information regarding their separation and methods of detection is still lacking. This chapter focused on different extraction and characterization techniques used for MPs in various food and marine systems. Extraction and detection of MPs is a crucial step due to their small size. Pre-extraction methods such as visual sorting are prone to error because MP resembles other organic matter. It is necessary to develop low-cost and high-efficiency methods for the detection and extraction of MPs. Techniques like floatation, membrane separation, chemical and enzymatic treatments, and extraction have been highlighted in this chapter. Identification techniques such as SEM-EDS, Raman spectroscopy, FTIR, and thermo-analytical methods were also discussed. Even though some basic research has been conducted, more investigation has to be done to design fast, convenient methods for extraction and identification of the type of MPs and on their implications on human health.

REFERENCES

[1] K. F. Drain, W. R. Murphy, and M. S. Otterburn, "Polymer waste – resource recovery," *Conserv. Recycl.,* vol. 4, pp. 201–218, 1981, doi: https://doi.org/10.1016/0361-3658(81)90025-4

[2] A. Dey, C. V. Dhumal, P. Sengupta, A. Kumar, N. K. Paramanik, and T. Alam, "Challenges and possible solutions to mitigate the problems of single-use plastics used for packaging food items: A review," *J. Food Sci. Technol.*, vol. 58, no. 9, pp. 3251–3269, 2021, doi: 10.1007/s13197-020-04885-6.

[3] A. Sivan, "New perspectives in plastic biodegradation," *Curr. Opin. Biotechnol.*, vol. 22, no. 3, pp. 422–426, 2011, doi: 10.1016/j.copbio.2011.01.013.

[4] M. Simon, N. Van Alst, and J. Vollertsen, "Quantification of microplastic mass and removal rates at wastewater treatment plants applying Focal Plane Array (FPA) -based Fourier Transform Infrared (FT-IR) imaging," *Water Res.*, vol. 142, pp. 1–9, 2018, doi: 10.1016/j.watres.2018.05.019.

[5] A. L. Andrady, "Microplastics in the marine environment,"*Mar. Pollut. Bull.*, vol. 62, no. 8, pp. 1596–1605, 2011, doi: 10.1016/j.marpolbul.2011.05.030.

[6] V. Hidalgo-ruz, L. Gutow, R. C. Thompson, and M. Thiel, "Microplastics in the marine environment: A review of the methods used for identification and quantification," *Environ. Sci. Technol.*, vol. 46, no. 6, pp. 3060–3075, 2012, doi: 10.1021/es2031505.

[7] H. Zhang, "Estuarine, Coastal and Shelf Science Transport of microplastics in coastal seas," *Estuar. Coast. Shelf Sci.*, vol. 199, pp. 74–86, 2017, doi: 10.1016/j.ecss.2017.09.032.

[8] Y. Qi, X. Yang, A. M. Pelaez, E. H. Lwanga, N. Beriot, H. Gertsen, P. Garbeva, and V. Geissen, "Science of the Total Environment Macro- and micro- plastics in soil-plant system: Effects of plastic mulch fi lm residues on wheat (Triticum aestivum) growth," *Sci. Total Environ.*, vol. 645, pp. 1048–1056, 2018, doi: 10.1016/j.scitotenv.2018.07.229.

[9] L. M. Hernandez, E. G. Xu, H. C. E. Larsson, R. Tahara, V. B. Maisuria, and N. Tufenkji, "Plastic teabags release billions of microparticles and nanoparticles into tea," *Environ. Sci. Technol.*, vol. 53, pp. 12300–12310, 2019, doi: 10.1021/acs.est.9b02540.

[10] G. Oliveri, M. Ferrante, M. Banni, C. Favara, I. Nicolosi, A. Cristaldi, M. Fiore, and P. Zuccarello, "Micro- and nano-plastics in edible fruit and vegetables. The first diet risks assessment for the general population," *Environ. Res.*, vol. 187, pp. 1–7, 2020, doi: 10.1016/j.envres.2020.109677.

[11] O. O. Fadare, B. Wan, L. Guo, and L. Zhao, "Chemosphere Microplastics from consumer plastic food containers: Are we consuming it?," *Chemosphere*, vol. 253, pp. 1–6, 2020, doi: 10.1016/j.chemosphere.2020.126787.

[12] V. Stock, C. Fahrenson, A. Thuenemann, M.H. Dönmez, L.Voss, L. Böhmert, A. Braeuning, A. Lampen, and H. Sieg, "Impact of artificial digestion on the sizes and shapes of microplastic particles," *Food Chem. Toxicol.*, 135, pp. 1–25, 2019, doi: 10.1016/j.fct.2019.111010.

[13] E. J. Carpenter, and K. L. Smith, "Plastics on the Sargasso Sea surface," *Science*, vol. 175, no. 4027, pp. 1240–1241, 1972, DOI: 10.1126/science.175.4027.1240.

[14] K. Ugwu, A. Herrera, and M. Gomez, "Microplastics in marine biota: A review," *Mar. Poll. Bull.*, vol. 169, pp. 112540–112550, 2021, doi: 10.1016/j.marpolbul.2021.112540.

[15] D. Gola, P. K. Tyagi, A. Arya, N. Chauhan, M. Agarwal, S. K. Singh, and S. Gola, "The impact microplastics on marine environment: A review," *Environmental Nanotechnology, Monitoring & Management*, vol. 16, pp. 100552–100557, doi: 10.1016/j.enmm.2021.100552.

[16] J. C. Anderson, B. J. Park, and V. P. Palace, "Microplastics in aquatic environments: Implications for Canadian ecosystem," *Environ. Pollut.*, vol. 218, pp. 269–280, 2016, doi: 10.1016/j.envpol.2016.06.074.

Detection and Extraction Techniques

[17] M. Smith, D. C. Love, C. M. Rochman, and R. A. Neff, "Microplastics in seafood and the implications for human health," *Curr. Environ. Health Rep.*, vol. 5, pp. 375–386, 2018, doi: 10.1007/s40572-018-0206-z.

[18] J. C. Prata, J. P. da Costa, I. Lopes, A. C. Duarte, and T. Rocha-Santos, "Environmental exposure to microplastics: An overview on possible human health effects," *Sci. Total Environ.*, vol. 702, pp. 134455–134486, 2020, doi: 10.1016/j.scitotenv.2019.134455.

[19] M. Xu, G. Halimu, Q. Zhang, Y. Song, X. Fu, and Y. Li, "Science of the Total Environment Internalization and toxicity: A preliminary study of effects of nanoplastic particles on human lung epithelial cell," *Sci. Total Environ.*, vol. 694, pp. 1–10, 2019, doi: 10.1016/j.scitotenv.2019.133794.

[20] R. C. Thompson, Y. Olsen, R. P. Mitchell, A. Davis, S. J. Rowland, G. John, D. Mcgonigle, and A.E. Russell., "Lost at sea: Where is all the plastic?," *Science*, vol. 304, no. 5672, pp. 838–838, 2004, doi: 10.1126/science.1094559.

[21] C. Alvarez-hernandez, C. Cairos, J. L. Darias, E. Mazzetti, C. H. Sanchez, J. G. Salamo, and J. H. Borges., "Microplastic debris in beaches of Tenerife (Canary Islands, Spain)," *Mar. Poll. Bull.*, vol. 146, pp. 26–32, 2019, doi: 10.1016/j.marpolbul.2019.05.064.

[22] M. Tiwari, T. D. Rathod, P. Y. Ajmal, R. C. Bhangare, and S. K. Sahu, "Distribution and characterization of microplastics in beach sand from three different Indian coastal environments," *Mar. Pollut. Bull.*, vol. 140, pp. 262–273, 2019, doi: 10.1016/j.marpolbul.2019.01.055.

[23] J. Wagner, Z. Wang, S. Ghosal, and C. Rochman, "Analytical methods novel method for the extraction and identification of microplastics in ocean trawl and fish gut," *Anal. Methods,* vol. 9, no. 9, pp. 1479–1490, 2017, doi: 10.1039/C6AY02396G.

[24] Z. Wang, J. Wagner, S. Ghosal, G. Bedi, and S. Wall, "Science of the Total Environment SEM/EDS and optical microscopy analyses of microplastics in ocean trawl and fish guts," *Sci. Total Environ.*, vol. 1, pp. 616–626, 2017, doi: 10.1016/j.scitotenv.2017.06.047.

[25] H. B. Jayasiri, C. S. Purushothaman, and A. Vennila, "Quantitative analysis of plastic debris on recreational beaches in Mumbai, India," *Mar. Pollut. Bull.*, vol. 77, no. 1–2, pp. 107–-112, 2013, doi: 10.1016/j.marpolbul.2013.10.024.

[26] A. Sridhar, M. Ponnuchamy, P. Senthil, K. Ashish, K. Dai, and V. N. Vo, "Techniques and modeling of polyphenol extraction from food: A review," *Environ. Chem. Lett.*, vol. 19, pp. 3409–3443, 2021, doi: 10.1007/s10311-021-01217-8.

[27] A. Sridhar, D. Kannan, A. Kapoor, and S. Prabhakar, "Extraction and detection methods of microplastics in food and marine systems: A critical review," *Chemosphere,* vol. 286, pp. 131653, 2022, doi: 10.1016/j.chemosphere.2021.131653.

[28] G. Kutralam-muniasamy, F. Pérez-guevara, I. Elizalde-martínez, and V. C. Shruti, "Branded milks – Are they immune from microplastics contamination?," *Sci. Total Environ.*, vol. 714, pp. 1–10, 2020, doi: 10.1016/j.scitotenv.2020.136823.

[29] Q. Li, J. Wu, X. Zhao, X. Gu, and R. Ji, "Separation and identification of microplastics from soil and sewage," *Environ. Pollut.*, vol. 254, pp. 1–9, 2019, doi: 10.1016/j.envpol.2019.113076.

[30] A. L. Lusher, N. A. Welden, P. Sobral, and M. Cole, "Sampling, isolating and identifying microplastics ingested by fish and invertebrates," *Anal. Methods*, vol. 9, pp. 1346–1360, 2017, doi: 10.1039/C6AY02415G.

[31] B. Beckingham, A. Apintiloaiei, C. Moore, and J. Brandes, " Hot or not: Systematic review and laboratory evaluation of the hor needle test for microscopic identification," *Microplast. Nanopast.*, vol 3, no. 8, pp. 1–13. 2023, doi: 10.1186/s43591-023-00056-4.

[32] G. Erni-cassola, M. I. Gibson, and R. C. Thompson, "Lost, but found with Nile red; a novel method to detect and quantify small microplastics (20 μm – 1 mm) in environmental samples," *Environ. Sci. Technol.*, vol. 51, no. 23, pp. 13641–13648, 2017, doi: 10.1021/acs.est.7b04512.

[33] F. Zhou, X. Wang, G. Wang, and Y. Zuo, "A rapid method for detecting microplastics based on Fluorescence Lifetime Imaging Technology (FLIM)," *Toxics*, vol. 10, pp. 118–135, 2022, doi: 10.3390/ toxics10030118.

[34] B. Quinn, F. Murphy, and C. Ewins, " Validation of density separation for the rapid recovery of microplastics from sediment," *Anal. Methods*, vol. 9, no. 9, pp. 1491–1498, 2016, doi: 10.1039/C6AY02542K.

[35] M. Nuelle, J. H. Dekiff, D. Remy, and E. Fries, "A new analytical approach for monitoring microplastics in marine sediments," *Environ. Pollut.*, vol. 184, pp. 161–169, 2014, doi: 10.1016/j.envpol.2013.07.027.

[36] X. Han, X. Lu, and R. D. Vogt, "An optimized density-based approach for extracting microplastics from soil and sediment samples," *Environ. Pollut.*, vol. 254, pp. 1–7, 2019, doi: 10.1016/j.envpol.2019.113009.

[37] M. Scheurer, and M. Bigalke, "Microplastics in Swiss floodplain soils," *Environ. Sci. Technol.*, vol. 52, no. 6, pp. 3591–3598, 2018, doi: 10.1021/acs.est.7b06003.

[38] A. Herrera, P. Garrido-amador, I. Martínez, M.D. Samper, J. Lopez-martínez, M. Gomez, and T. T. Packard, "Novel methodology to isolate microplastics from vegetal-rich samples," *Mar. Pollut. Bull.*, vol. 129, pp. 61--69, 2018, doi:10.1016/j.marpolbul.2018.02.015.

[39] S. Felsing, C. Kochleus, S. Buchinger, N. Brennholt, F. Stock, and G. Reifferscheid, "A new approach in separating microplastics from environmental samples based on their electrostatic behavior," *Environ. Pollut.* vol. 234, pp. 20–28, 2018, doi: 10.1016/j. envpol.2017.11.013.

[40] J. Grbic, B. Nguyen, E. Guo, J. B. You, D. Sinton, and C. M. Rochman, "Magnetic extraction of microplastics from environmental samples," *Environ. Sci. Technol. Lett.*, vol. 6, pp. 68–72, 2019, doi: 10.1021/acs.estlett.8b00671.

[41] N. Fraunholcz, "Separation of waste plastics by froth flotation — a review, part I," *Miner. Eng.*, vol. 17, pp. 261–268, 2004, doi: 10.1016/j.mineng.2003.10.028.

[42] W. Fu, J. Min, W. Jiang, Y. Li, and W. Zhang, "Science of the Total Environment Separation, characterization and identification of microplastics and nanoplastics in the environment," *Sci. Total Environ.*, vol. 721, p. 1–26, 2020, doi: 10.1016/j.scitotenv.2020.137561.

[43] L. Huang, H. Wang, C. Wang, J. Zhao, and B. Zhang, "Microwave-assisted surface modification for the separation of polycarbonate from polymethylmethacrylate and polyvinyl chloride waste plastics by flotation," *Manage. Res.*, vol. 35, no. 3, pp. 294–300, 2017, doi: 10.1177/0734242X16682078.

[44] E.M. Crichton, M. Noël, E.A. Gies, and P.S. Ross, " "A novel, density-independent and FTIR-compatible approach for the rapid extraction of microplastics from aquatic sediments" *Anal. Methods*, vol. 9, no. 9, pp. 1419–1428, 2017, doi: 10.1039/C6AY02733D.

[45] C. Scopetani, D. Chelazzi, J. Mikola, V. Leiniö, R. Heikkinen, and A. Cincinelli, "Science of the Total Environment Olive oil-based method for the extraction, quantification and identification of microplastics in soil and compost samples," *Sci. Total Environ.*, vol. 733, pp. 139338–139344, 2020, doi: 10.1016/j.scitotenv.2020.139338.

[46] C. Junhao, Z. Xining, G. Xiaodong, Z. Li, H. Qi, H. Kadambot, and M. Siddique, "Extraction and identification methods of microplastics and nanoplastics in agricultural

soil: A review," *J. Environ. Manage.*, vol. 294, pp. 112997–113009, 2021, doi: 10.1016/j.jenvman.2021.112997.

[47] Q. Qiu, Z. Tan, J. Wang, J. Peng, M. Li, and Z. Zhan, "Extraction, enumeration and identification methods for monitoring microplastics in the environment," *Estuar. Coast. Shelf Sci.*, vol. 176, pp. 102–109, 2016, doi: 10.1016/j.ecss.2016.04.012.

[48] E. M. Tuuri, J. R. Gascookem, and S. C. Leterme, "Efficacy of chemical digestion methods to reveal undamaged microplastics from planktonic samples," *Sci. Total Environ.*, vol. 947, pp. 174279, 2024, doi: 10.1016/j.scitotenv.2024.174279.

[49] M. Claessens, L. Van Cauwenberghe, M. B. Vandegehuchte, and C. R. Janssen, "New techniques for the detection of microplastics in sediments and field collected organisms," *Mar. Pollut. Bull.*, vol. 70, no. 1–2, pp. 227–233, 2013, doi: 10.1016/j.marpolbul.2013.03.009.

[50] A. Dehaut, A. Cassone, L. Frere, L. Hermabessiere, C. Himber, E. Rinnert, G. Riviere, C. Lambert, P. Soudant, A. Huvet, G. Duflos, and I. Paul-Pont, "Microplastics in seafood: Benchmark protocol for their extraction and," *Environ. Poll.*, vol. 215, pp. 223–233, 2016, doi: 10.1016/j.envpol.2016.05.018.

[51] T. Maes, R. Jessop, N. Wellner, K. Haupt, and A. G. Mayes, "A rapid-screening approach to detect and quantify microplastics based on fluorescent tagging with Nile Red," *Sci. Rep.*, vol. 7, no. 1, pp. 1–10, 2017, doi: 10.1038/srep44501.

[52] T. Naidoo, K. Goordiyal, and D. Glassom, "Are Nitric Acid (HNO_3) digestions efficient in isolating microplastics from Juvenile Fish?," *Water Air Soil Poll.*, vol. 228, pp. 1–11. 2017, doi: 10.1007/s11270-017-3654-4.

[53] A. Karami, A. Golieskardi, C. Keong, N. Romano, Y. Bin, and B. Salamatinia, "Science of the Total Environment: A high-performance protocol for extraction of microplastics in fish," *Sci. Total Environ.*, vol. 578, pp. 485–494, 2016, doi: 10.1016/j.scitotenv.2016.10.213.

[54] M. Cole, H. Webb, P. K. Lindeque, E. S. Fileman, C. Halsband, and T. S. Galloway, "Seawater samples and marine organisms," *Sci. Rep.*, vol. 4, no. 1, pp. 1–8, 2014, doi: 10.1038/srep04528.

[55] E. M. Foekema, C. De Gruijter, M. T. Mergia, J. A. Van Franeker, A. J. Murk, and A. A. Koelmans, "Plastic in north sea fish," *Environ. Sci. Technol.*, vol. 47, no. 15, pp. 8818–8824, 2013, doi: 10.1021/es400931b.

[56] J. P. W. Desforges, M. Galbraith, and P. S. Ross, "Ingestion of microplastics by Zooplankton in the Northeast Pacific Ocean," *Arch. Environ. Contam. Toxicol.*, vol. 69, pp. 320–330, 2015, doi: 10.1007/s00244-015-0172-5.

[57] S. Kühn, B. Van Werven, A. Van Oyen, A. Meijboom, E. L. Bravo, and J. A. Van Franeker, "The use of potassium hydroxide (KOH) solution as a suitable approach to isolate plastics ingested by marine organisms," *Mar. Poll. Bull.*, vol. 115, no. 1–2, pp. 86–90, 2016, doi: 10.1016/j.marpolbul.2016.11.034.

[58] O. Mbachu, G. Jenkins, C. Pratt, and P. Kaparaju, "Enzymatic purification of microplastics in soil" *MethodsX*, vol. 8, pp. 101254–101265, 2021, doi: 10.1016/j.mex.2021.101254.

[59] T. M. Karlssona, A. D. Vethaaka, B. C. Almrothd, F. Ariesee, M. Velzena, M. Hasselllövb, and H. A. Lesliea, "Screening for microplastics in sediment, water, marine invertebrates and fish: Method development and microplastic accumulation," *Mar. Pollut. Bull.*, vol. 122, no. 1-2, pp. 0–1, 2017, doi: 10.1016/j.marpolbul.2017.06.081.

[60] W. Courtene-jones, B. Quinn, F. Murphy, and F. Gary, "Analytical Methods Optimisation of enzymatic digestion and validation of specimen preservation methods for the analysis of ingested microplastics," *Anal. Methods*, vol. 9, pp. 1437–1445, 2017, doi: 10.1039/C6AY02343F.

[61] M. G. J. Lo Der, H. K. Imhof, M. Ladehoff, L. A. Löschel, C. Lorenz, S. Mintenig, S. Piehl, S. Primpke, I. Schrank, C. Laforsch, and G. Gerdts, "Enzymatic purification of microplastics in environmental samples," *Environ. Sci Technol.*, vol. 51, no. 24, pp. 14283–14292, 2017, doi: 10.1021/acs.est.7b03055.

[62] M. Kumar, X. Xiong, M. He, D. C. W. Tsang, J. Gupta, E. Khan, S. Harrad, D. Hou, Y. Sik Ok, and N. S. Bolan, "Microplastics as pollutants in agricultural soils," *Environ. Pollut.*, vol. 265, p. 1–11, 2020, doi: 10.1016/j.envpol.2020.114980.

[63] S. Zhao, M. Danley, J. E. Ward, D. Li, and T. J. Mincer, "An approach for extraction, characterization and quantitation of microplastic in natural marine snow using Raman microscopy," *Anal. Methods*, vol. 9, no. 9, pp. 1–9, 2016, doi: 10.1039/C6AY02302A.

[64] A. S. Tagg, J. P. Harrison, Y. Ju-nam, M. Sapp, E. L. Bradley, and C. J. Sinclair, "Fenton's reagent for the rapid and efficient isolation of microplastics from wastewater," *Chem. Commun.*, vol. 53, no. 2, pp. 372–375, 2017, doi: 10.1039/c6cc08798a.

[65] R. R. Hurley, A. L. Lusher, M. Olsen, and L. Nizzetto, "Validation of a method for extracting microplastics from complex, organic-rich, environmental matrices," *Environ. Sci. Technol.*, vol. 52, pp. 7409–7417, 2018, doi: 10.1021/acs.est.8b01517.

[66] R. M. Blair, S. Waldron, and C. Gauchotte-lindsay, "Microscopy and elemental analysis characterisation of microplastics in sediment of a freshwater urban river in Scotland, UK," *Environ. Sci. Poll. Res.*, vol. 26, pp. 12491–12504, 2019, doi: 10.1007/s11356-019-04678-1.

[67] D. A. Cooper, and P. L. Corcoran, "Effects of mechanical and chemical processes on the degradation of plastic beach debris on the island of Kauai, Hawaii," *Mar. Pollut. Bull.*, vol. 60, no. 5, pp. 650–654, 2010, doi: 10.1016/j.marpolbul.2009.12.026.

[68] A. Käppler, M. Fischer, B. M. Scholz-Böttcher, S. Oberbeckmann, M. Labrenz, D. Fischer, K. J. Eichhorn, and B.Voit, "Comparison of μ -ATR-FTIR spectroscopy and py-GCMS as identification tools for microplastic particles and fibers isolated from river sediments," *Anal. Bioanal. Chem.*, vol. 410, pp. 5313–5327, 2018, doi: 10.1007/s00216-018-1185-5.

[69] I. W. Levin, and R. Bhargava, "Fourier transform infrared vibrational spectroscopic imaging: Integrating microscopy," *Annu. Rev. Phys. Chem.*, vol. 56, no. 1, pp. 429–474, 2005, doi: 10.1146/annurev.physchem.56.092503.141205.

[70] S. M. Mintenig, M. G. J. Löder, S. Primpke, and G. Gerdts, "Science of the Total Environment low numbers of microplastics detected in drinking water from ground water sources," *Sci. Total Environ.*, vol. 648, pp. 631–635, 2019, doi: 10.1016/j.scitotenv.2018.08.178.

[71] S. A. Mason, V. G. Welch, and J. Neratko, "Synthetic polymer contamination in bottled water," *Front. Chem.,* vol. 6, pp. 1–11, 2018, doi: 10.3389/fchem.2018.00407.

[72] M. R. Junga, F. D. Horgena, S. V. Orski, V. Rodriguez, K. L. Beers, G. H. Balazs, T. T. Jones, T. M. Work, K. C. Brignac, S. J. Royer, K. D. Hyrenbach, B. A. Jensen, and J. M. Lynch, "Validation of ATR FT-IR to identify polymers of plastic marine debris, including those ingested by marine organisms," *Mar. Pollut. Bull.*, vol. 127, pp. 704–716, 2018, doi: 10.1016/j.marpolbul.2017.12.061.

[73] C. F. Araujo, M. M. Nolasco, A. M. P. Ribeiro, and P. J. A. Ribeiro-claro, "Identification of microplastics using Raman spectroscopy: Latest developments and future prospects," *Water Res.*, vol. 142, pp. 426–440, 2018, doi: 10.1016/j.watres.2018.05.060.

[74] A. M. Elert, R. Becker, E. Duemichen, P. Eisentraut, J. Falkenhagen, H. Sturm, and U. Braun, "Comparison of different methods for MP detection: What can we learn from them, and why asking the right question before measurements matters? *," *Environ. Pollut.*, 231, pp. 1–9, 2017, doi: 10.1016/j.envpol.2017.08.074.

[75] I. Gambino, C. Malitesta, F. Bagordo, T. Grassi, A. Panico, S. Fraissinet, A. De Donno, and G. E. De Benedetto, "Characterization of microplastics in water bottled in different packaging by Raman spectroscopy" *Environ. Sci. Water Res. Technol.*, vol. 9, pp. 3391–3397, 2023, doi: 1 10.1039/D3EW00197K

[76] M. Bläsing, and W. Amelung, "Science of the Total Environment plastics in soil: Analytical methods and possible sources," *Sci. Total Environ.*, vol. 612, pp. 422–435, 2018, doi: 10.1016/j.scitotenv.2017.08.086.

[77] S. Ghosal, M. Chen, J. Wagner, Z. Wang, and S. Wall, "Molecular identification of polymers and anthropogenic particles extracted from oceanic water and fish stomach: A Raman micro- spectroscopy study," *Environ. Pollut.*, vol. 233, pp. 1–12, 2017, doi: 10.1016/j.envpol.2017.10.014.

[78] S. K. Bharti, and R. Roy, "Quantitative H NMR spectroscopy," *Trends Anal. Chem.*, vol. 35, pp. 5–26, 2012, doi: 10.1016/j.trac.2012.02.007.

[79] N. Peez, J. Becker, S. M. Ehlers, M. Fritz, C. B. Fischer, J. H. Koop, C. Winkelmann, and W. Imhof, "Correction to: Quantitative analysis of PET microplastics in environmental model samples using quantitative 1 H-NMR spectroscopy: Validation of an optimized and consistent sample clean-up method," *Anal. Bioanal. Chem.*, vol. 411, pp. 7409–7418, 2019, doi: 10.1007/s00216-019-02089-2.

[80] F. Watteau, M. Dignac, A. Bouchard, and A. Revallier, "Microplastic detection in soil amended with municipal solid waste composts as revealed by Transmission Electronic Microscopy and Pyrolysis/GC/MS," *Front. Sustain. Food Syst.*, vol. 2, pp. 1–14, 2018, doi: 10.3389/fsufs.2018.00081.

3 Adsorptive Techniques for the Remediation of Microplastics

Manisha Bhardwaj, Anjana Bhardwaj,
Sukhdev Prajapati, Rajendra Vishwakarma,
Jaya Dwivedi, and Swapnil Sharma

3.1 INTRODUCTION

Plastic waste, particularly microplastics (MPs), has a pervasive environmental challenge, infiltrating aquatic ecosystems worldwide and posing a threat to both public health and the environment [1]. MPs are microscopic particles of plastic materials measuring less than 5mm in size. They can originate from a variety of sources, including textile synthetic fibers, microbeads from personal care products, and the fragmentation of larger plastic objects [2]. Four different sizes of plastic waste can be found in environments: macroplastics, which are larger than 25 mm; mesoplastics, which are between 5 and 25 mm; MPs, which are between 100 and 5 mm; and nanoplastics, which are smaller than 100 nm [3, 4]. Plastic output surged drastically to 348 million tons in 2017, and predictions show that by 2050, there will be almost 12,000 million tons of plastic trash globally [5]. Since understanding how MPs interact with enzymes within organisms is crucial for assessing the broader effects of MP contamination on ecosystems and human well-being, researchers are actively investigating the biological effects of MPs, particularly their influence on enzyme function [6].

Joanne and Gopinath *et al.* (2021) investigated the effects of MPs on inflammation using sophisticated *in vitro* models of the stomach and lungs, as well as MPs variations such as polyethylene and polystyrene. The results showed that certain MNP (micronanoplastic) caused inflammation, lowered tissue function, and damaged epithelial barriers with SEM images and anti-oxidant enzyme activity displayed in Figure 3.1 and Figure 3.2, respectively [7, 8].

The utilization of adsorptive strategies has appeared to guarantee the evacuation of MPs from different natural components. These strategies such as filtration, coagulation, adsorption biodegradation, chemical corruption, and oxidation take after the adsorption standards, which start the adsorption of MPs. By utilizing this strategy, MPs can be successfully removed from soils and water bodies, diminishing the conceivable negative impacts they may have on the economy and human condition.

42   DOI: 10.1201/9781003486947-3

Adsorptive Techniques 43

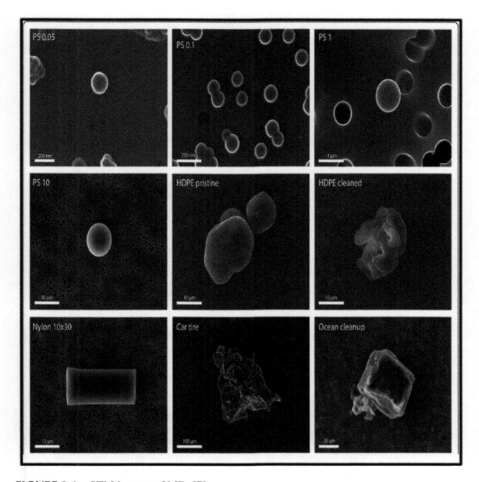

FIGURE 3.1 SEM images of MPs [7].

Consequently, there is a crucial need to develop technologies for removing MPs from the environment [9, 10]. This chapter highlights sources, environmental impact, and removal techniques for the elimination of MPs.

3.2 MPS: SOURCES AND ENVIRONMENTAL IMPACT

3.2.1 Sources

Ecosystems are exposed to MPs through a variety of sources and mechanisms. These include agricultural runoff, emissions from sewage and industrial sites, and the inclusion of MPs in goods like personal hygiene products, etc. Secondary MPs are formed when bigger plastic items, like tyres and synthetic textiles, break down primary MPs

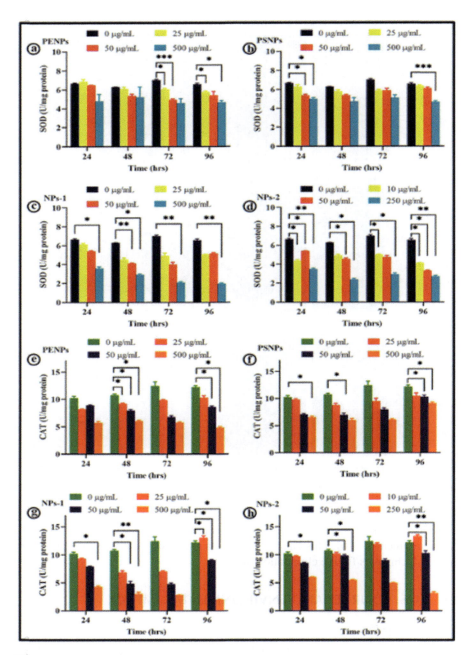

FIGURE 3.2 Anti-oxidant enzyme activity of MPs [8].

Adsorptive Techniques

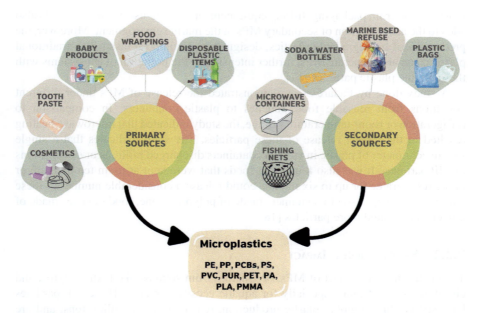

FIGURE 3.3 Sources of MPs.

[11, 12]. Plastics and MPs are released into the environment continuously through freshwater, marine, and terrestrial ecosystems.

MPs can be divided into primary and secondary sources, each playing a distinct role in environmental pollution, as shown in Figure 3.3.

3.2.1.1 Primary Source

Primary MPs originate from household and industrial products like personal care items, cosmetics, medication delivery systems, etc. These MPs are intentionally produced for use in products such as exfoliants and resin pellets [13]. They enter the environment through various pathways, including effluent discharge from wastewater treatment plants, significantly contributing to MP contamination in aquatic ecosystems [14]. Primary MPs vary in size, shape, and composition based on the products they are used in, and they are dispersed through mechanisms like wind and ocean currents. Additionally, city dust, tyre dust, and road markings are notable primary sources of MPs, further increasing their presence in the environment [15].

3.2.1.2 Secondary Source

Secondary MPs are generated when larger plastic items degrade over time due to mechanical processes or exposure to UV light. This degradation causes the release of MPs into the environment. Land-based sources, such as beaches undergoing weathering, are significant contributors to the production of secondary MPs [16].

Plastic waste from packaging, fishing equipment, and single-use consumer goods also adds to the accumulation of secondary MPs in the marine environment. Moreover, the production of biodegradable plastics, designed to decompose faster than traditional plastics, can also generate MPs, further intensifying the pollution of ecosystems with these minute plastic particles [17].

A study done by Fadare *et al.* demonstrated the release of MPs under different conditions from reusable food pouches to plastic containers. In comparison to refrigeration or room-temperature storage, the study indicated that microwave heating resulted in the largest release of these particles. The study indicates that a single square centimeter of plastic in various containers discharged millions or even billions of MP particles. It was also found that foods that were kept at room temperature or in the refrigerator for up to six months could release a considerable number of these particles. In comparison to containers made of polypropylene, food pouches made of polyethene emitted more particles [18].

3.2.2 ENVIRONMENTAL IMPACT

The environmental impact of MPs, originating from sources like textiles, tyres, and city dust, is significant, especially for aquatic ecosystems [19]. These MP particles have surged due to global plastic production, reaching 359.0 million tons, and are projected to hit 500.0 million tons by 2025 [20]. In 2013, China was the top plastic producer with 114.0 million tons. With about 50.0 million tons produced, the European Union surpassed China to become the second-largest plastic-producing region. North America produced 49.0 million tons of plastic, which was a significant contribution as well. The Mediterranean Sea region has become one of the most plastic- and MP-polluted regions in the world, even though it is acknowledged as a crucial resource for human survival. It is predicted that these concentrations will quadruple in the upcoming years in the absence of sufficient intervention [21]. Significant environmental effects result from the spread of these microscopic particles, especially in aquatic ecosystems. Toxic organic and inorganic pollutants, including heavy metals and persistent organic pollutants (POPs), can build up in MPs and be released. When primary or secondary MPs are introduced into terrestrial and aquatic environments, the life cycle of MPs usually starts. This cycle demonstrates the widespread impact of MP pollution and the urgent need for practical approaches to reduce the adverse impacts on marine ecosystems [22].

3.3 ADSORBENT MATERIALS

The efficiency of various adsorbent materials in removing MPs from the environment has been studied. Typical adsorbents consist of the following.

3.3.1 ACTIVATED CARBON

Activated carbon is useful for adsorbing organic pollutants, such as MPs, from water and sediment because of its large surface area and pore structure [23]. A study done by Xing *et al.* (2023) demonstrates activated carbon (modified by zinc chloride) can adsorb polystyrene MPs and also observed several interactions, including π-π, n-π, electrostatic, and pore-filling interactions. The n-π and π-π interactions did not prove to be decisive, even though all interactions were significant. Furthermore, adsorption capacity was found to be enhanced by common anions in water, though lake water experienced a decrease in this regard. The study also showed that high-temperature calcination could completely restore activated carbon's adsorption capacity, providing a viable method for removing MPs from drinking water treatment facilities and raising the possibility of improving MPs adsorption through material modifications [24].

3.3.2 Magnetic Nanoparticles

Due to the large surface area and special physicochemical characteristics, nanomaterials like carbon nanotubes, graphene oxide, and magnetic nanoparticles have been studied for their potential to adsorb MPs [25].

In a study investigated by Heo *et al.* (2022), the effectiveness of magnetic iron oxide (Fe_3O_4) nanoparticles in the adsorption of micron-sized polystyrene MP particles from water was observed, whereas when combined with iron oxide nanoparticles the microparticles were removed from the water in a matter of minutes by using a magnet. The formation of Fe_3O_4 complexes was revealed by analysis using transmission electron microscopy, indicating the adsorption of particles onto the iron oxide nanoparticles. The iron oxide accumulation with particles was driven by hydrophobic intuition and encouraged to come about in high concentrations of iron oxide particles, though particles in freshwater anticipated adsorption in addition, Fe_3O_4 complexes might abdicate press oxide particles for recuperation [26].

3.3.3 Clay Minerals

Clay minerals such as bentonite and kaolinite have been studied for their ability to adsorb MPs due to their vast surface zone and negatively charged surfaces, which can pull in and hold onto strongly charged MPs [27]. Wang *et al.* (2023) studied the total and association of fresh and matured polyethylene (PE) MPs with various clay minerals under a variety of hydrochemical circumstances. It appears that the kind of cations present is important for PE aggregation, with ancient PEs being more stable in arrangements with monovalent electrolytes than divalent ones. The behavior of PEs in contact with clay minerals is typically determined by electrostatic shock, while the precise type of clay mineral has little influence on the settling percentage. Furthermore, elevated NaCl concentrations cause PEs to settle with clay particles, which promotes these observations about the movement of MPs in aquatic environments [28].

3.3.4 Biochar-Zeolites

Zeolites are crystalline aluminosilicates that can absorb MPs based on their estimation and charge due to their porous structure [29]. Babalaret *et al.* (2024) used PEG (polyethylene glycol) and PEI (polyethyleneimine) to enhance the electrostatic properties of MACZ (magnetic-activated biochar-zeolite composite), allowing it to adsorb MPs at distances of 2 μm and 15 μm. The arrangement of MACZ resulted in maximum adsorption capacities of 736 mg/g and 769 mg/g for MPs measuring 2 μm and 15 μm, respectively. The optimization protocol suggested a pH of 4 and temperatures of 28 °C and 24 °C, and the fabric was found to be recovered after four cycles [30].

3.3.5 HYDROGELS

Hydrogels have appeared as extraordinary guarantees as adsorbent materials for MP adsorption because of their permeable structure and high water substance. As these hydrophilic polymer systems swell and adsorb toxins, they show a solid partiality for water and can viably capture MPs. Functionalized hydrogels can move forward the proficiency of adsorption by authoritative MPs to particular authoritative locales. Hydrogel synthesis's flexibility also makes it conceivable to alter its properties to particularly target different sorts and sizes of MPs, which makes them valuable and effective for the remediation of MPs in sea-going situations [31].

Gua *et al.* (2022) investigated a light-driven keen hydrogel actuator created with a complex structure composed of graphene oxide nanosheets, poly(N-isopropyl acrylamide) hydrogels, and copolymers of polyethyleneimine and polydopamine that are covalently reinforced. This novel design has specific functions: it acts as a photothermal converter, adsorbs MCPs (microplastic compound pollutants), and provides an actuation matrix. As a consequence, the actuator can simultaneously detect and absorb MCPs while remaining responsive, much like a soft swimming robot. This actuator is notable for its high rates of adsorption (94.63%) and desorption efficiency (99.12%), remarkable adsorbing selectivity (97.09% for ferric ion-adsorbed MCPs), remarkably low detection limit (0.98 μm for ferric ion), and adaptable untethered photothermal actuation capabilities [32].

Other materials such as bentonite, chitosan, and natural fibers (cellulose, cotton, and wool) offer a potential way to remove MPs from water systems and are listed in Table 3.1. Due to their porous structure and electrostatic interactions, low cost, broad availability, and environmentally beneficial qualities, they are an ideal option for extensive MP adsorption in aquatic environments.

The choice of an effective adsorbent material is influenced by several aspects, including MP properties, cost, availability, ease of regeneration, and environmental conditions. It is crucial to thoroughly assess an adsorbent's effectiveness, selectivity, and environmental impact before mass application for the adsorption of MPs [33].

3.4 ADSORPTION MECHANISMS

MPs adsorb through a variety of mechanisms, each of which influences how MPs and pollutants interact. Numerous mechanisms, such as hydrophobic interactions, partitioning, electrostatic interactions, and other noncovalent interactions, drive the adsorption of MPs as demonstrated in Figure 3.4. The adsorption of MPs involves several mechanisms, reflecting the complex interactions between the plastic particles and the surrounding environment. One primary mechanism is hydrophobic interaction, stemming from the hydrophobic nature of the resin, which is the main component of

TABLE 3.1
Material, Sources, Size, and Removal Efficiency Removal of MPs

S.No	Material	Size of MPs (µm)	Removal efficiency (%)	References
1.	Activated Carbon	20–50	92.8	[34]
2.	Zeolites	10	>96	[35]
3.	Silica Gel	10	96	[36]
4.	Chitosan	10-15	71.6–92.1	[37]
5.	Graphene Oxide	20-30	89.8	[38]
6.	Bentonite	12	60	[39]
7.	Cellulose	5	94	[40]
8.	Clay	20	78	[41]
9.	Polymeric Resins	100–355	90	[42]
10.	Carbon Nanotubes	20-30	80	[43]
11.	Magnesium Oxide	10-15	10	[44]
12.	Alumina	>5	69	[45]
13.	Lignin	90	80	[46]
14.	Cyclodextrins	100	90	[47]
15.	Jute Fiber	100	95	[48]
16.	Citrus Peel Extract	79	90	[49]
17.	Hydrogels	88	90	[50]
18.	Nanocellulose	79	80-90	[51]
19.	Grape Pomace Extract	100	85	[52]
20.	Neem Powder	>5	80	[53]
21.	Mushroom Mycelium	100	90	[54]
22.	Bamboo Charcoal	89	70-90	[55]
23.	Oat Hulls	>50	70	[11]
24.	Polydimethylsiloxane	25	80	[56]
25.	Titanium Dioxide	20-60	95	[57]
26.	Manganese Dioxide	100	80	[58]
27.	Iron Oxides	90	50-90	[59]
28.	Silver Nanoparticles	1-5	80	[60]
29.	Copper Nanoparticles	100	85	[61]
30.	Zinc Oxide Nanoparticles	1-5	90	[62]
31.	Pectin	100	80	[57]
32.	Tannins	1-5	90	[63]
33.	Water Hyacinth	100	80	[64]
34.	Algae Biomass	1-5	90	[65]
35.	Rice Husk	5-100	90	[66]
36.	Sugarcane Bagasse	1-5	80	[67]
37.	Orange Peel	100	90	[68]
38.	Shrimp Shells	100	80	[69]
39.	Seaweed Extracts	5-90	95	[70]
40.	Pine Bark	5-100	90	[71]

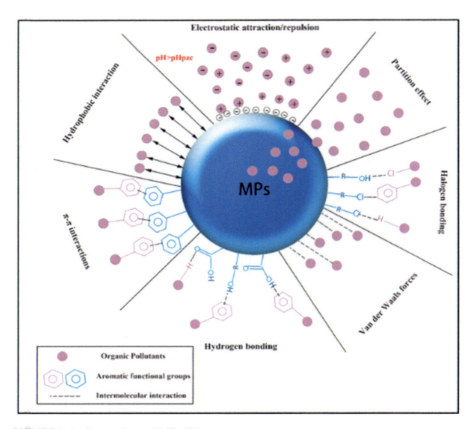

FIGURE 3.4 Interactions of MPs [9].

plastics [72]. This characteristic causes MPs to have hydrophobic surfaces, making them highly attractive to organic pollutants characterized by high-fat solubility and low water solubility. The water partition coefficient (Log K_{ow}) serves as a basic pointer of a substance's hydrophobicity, with higher partition coefficient values demonstrating more noteworthy partiality for adsorption onto MPs. Subsequently, natural compounds with raised partition coefficient values are more likely to follow to MPs [73].

MPs regularly carry a net negative charge due to the nearness of useful bunches on their surfaces. This charge encourages intelligence with emphatically charged particles or polar particles displayed within the surrounding media. Moreover, dividing plays a significant part, wherein natural toxins are drawn to the surface of MPs due to contrasts in their dissolvability between the plastic surface and the surrounding water. This handle is affected by variables such as the chemical composition of the plastic, the sort and concentration of poisons, and natural conditions such as pH and temperature. Other noncovalent intelligence, such as van der Waals strengths and hydrogen holding, contribute to the adsorption of natural toxins onto MPs. These different components frequently act synergistically, with their relative significance

Adsorptive Techniques

depending on variables such as the particular properties of the MPs and the nature of the poisons and encompassing environment. By and large, the adsorption of MPs may be a complex wonder driven by different interrelated instruments, highlighting the need for better understanding of its workings [10].

3.5 REDUCTION STRATEGIES

One of the most successful ways to reestablish the environment and diminish MP contamination is to decrease inputs at the source through reusing and efficient waste management strategies. Integrated waste management systems follow the three R's (reduce, reuse, and recycle) and prioritize improving the lifespan of plastics and MPs [74]. This methodology not only helps reduce energy and resource consumption, but it also reduces harmful emissions and keeps huge sums of unmanaged MPs waste out of the environment and, inevitably, the food chain.

Furthermore, there is a growing emphasis on technological advancements in both the scientific and economic fields to address MP pollution. This includes the development of innovative technological approaches aiming at reducing plastic waste and enhancing recycling capabilities. In addition to technological solutions, political measures and regulations play a vital role in combating plastic pollution [75-76].

3.5.1 TECHNOLOGICAL STRATEGIES FOR REMOVAL OF MPs

One of the main sources of environmental MPs is wastewater treatment plants. To successfully remove MPs from wastewater and avoid contaminating effluent with plastic particles, innovative wastewater treatment methods must be implemented. Thus, wastewater treatment plants are the major focus where technological approaches for MP removal are developed and practiced. Modern technologies and their use for removing MPs are the main focus of existing solutions. According to the reports, these systems were reported to filter out approximately 95% to 99.9% of MPs from wastewater [77].

Figure 3.5 demonstrates several methods for lowering the number of MP particles in wastewater treatment facilities and other water-using processes. Membranes, electrodeposition, and coagulation are a few of the frequently mentioned advanced wastewater technologies for the removal of MPs. The sections that follow will include a list of techniques for removing MPs [78].

3.5.1.1 Filtration

The removal of MPs from different aqueous mediums has been thoroughly investigated for sand, cloth, disc, drum, and biologically active filters. For example, a 95% removal rate of MPs during sewage treatment was demonstrated utilizing a sand filter. In research, cloth filtering successfully removed MPs at a remarkable 97% clearance rate. The effectiveness of disc filters and drum filters in eliminating particles, including MPs, from wastewater was reported [79]. A study by Talvitie *et al.* (2017) found that particles smaller than 25 μm were effectively removed by these filters, and further, the drum filter achieved an 80% removal rate, while the disc filter maintained a

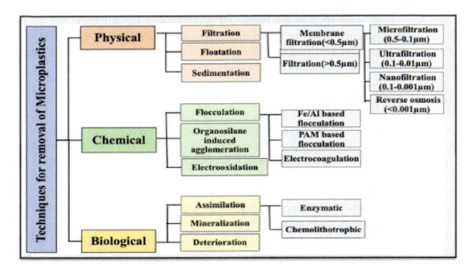

FIGURE 3.5 Techniques for removal of MPs.

removal effectiveness of over 90% for particles in the 50–25 μm range. However, after four months of use, the disc filter's flow was reduced by scaling and fouling, while the drum filter was spared this problem because it used a specific backwashing and filter material extraction mechanism. These filters have proven to be effective tools in tackling MP pollution in wastewater treatment processes [80]. Micropollutants were successfully removed from wastewater through a series of treatment steps involving mechanical, chemical, and biological processes (such as activated sludge treatment) followed by Biological Activated Carbon Filtration (BAF), achieving an impressive retention capacity of over 99% in a large, advanced sewage treatment plant that serves about 800,000 inhabitants. Moreover, filtration techniques have their limitations, especially when it comes to pore size reduction, which raises the possibility of clogs. Up to a certain size, filtration is an effective way to eliminate MPs [81].

3.5.1.2 Membrane Bioreactor and Membrane Filtration

One of the most promising methods for removing MPs is the Membrane Bioreactor (MBR), which combines biological wastewater treatment with a membrane technique such as microfiltration or ultrafiltration. Although MBRs are widely used in wastewater treatment facilities, demonstrating their commercial feasibility, their cost is usually higher than that of other traditional tertiary treatments. The ability of MBR systems to remove MPs has been compared to sewage treatment plants that use granular sand filtration or activated sludge for secondary or tertiary treatment. Furthermore, MBRs have been used against other cutting-edge tertiary treatments such as flotation with dissolved air and fast-gravity sand filters, which also accomplish high separation rates (>95%) for MPs from primary and secondary wastewater streams. These comparisons demonstrate the effectiveness of MBR technology and the elimination of MPs and its versatility in wastewater treatment procedures [82].

MBR demonstrates an amazing removal efficiency of 100%, showing an improvement over a wastewater treatment system with three stages that could only remove 60% of the MPs investigated by Baresel *et al.* (2019) and Lares *et al.* (2018). Along with MBR, dynamic membranes have been proposed as a solution for removing MPs from wastewater treatment. Dynamic membranes provide various advantages over traditional membranes, including lower system and material costs, as well as decreased energy usage [83, 84].

3.5.1.3 Flocculation

The flocculation preparation is a principal strategy to treat MPs in water due to its economy, high proficiency, and convenience. Ma *et al.* (2019) showed that salts based on aluminum (Al) are superior in water treatment methods compared to salts based on iron (Fe) at dispensing with particulate matter. There is a connection between molecule measure and evacuation productivity, as seen by the increment in PE evacuation effectiveness with diminishing molecule measure. The disposal of PE is much moved forward by the expansion of polyacrylamide (PAM), especially the anionic shape. Indeed, although coagulation and ultrafiltration have appeared to guarantee evacuating PE particles from drinking water, it is pivotal to keep in mind that utilizing conventional coagulants based on Al may cause some film defilement amid ultrafiltration. Despite this, it is clear that coagulation and ultrafiltration strategies may be common choices for the expulsion of MPs from drinking water based on laboratory-scale trials. Additional investigation is required to affirm their pertinence in genuine water treatment plants [85].

3.5.1.4 Electrooxidation

Electrooxidation treatment has been investigated as a potential solution to this environmental issue since washing machine wastewater is a major source of fibers and MPs in surface and municipal waters. According to the research done by Duran *et al.* (2018), addition of Na_2SO_4 to MPs water resulted in the elimination of MPs and COD (chemical oxygen demand).

Duran's results also show promising effective ways to reduce MPs and other contaminants in wastewater from washing machines, particularly when the BDD (boron-doped diamond) anode is used. These results are predicated on electrochemical pilot systems, meaning that more investigation and advancement are required to confirm the scalability and practical application of EO (electrochemical oxidation) treatment in bigger wastewater treatment plants [86–91].

3.5.1.5 Degradation by Microorganisms

MPs polymers can break down naturally by a variety of methods, including mechanical stress, UV radiation, heat, chemicals, and biological activity, into smaller molecules like oligomers and monomers, and eventually into carbon dioxide and water [92, 93]. The breakdown of MPs is significantly influenced by microbial activity, which involves several mechanisms. The polymeric structure is first weakened by abiotic elements, which then catalyze additional biodegradation processes [94]. These abiotic chemicals not only promote degradation but also serve as triggers for microbial degradation processes.

The initiation of biological deterioration occurs when a microbial biofilm forms on the plastic surface, partially piercing the material. The plastic's surface properties, such as hydrophobicity or rising hydrophilicity, influence the creation of this biofilm. The production of biofilms and the ensuing processes of biodegradation are also influenced by environmental conditions such as salt, pH levels, temperature, and UV exposure.

Extracellular polymeric substances (EPS) secreted by microorganisms during biofilm formation are essential for enhancing the biofilm's cohesiveness and adherence to plastic surfaces, which causes physical degradation. Due to the existence and activity of the biofilm, the plastic material is physically broken down throughout this process [95]. On the other hand, when various microbial populations grow on the MP surface, chemical degradation takes place [96]. Under such circumstances, some bacteria, such as chemolithotrophic bacteria, may emit acid molecules, such as sulfuric acid (*Thiobacillus*), nitric acid (*Nitrobacter*), or nitrous acid (*Nitrosomonas*). These acidic substances can change the MP pore pH, which can cause gradual deterioration and change the microstructure of the plastic matrix. However, the intrinsic durability of plastic polymers, which are made up of lengthy chains of hydrogen and carbon atoms with evenly distributed charges, limits the amount of bio-fragmentation. The complete breakdown is hampered by this structural stability, despite the microbial activity and chemical alterations brought on by biofilms.

Enzymes produced by bacteria, including lipases, dioxygenases, endopeptidases, and mono-and dioxygenases, are essential for destabilizing the local electric charge of plastic polymers. For instance, these enzymes can oxidize and introduce one or two oxygen molecules, which can lead to breakdown and the formation of amides or carboxylic acids. A variety of plastics, such as polyethylene, PVC (polyvinyl chloride), PHB (polyhydroxybutyrate), poly(3-hydroxybutyrate-co-3-mercaptopropionate), and poly(3-hydroxy propionate), can be broken down by some bacteria, such as *Pseudomonas* [97]. *Streptomyces* is capable of breaking down polyester, whereas *Bacillus brevis* is majorly at breaking down polycaprolactone. However, the simple fact that MP degradation produces monomers does not ensure that microbes would assimilate them. These plastic monomers are oxidized inside microbial cells via a variety of catabolic processes, including fermentation, anaerobic respiration, and aerobic respiration. In addition to producing energy, this process helps maintain the structure of cells and makes it easier to produce new biomass. The absorption capacity of microorganisms is determined by their ability to grow under aerobic or anaerobic circumstances [98].

3.6 MP TRANSPORT IN FOOD WEBS: IMPLICATIONS FOR HUMAN HEALTH

MPs have been found in a variety of organisms at varying trophic levels, indicating their movement through the food chain over time. After entering aquatic environments, MPs may go up and down the food chain, moving from plankton to bigger predators like *Mytilus edulis* to *Carcinus maenas*. The density of MPs in the water column plays a role in its availability to planktivorous organisms and also to larger predators, potentially facilitating its transfer between trophic levels. Research has demonstrated

Adsorptive Techniques 55

that MPs build up continuously in aquatic food systems having higher detection rates of MPs over time in seabirds, marine animals, turtles, etc. Furthermore, MPs can absorb and retain hazardous substances like viruses, worsening health issues, which emphasizes that it is crucial to carry out extensive research to fully evaluate the long-term health effects of MP exposure [99, 100].

3.6.1 SEAFOOD

The water-based environment is being thoroughly investigated for MP contamination, as sea products constitute a key component of human diets. MPs may enter fish organs by active capture, such as gill water filtering, or by ingestion of infected prey. MP contamination is frequent in shellfish and other fish species, including Atlantic cod, European hake, red mullet, and European pilchard. MPs have been found in the digestive systems of some edible fish species, including Rhizoprionodon Land II and Scomberomorus Cavalla. Several variables, particularly salinity, may influence the quantity of MPs the marine environment consumes. Seafood is an important source of MP exposure, and its consumption by marine life raises concerns for human health. MP pollution has an influence on the freshwater fishing and aquaculture sectors as well as on marine environments. Stable isotope analysis has grown in popularity as a method for measuring the amount of MP contamination in freshwater food systems and providing information about eating habits. These studies are crucial for understanding how MPs propagate throughout marine habitats, as well as for safeguarding and sustaining seafood fisheries [101, 102].

3.6.2 SALT FOR HUMAN CONSUMPTION

In recent years, several studies have been conducted to investigate MP contamination in salt meant for human consumption. Salt is widely utilized in cosmetics, medications, and food preservation, and it is also essential for optimal nutrition.

Table salt has been shown to include MPs all over the world; sea salt has the greatest levels of contamination since it is made from saltwater, which frequently contains MPs. Both direct industrial inputs and environmental deterioration are sources of contamination. There have been attempts to create cleaner sea salt, such as coagulation procedures to get rid of MPs. The maximum amount of MPs humans might be exposed to each year via salt intake is 6110 particles [103].

3.6.3 DRINKING WATER

Worldwide, MPs have been found in drinking water as well as tap water, raising questions about possible health hazards from ingestion. Studies on bottled water with various packaging styles revealed variable MP concentrations; no statistically significant variations were found between the samples. On the other hand, compared to glass and single-use PET (polyethylene terephthalate bottles), mineral water from reusable PET bottles showed a greater MP level. The amounts of MP in treated water are still notable. According to reports from the World Health Organization, tap water includes an estimated 5 particles of MPs per liter, which means that an individual may consume up to 10 particles each day. People who drink most of their water from plastic bottles could be consuming a lot more MPs than people who just drink tap water. This concern emphasizes the need for more study and preventative steps to deal with this problem [104].

3.6.4 POLITICAL MEASURES AND REGULATIONS

Various laws, rules, and voluntary commitments are being implemented globally to tackle the issue of environmental pollution resulting from MPs [105]. Various stages, such as plastic production and disposal, should be the focus of these strategies to reduce the environmental impact. On a global scale, organizations like the United Nations Environment Programme (UNEP) and forums such as the G7 meetings play a crucial role as leaders in initiatives aimed at reducing marine pollution, including plastic pollution [106]. In Europe, the Marine Strategy Framework Directive mandates Member States to take proactive measures to clean up and reduce pollution in oceans, highlighting the international commitment to addressing plastic pollution and its impacts on marine ecosystems [107]. These concerted efforts at both national and international levels are vital in combating plastic pollution and promoting sustainable practices to protect the environment. To decrease MP pollution in the environment, this strategy focuses on lowering the manufacture and use of plastic, MPs, and aims to recycle all plastic packaging by 2030. This vision advocates for an intelligent, innovative, and sustainable plastics industry that prioritizes reuse, repair, and recycling aspects in design and manufacturing processes to increase environmental sustainability and minimize MP waste [108]. Waste reduction also requires promoting the use of reusable products like deposit bottles and reusable plastic bags. Reusable product substitutions can drastically reduce plastic waste by adhering to the "Reduce, Reuse, Recycle" philosophy. To prevent plastics from entering the environment, a genuinely sustainable recycling strategy needs a strong garbage-collecting infrastructure.

Additionally, there is a need for a comprehensive take-back system for clothing, which is still lacking in many countries. Recycling also plays a crucial role in extending the lifespan of synthetic fibers. As these fibers are reused, their quality diminishes, eventually leading to disposal. Sustainable consumption practices are vital in addressing environmental and health risks associated with MP pollution. The textile industry plays a significant role in developing strategies to reduce MP pollution. For instance, there has been a notable increase in global consumption of recycled polyester, growing by 58% in just one year from 2015 to 2016 [109]. Innovative approaches include the use of bio-based fibers and transforming biological materials like crayfish peel, trees, hemp, nettle, and flax preferably sourced locally into textile fibers. These new approaches aim to reduce reliance on synthetic fibers and promote eco-friendly alternatives that can contribute to a more sustainable textile industry. Several states in the United States, Canada, and the United Kingdom have implemented laws prohibiting the use of MPs in personal care and cosmetic products to prevent their direct discharge into the environment. A ban should be imposed on single-use plastic products like straws and cups. Similar prohibitions are being pursued in India, the UK, and the EU [110].

3.7 CONCLUSION

MPs are a significant environmental pollutant, causing toxic effects in humans due to their long-lasting properties and lack of biodegradability. Techniques like filtration, coagulation, biodegradation, and adsorption have been developed to eliminate MPs, with adsorption techniques being the most preferred due to their efficiency, low maintenance, and cost-effectiveness. Adsorbent materials include activated carbon, biodegradable polymers, magnetic nanoparticles, bio-based materials, etc. MPs are introduced into ecosystems through various sources, including farm run-off, sewage and industrial emissions, and intentional inclusion in personal hygiene products. Secondary MPs are formed when larger plastic items break down over time, contributing to environmental pollution. Land-based sources, plastic waste from packaging, fishing equipment, and single-use consumer goods also contribute to the production of secondary MPs. Global plastic production is projected to reach 500.0 million tons by 2025, with China being the top producer. These MPs threaten marine life by bioaccumulating in tissues and entering the food chain. Adsorption of MPs from the environment has been studied using various materials, such as activated carbon, magnetic nanoparticles, clay minerals, biochar-zeolites, and hydrogels, which further includes the process of MPs involving hydrophobic interactions, partitioning, electrostatic interactions, partitioning, and noncovalent interactions. Efficient waste management practices and a waste management system that adheres to the four Rs hierarchy can help reduce energy and resource consumption, mitigate emissions, and prevent mismanaged MPs waste from entering ecosystems and the food chain. Technological advancements, political measures, and regulations are essential in combating plastic pollution. Water treatment plants are also major sources of environmental MPs, and modern technologies like membranes, electrodeposition, and coagulation are used to remove MPs. Other methods include filtration techniques, membrane bioreactors, flocculation, dissolved air flotation, electrocoagulation, agglomeration fixation, electrooxidation treatment, and BDD anodes. Degradation by microorganisms is another method for removing MPs, but the intrinsic durability of plastic polymers limits bio-fragmentation. Thus, the development of new materials and technologies is an urgent requirement to eliminate MPs from the environment.

REFERENCES

[1] Campanale, C., Galafassi, S., Savino, I., Massarelli, C., Ancona, V., Volta, P., & Uricchio, V. F. (2022). Microplastics pollution in the terrestrial environments: Poorly known diffuse sources and implications for plants. *Science of the Total Environment*, 805, 150431. https://doi.org/10.1016/j.scitotenv.2021.150431

[2] Costigan, E., Collins, A., Hatinoglu, M. D., Bhagat, K., MacRae, J., Perreault, F., & Apul, O. (2022). Adsorption of organic pollutants by microplastics: Overview of a dissonant literature. *Journal of Hazardous Materials Advances*, 6, 100091. https://doi.org/10.1016/j.hazadv.2022.100091

[3] Sun, J., Peng, Z., Zhu, Z. R., Fu, W., Dai, X., & Ni, B. J. (2022). The atmospheric microplastics deposition contributes to microplastic pollution in urban waters. *Water Research*, 225, 119116. https://doi.org/10.1016/j.watres.2022.119116

[4] Prata, J. C., da Costa, J. P., Lopes, I., Duarte, A. C., & Rocha-Santos, T. (2020). Environmental exposure to microplastics: An overview on possible human health effects. *Science of the Total Environment*, 702, 134455. https://doi.org/10.1016/j.scitot env.2019.134455

[5] Bäuerlein, P. S., Hofman-Caris, R. C., Pieke, E. N., & Ter Laak, T. L. (2022). Fate of microplastics in the drinking water production. *Water Research*, 221, 118790. https://doi.org/10.1016/j.watres.2022.118790

[6] Othman, A. R., Hasan, H. A., Muhamad, M. H., Ismail, N. I., & Abdullah, S. R. S. (2021). Microbial degradation of microplastics by enzymatic processes: A review. *Environmental Chemistry Letters*, 19, 3057–3073. https://doi.org/10.1007/s10 311-021-01197-9

[7] Donkers, J. M., Höppener, E. M., Grigoriev, I., Will, L., Melgert, B. N., van der Zaan, B., ... & Kooter, I. M. (2022). Advanced epithelial lung and gut barrier models demonstrate passage of microplastic particles. *Microplastics and Nanoplastics*, 2(1), 6. https://doi.org/10.1186/s43591-021-00024-w

[8] Gopinath, P. M., Twayana, K. S., Ravanan, P., Thomas, J., Mukherjee, A., Jenkins, D. F., & Chandrasekaran, N. (2021). Prospects on the nano-plastic particles internalization and induction of cellular response in human keratinocytes. *Particle and Fibre Toxicology*, 18, 1–24. https://doi.org/10.1186/s12989-021-00428-9

[9] Jiménez-Skrzypek, G., Hernández-Sánchez, C., Ortega-Zamora, C., González-Sálamo, J., González-Curbelo, M. Á., & Hernández-Borges, J. (2021). Microplastic-adsorbed organic contaminants: analytical methods and occurrence. *TrAC Trends in Analytical Chemistry*, 136, 116186. https://doi.org/10.1016/j.trac.2021.116186

[10] Reineccius, J., Bresien, J., & Waniek, J. J. (2021). Separation of microplastics from mass-limited samples by an effective adsorption technique. *Science of The Total Environment*, 788, 147881. https://doi.org/10.1016/j.scitotenv.2021.147881

[11] Auta, H. S., Emenike, C. U., & Fauziah, S. H. (2017). Distribution and importance of microplastics in the marine environment: a review of the sources, fate, effects, and potential solutions. *Environment International*, 102, 165–176. https://doi.org/10.1016/ j.envint.2017.02.013

[12] Andrady, A. L. (2017). The plastic in microplastics: A review. *Marine Pollution Bulletin*, 119(1), 12–22. https://doi.org/10.1016/j.marpolbul.2017.01.082

[13] Anderson, P. J., Warrack, S., Langen, V., Challis, J. K., Hanson, M. L., & Rennie, M. D. (2017). Microplastic contamination in lake Winnipeg, Canada. *Environmental Pollution*, 225, 223–231. https://doi.org/10.1016/j.envpol.2017.02.072

[14] Caron, A. G., Thomas, C. R., Berry, K. L., Motti, C. A., Ariel, E., & Brodie, J. E. (2018). Ingestion of microplastic debris by green sea turtles (Chelonia mydas) in the Great Barrier Reef: Validation of a sequential extraction protocol. *Marine Pollution Bulletin*, 127, 743–751. https://doi.org/10.1016/j.marpolbul.2017.12.062

[15] Kole, P. J., Löhr, A. J., Van Belleghem, F. G., & Ragas, A. M. (2017). Wear and tear of tyres: a stealthy source of microplastics in the environment. *International Journal of Environmental Research and Public Health*, 14(10), 1265. https://doi.org/10.3390/ije rph14101265

[16] Duis, K., & Coors, A. (2016). Microplastics in the aquatic and terrestrial environment: sources (with a specific focus on personal care products), fate and effects. *Environmental Sciences Europe*, 28(1), 2. https://doi.org/10.1186/s12 302-015-0069-y

Adsorptive Techniques

[17] Good, T. P., June, J. A., Etnier, M. A., & Broadhurst, G. (2010). Derelict fishing nets in Puget sound and the Northwest straits: Patterns and threats to marine fauna. *Marine Pollution Bulletin*, 60(1), 39–50. https://doi.org/10.1016/j.marpolbul.2009.09.005

[18] Fadare, O. O., & Okoffo, E. D. (2020). Covid-19 face masks: A potential source of microplastic fibers in the environment. *Science of the Total Environment*, 737, 140279. https://doi.org/10.1016%2Fj.scitotenv.2020.140279

[19] Bui, X. T., Nguyen, P. T., Nguyen, V. T., Dao, T. S., & Nguyen, P. D. (2020). Microplastics pollution in wastewater: Characteristics, occurrence and removal technologies. *Environmental Technology & Innovation*, 19, 101013. https://doi.org/10.1016/j.eti.2020.101013

[20] Huang, D., Tao, J., Cheng, M., Deng, R., Chen, S., Yin, L., & Li, R. (2021). Microplastics and nanoplastics in the environment: Macroscopic transport and effects on creatures. *Journal of Hazardous Materials*, 407, 124399. https://doi.org/10.1016/j.jhazmat.2020.124399

[21] Ryan, P.G. (2015) A Brief History of Marine Litter Research. In: Bergmann, M., Gutow, L. and Klages, M., Eds., Marine Anthropogenic Litter, Springer, Berlin, Open Chpt. 1, 1-28. https://doi.org/10.1016/j.jhazmat.2020.124399

[22] Wang, Q., Bai, J., Ning, B., Fan, L., Sun, T., Fang, Y., ... & Gao, Z. (2020). Effects of bisphenol A and nanoscale and microscale polystyrene plastic exposure on particle uptake and toxicity in human Caco-2 cells. *Chemosphere*, 254, 126788. https://doi.org/10.1016/j.chemosphere.2020.126788

[23] Xing, X., Zhang, Y., Zhou, G., Zhang, Y., Yue, J., Wang, X., ... & Zhang, J. (2023). Mechanisms of polystyrene nanoplastics adsorption onto activated carbon modified by ZnCl2. *Science of The Total Environment*, 876, 162763. https://doi.org/10.1016/j.scitotenv.2023.162763

[24] Arenas, L. R., Gentile, S. R., Zimmermann, S., & Stoll, S. (2021). Nanoplastics adsorption and removal efficiency by granular activated carbon used in drinking water treatment process. *Science of the Total Environment*, 791, 148175. https://doi.org/10.1016/j.scitotenv.2021.148175

[25] Wang, H. P., Huang, X. H., Chen, J. N., Dong, M., Nie, C. Z., & Qin, L. (2023). Modified superhydrophobic magnetic Fe_3O_4 nanoparticles for removal of microplastics in liquid foods. *Chemical Engineering Journal*, 476, 146562. https://doi.org/10.1016/j.cej.2023.146562

[26] Heo, Y., Lee, E. H., & Lee, S. W. (2022). Adsorptive removal of micron-sized polystyrene particles using magnetic iron oxide nanoparticles. *Chemosphere*, 307, 135672. https://doi.org/10.1016/j.chemosphere.2022.135672

[27] Ding, L., Yu, X., Guo, X., Zhang, Y., Ouyang, Z., Liu, P., ... & Zhu, L. (2022). The photodegradation processes and mechanisms of polyvinyl chloride and polyethylene terephthalate microplastic in aquatic environments: Important role of clay minerals. *Water Research*, 208, 117879. https://doi.org/10.1016/j.watres.2021.117879

[28] Wang, Y., Chen, X., Wang, F., & Cheng, N. (2023). Influence of typical clay minerals on aggregation and settling of pristine and aged polyethylene microplastics. *Environmental Pollution*, 316, 120649. https://doi.org/10.1016/j.envpol.2022.120649

[29] Hanif, M. A., Ibrahim, N., Dahalan, F. A., Md. Ali, U. F., Hasan, M., Azhari, A. W., & Jalil, A. A. (2023). Microplastics in facial cleanser: Extraction, identification, potential toxicity, and continuous-flow removal using agricultural waste–based biochar. *Environmental Science and Pollution Research*, 30(21), 60106–60120. https://doi.org/10.1007/s11356-023-26741-8

[30] Babalar, M., Siddiqua, S., & Sakr, M. A. (2024). A novel polymer coated magnetic activated biochar-zeolite composite for adsorption of polystyrene

microplastics: Synthesis, characterization, adsorption and regeneration performance. *Separation and Purification Technology*, 331, 125582. https://doi.org/10.1016/j.seppur.2023.125582

[31] Mendonça, I., Sousa, J., Cunha, C., Faria, M., Ferreira, A., & Cordeiro, N. (2023). Solving urban water microplastics with bacterial cellulose hydrogels: Leveraging predictive computational models. *Chemosphere*, 314, 137719. https://doi.org/10.1016/j.chemosphere.2022.137719

[32] Guo, Q., Liu, Y., Liu, J., Wang, Y., Cui, Q., Song, P., ... & Zhang, C. (2022). Hierarchically structured hydrogel actuator for microplastic pollutant detection and removal. *Chemistry of Materials*, 34(11), 5165–5175. https://doi.org/10.1021/acs.chemmater.2c00625

[33] Chen, Z., Fang, J., Wei, W., Ngo, H. H., Guo, W., & Ni, B. J. (2022). Emerging adsorbents for micro/nanoplastics removal from contaminated water: advances and perspectives. *Journal of Cleaner Production*, 371, 133676. https://doi.org/10.1016/j.jclepro.2022.133676

[34] Kim, K. T., & Park, S. (2021). Enhancing microplastics removal from wastewater using electro-coagulation and granule-activated carbon with thermal regeneration. *Processes*, 9(4), 617. https://doi.org/10.3390/pr9040617

[35] Shen, M., Hu, T., Huang, W., Song, B., Zeng, G., & Zhang, Y. (2021). Removal of microplastics from wastewater with aluminosilicate filter media and their surfactant-modified products: Performance, mechanism and utilization. *Chemical Engineering Journal*, 421, 129918. https://doi.org/10.1016/j.cej.2021.129918

[36] Sturm, M. T., Horn, H., & Schuhen, K. (2021). Removal of microplastics from waters through agglomeration-fixation using organosilanes—effects of polymer types, water composition and temperature. *Water*, 13(5), 675. https://doi.org/10.3390/w13050675

[37] Sun, C., Wang, Z., Zheng, H., Chen, L., & Li, F. (2021). Biodegradable and re-usable sponge materials made from chitin for efficient removal of microplastics. *Journal of Hazardous Materials*, 420, 126599. https://doi.org/10.1016/j.jhazmat.2021.126599

[38] Peng, G., Xiang, M., Wang, W., Su, Z., Liu, H., Mao, Y., ... & Zhang, P. (2022). Engineering 3D graphene-like carbon-assembled layered double oxide for efficient microplastic removal in a wide pH range. *Journal of Hazardous Materials*, 433, 128672. https://doi.org/10.1016/j.jhazmat.2022.128672

[39] Spacilova, M., Dytrych, P., Lexa, M., Wimmerova, L., Masin, P., Kvacek, R., & Solcova, O. (2023). An innovative sorption technology for removing microplastics from wastewater. *Water*, 15(5), 892. https://doi.org/10.3390/w15050892

[40] Egea-Corbacho, A., Martín-García, A. P., Franco, A. A., Albendín, G., Arellano, J. M., Rodríguez, R., ... & Coello, M. D. (2022). A method to remove cellulose from rich organic samples to analyse microplastics. *Journal of Cleaner Production*, 334, 130248. https://doi.org/10.1016/j.jclepro.2021.130248

[41] Nakazawa, Y., Abe, T., Matsui, Y., Shinno, K., Kobayashi, S., Shirasaki, N., & Matsushita, T. (2021). Differences in removal rates of virgin/decayed microplastics, viruses, activated carbon, and kaolin/montmorillonite clay particles by coagulation, flocculation, sedimentation, and rapid sand filtration during water treatment. *Water Research*, 203, 117550. https://doi.org/10.1016/j.watres.2021.117550

[42] Yang, L., Li, K., Cui, S., Kang, Y., An, L., & Lei, K. (2019). Removal of microplastics in municipal sewage from China's largest water reclamation plant. *Water Research*, 155, 175–181. https://doi.org/10.1016/j.watres.2019.02.046

[43] Tang, Y., Zhang, S., Su, Y., Wu, D., Zhao, Y., & Xie, B. (2021). Removal of microplastics from aqueous solutions by magnetic carbon nanotubes. *Chemical Engineering Journal*, 406, 126804. https://doi.org/10.1016/j.cej.2020.126804

Adsorptive Techniques

[44] Zhang, Y., Zhao, J., Liu, Z., Tian, S., Lu, J., Mu, R., & Yuan, H. (2021). Coagulation removal of microplastics from wastewater by magnetic magnesium hydroxide and PAM. *Journal of Water Process Engineering*, 43, 102250. https://doi.org/10.1016/j.jwpe.2021.102250

[45] Skaf, D. W., Punzi, V. L., Rolle, J. T., & Kleinberg, K. A. (2020). Removal of micronsized microplastic particles from simulated drinking water via alum coagulation. *Chemical Engineering Journal*, 386, 123807. https://doi.org/10.1016/j.cej.2019.123807

[46] Găgeanu, I., Carvalheiro, F., Ekielski, A., & Duarte, L. C. (2023). Lignin utilization for the removal of microplastic particles from water. *Inmateh-Agricultural Engineering*, 71, 511–521. https://doi.org/10.35633/inmateh-71-44

[47] Tian, B., Hua, S., Tian, Y., & Liu, J. (2021). Cyclodextrin-based adsorbents for the removal of pollutants from wastewater: A review. *Environmental Science and Pollution Research*, 28(2), 1317–1340. https://doi.org/10.1007/s11356-020-11168-2

[48] Shahinur, S., Hasan, M., Ahsan, Q., Sultana, N., Ahmed, Z., & Haider, J. (2021). Effect of rot-, fire-, and water-retardant treatments on jute fiber and their associated thermoplastic composites: A study by FTIR. *Polymers*, 13(15), 2571. https://doi.org/10.3390/polym13152571

[49] Gómez-Aguilar, D. L., Rodríguez-Miranda, J. P., & Salcedo-Parra, O. J. (2022). Fruit peels as a sustainable waste for the biosorption of heavy metals in wastewater: A review. *Molecules*, 27(7), 2124. https://doi.org/10.3390/molecules27072124

[50] Verster, L. (2021). *Hydrogel Soft Dendritic Colloids as Active Marine Microplastic Cleaners*. North Carolina State University.

[51] Leppänen, I., Lappalainen, T., Lohtander, T., Jonkergouw, C., Arola, S., & Tammelin, T. (2022). Capturing colloidal nano-and microplastics with plant-based nanocellulose networks. *Nature Communications*, 13(1), 1814. https://doi.org/10.1038/s41467-022-29446-7

[52] Berger, C., Mattos, B. D., Amico, S. C., de Farias, J. A., Coldebella, R., Gatto, D. A., & Missio, A. L. (2020). Production of sustainable polymeric composites using grape pomace biomass. *Biomass Conversion and Biorefinery*, 12, 1–12. https://doi.org/10.1007/s13399-020-00966-w

[53] Khan, Q., Imran, U., Ullman, J. L., & Khokhar, W. A. (2023). Turbidity removal through the application of powdered azadirachta indica (neem) seeds. *Mehran University Research Journal of Engineering and Technology*, 42(1), 1–8. https://doi.org/10.22581/muet1982.2301.01

[54] Fu, X., Zhang, S., Zhang, X., Zhang, Y., Li, B., Jin, K., ... & Li, Q. (2023). Sustainable microplastic remediation with record capacity unleashed via surface engineering of natural fungal mycelium framework. *Advanced Functional Materials*, 33(27), 2212570. https://doi.org/10.1002/adfm.202212570

[55] Chaturvedi, K., Singhwane, A., Dhangar, M., Mili, M., Gorhae, N., Naik, A., ... & Verma, S. (2023). Bamboo for producing charcoal and biochar for versatile applications. *Biomass Conversion and Biorefinery*, 14, 1–27. https://doi.org/10.1007/s13399-022-03715-3

[56] Rius-Ayra, O., Biserova-Tahchieva, A., Sansa-López, V., & Llorca-Isern, N. (2022). Superhydrophobic PDMS coated 304 stainless-steel mesh for the removal of HDPE microplastics. *Progress in Organic Coatings*, 170, 107009. https://doi.org/10.1016/j.porgcoat.2022.107009

[57] Rozman, U., Klun, B., Marolt, G., Imperl, J., & Kalčíková, G. (2023). A study of the adsorption of titanium dioxide and zinc oxide nanoparticles on polyethylene microplastics and their desorption in aquatic media. *Science of The Total Environment*, 888, 164163. https://doi.org/10.1016/j.scitotenv.2023.164163

[58] Ye, H., Wang, Y., Liu, X., Xu, D., Yuan, H., Sun, H., ... & Ma, X. (2021). Magnetically steerable iron oxides-manganese dioxide core–shell micromotors for organic and microplastic removals. *Journal of Colloid and Interface Science*, 588, 510–521. https://doi.org/10.1016/j.jcis.2020.12.097

[59] Zandieh, M., & Liu, J. (2022). Removal and degradation of microplastics using the magnetic and nanozyme activities of bare iron oxide nanoaggregates. *Angewandte Chemie*, 134(47), e202212013. https://doi.org/10.1002/ange.202212013

[60] Li, P., Zou, X., Wang, X., Su, M., Chen, C., Sun, X., & Zhang, H. (2020). A preliminary study of the interactions between microplastics and citrate-coated silver nanoparticles in aquatic environments. *Journal of Hazardous Materials*, 385, 121601. https://doi.org/10.1016/j.jhazmat.2019.121601

[61] Mozafarjalali, M., Hamidian, A. H., & Sayadi, M. H. (2023). Microplastics as carriers of iron and copper nanoparticles in aqueous solution. *Chemosphere*, 324, 138332. https://doi.org/10.1016/j.chemosphere.2023.138332

[62] De Falco, F., Gentile, G., Avolio, R., Errico, M. E., Di Pace, E., Ambrogi, V., ... & Cocca, M. (2018). Pectin based finishing to mitigate the impact of microplastics released by polyamide fabrics. *Carbohydrate Polymers*, 198, 175–180. https://doi.org/10.1016/j.carbpol.2018.06.062

[63] Park, J. W., Lee, S. J., Hwang, D. Y., & Seo, S. (2021). Removal of microplastics via tannic acid-mediated coagulation and in vitro impact assessment. *Rsc Advances*, 11(6), 3556–3566. https://doi.org/10.1039/D0RA09645H

[64] Janssen, T. W. (2023). *Floating Water Hyacinths Consistently Trap Surface Macroplastics Along the River Course* (Doctoral dissertation, MSc thesis, Wageningen University and Research).

[65] Cheng, Y. R., & Wang, H. Y. (2022). Highly effective removal of microplastics by microalgae Scenedesmus abundans. *Chemical Engineering Journal*, 435, 135079. https://doi.org/10.1016/j.cej.2022.135079

[66] Yogarathinam, L. T., Usman, J., Othman, M. H. D., Ismail, A. F., Goh, P. S., Gangasalam, A., & Adam, M. R. (2022). Low-cost silica based ceramic supported thin film composite hollow fiber membrane from guinea corn husk ash for efficient removal of microplastic from aqueous solution. *Journal of Hazardous Materials*, 424, 127298. https://doi.org/10.1016/j.jhazmat.2021.127298

[67] Dantas, T. N. C., Dantas Neto, A. A., Oliveira, A. C., & Moura, M. C. P. D. A. Use of surfactant-modified adsorbents in the removal of microplastics from wastewater. Available at SSRN 4456514. https://dx.doi.org/10.2139/ssrn.4456514

[68] Al-sareji, O. J., Abdulzahra, M. A., Hussein, T. S., Shlakaa, A. S., Karhib, M. M., Meiczinger, M., ... & Hashim, K. S. (2023). Removal of pharmaceuticals from water using laccase immobilized on orange peels waste-derived activated carbon. *Water*, 15(19), 3437. https://doi.org/10.3390/w15193437

[69] Eamrat, R., Rujakom, S., Pussayanavin, T., Taweesan, A., Witthayaphirom, C., & Kamei, T. (2024). Optimizing biocoagulant aid from shrimp shells (Litopenaeus vannamei) for enhancing microplastics removal from aqueous solutions. *Environmental Technology & Innovation*, 33, 103457. https://doi.org/10.1016/j.eti.2023.103457

[70] Sundbæk, K. B., Koch, I. D. W., Villaro, C. G., Rasmussen, N. S., Holdt, S. L., & Hartmann, N. B. (2018). Sorption of fluorescent polystyrene microplastic particles to edible seaweed Fucus vesiculosus. *Journal of Applied Phycology*, 30, 2923–2927. https://doi.org/10.1007/s10811-018-1472-8

[71] Subair, A., Krishnamoorthy Lakshmi, P., Chellappan, S., & Chinghakham, C. (2024). Removal of polystyrene microplastics using biochar-based continuous flow fixed-bed

column. *Environmental Science and Pollution Research*, 31, 1–13. https://doi.org/10.1007/s11356-024-32088-5

[72] Fu, L., Li, J., Wang, G., Luan, Y., & Dai, W. (2021). Adsorption behavior of organic pollutants on microplastics. *Ecotoxicology and Environmental Safety*, 217, 112207. https://doi.org/10.1016/j.ecoenv.2021.112207

[73] Wu, P., Cai, Z., Jin, H., & Tang, Y. (2019). Adsorption mechanisms of five bisphenol analogues on PVC microplastics. *Science of the Total Environment*, 650, 671–678. https://doi.org/10.1016/j.scitotenv.2018.09.049

[74] Prata, J. C., Silva, A. L. P., Da Costa, J. P., Mouneyrac, C., Walker, T. R., Duarte, A. C., & Rocha-Santos, T. (2019). Solutions and integrated strategies for the control and mitigation of plastic and microplastic pollution. *International Journal of Environmental Research and Public Health*, 16(13), 2411. https://doi.org/10.3390/ijerph16132411

[75] Calcott, P., & Walls, M. (2000). Can downstream waste disposal policies encourage upstream "design for environment"? *American Economic Review*, 90(2), 233–237. https://doi.org/10.1257/aer.90.2.233

[76] Schneider, D. R., & Ragossnig, A. (2015). Recycling and incineration, contradiction or coexistence? *Waste Management & Research*, 33(8), 693–695. https://doi.org/10.1177/0734242X15593421

[77] Habib, R. Z., Thiemann, T., & Al Kendi, R. (2020). Microplastics and wastewater treatment plants—a review. *Journal of Water Resource and Protection*, 12(01), 1. http://www.scirp.org/journal/Paperabs.aspx?PaperID=97637

[78] Padervand, M., Lichtfouse, E., Robert, D., & Wang, C. (2020). Removal of microplastics from the environment. A review. *Environmental Chemistry Letters*, 18(3), 807–828. https://doi.org/10.1007/s10311-020-00983-1

[79] Talvitie, J., Mikola, A., Koistinen, A., & Setälä, O. (2017). Solutions to microplastic pollution–Removal of microplastics from wastewater effluent with advanced wastewater treatment technologies. *Water Research*, 123, 401–407. https://doi.org/10.1016/j.watres.2017.07.005

[80] Talvitie, J., Mikola, A., Setälä, O., Heinonen, M., & Koistinen, A. (2017). How well is microlitter purified from wastewater?–A detailed study on the stepwise removal of microlitter in a tertiary level wastewater treatment plant. *Water Research*, 109, 164–172. https://doi.org/10.1016/j.watres.2016.11.046

[81] Enfrin, M., Dumée, L. F., & Lee, J. (2019). Nano/microplastics in water and wastewater treatment processes–origin, impact and potential solutions. *Water Research*, 161, 621–638. https://doi.org/10.1016/j.watres.2019.06.049

[82] Poerio, T., Piacentini, E., & Mazzei, R. (2019). Membrane processes for microplastic removal. *Molecules*, 24(22), 4148. https://doi.org/10.3390/molecules24224148

[83] Baresel, C., Harding, M., & Fang, J. (2019). Ultrafiltration/granulated active carbon-biofilter: efficient removal of a broad range of micropollutants. *Applied Sciences*, 9(4), 710. https://doi.org/10.3390/app9040710

[84] Lares, M., Ncibi, M. C., Sillanpää, M., & Sillanpää, M. (2018). Occurrence, identification and removal of microplastic particles and fibers in conventional activated sludge process and advanced MBR technology. *Water Research*, 133, 236–246. https://doi.org/10.1016/j.watres.2018.01.049

[85] Ma, B., Xue, W., Hu, C., Liu, H., Qu, J., & Li, L. (2019). Characteristics of microplastic removal via coagulation and ultrafiltration during drinking water treatment. *Chemical Engineering Journal*, 359, 159–167. https://doi.org/10.1016/j.cej.2018.11.155

[86] Wang, Y., Li, Y. N., Tian, L., Ju, L., & Liu, Y. (2021). The removal efficiency and mechanism of microplastic enhancement by positive modification dissolved air flotation. *Water Environment Research*, 93(5), 693–702. https://doi.org/10.1002/wer.1352

[87] Perren, W., Wojtasik, A., & Cai, Q. (2018). Removal of microbeads from wastewater using electrocoagulation. *ACS Omega*, 3(3), 3357–3364. https://doi.org/10.1021/acsomega.7b02037

[88] Herbort, A. F., Sturm, M. T., & Schuhen, K. (2018). A new approach for the agglomeration and subsequent removal of polyethylene, polypropylene, and mixtures of both from freshwater systems–a case study. *Environmental Science and Pollution Research*, 25, 15226–15234. https://doi.org/10.1007/s11356-018-1981-7

[89] Herbort, A. F., Sturm, M. T., Fiedler, S., Abkai, G., & Schuhen, K. (2018). Alkoxy-silyl induced agglomeration: a new approach for the sustainable removal of microplastic from aquatic systems. *Journal of Polymers and the Environment*, 26, 4258–4270. https://doi.org/10.1007/s10924-018-1287-3

[90] Schuhen, K., Sturm, M. T., & Herbort, A. F. (2019). Technological approaches for the reduction of microplastic pollution in seawater desalination plants and for sea salt extraction. *Plastics in the Environment*, 1–16.

[91] Duran, F. E., de Araujo, D. M., do Nascimento Brito, C., Santos, E. V., Ganiyu, S. O., & Martinez-Huitle, C. A. (2018). Electrochemical technology for the treatment of real washing machine effluent at pre-pilot plant scale by using active and non-active anodes. *Journal of Electroanalytical Chemistry*, 818, 216–222. https://doi.org/10.1016/j.jelechem.2018.04.029

[92] Shimao, M. (2001). Biodegradation of plastics. *Current Opinion in Biotechnology*, 12(3), 242–247. https://doi.org/10.1016/S0958-1669(00)00206-8

[93] Singh, B., & Sharma, N. (2008). Mechanistic implications of plastic degradation. *Polymer Degradation and Stability*, 93(3), 561–584. https://doi.org/10.1016/j.polymdegradstab.2007.11.008

[94] Jakubowicz, I., Yarahmadi, N., & Petersen, H. (2006). Evaluation of the rate of abiotic degradation of biodegradable polyethylene in various environments. *Polymer Degradation and Stability*, 91(7), 1556–1562. https://doi.org/10.1016/j.polymdegradstab.2005.09.018

[95] Bonhomme, S., Cuer, A., Delort, A. M., Lemaire, J., Sancelme, M., & Scott, G. (2003). Environmental biodegradation of polyethylene. *Polymer Degradation and Stability*, 81(3), 441–452. https://doi.org/10.1016/S0141-3910(03)00129-0

[96] Zettler, E. R., Mincer, T. J., & Amaral-Zettler, L. A. (2013). Life in the "plastisphere": microbial communities on plastic marine debris. *Environmental Science & Technology*, 47(13), 7137–7146. https://doi.org/10.1021/es401288x

[97] Ghosh, S. K., Pal, S., & Ray, S. (2013). Study of microbes having potentiality for biodegradation of plastics. *Environmental Science and Pollution Research*, 20, 4339–4355. https://doi.org/10.1007/s11356-013-1706-x

[98] Al Mamun, A., Prasetya, T. A. E., Dewi, I. R., & Ahmad, M. (2023). Microplastics in human food chains: Food becoming a threat to health safety. *Science of the Total Environment*, 858, 159834. https://doi.org/10.1016/j.scitotenv.2022.159834

[99] Pironti, C., Ricciardi, M., Motta, O., Miele, Y., Proto, A., & Montano, L. (2021). Microplastics in the environment: Intake through the food web, human exposure and toxicological effects. *Toxics*, 9(9), 224. https://doi.org/10.3390/toxics9090224

[100] Mercogliano, R., Avio, C. G., Regoli, F., Anastasio, A., Colavita, G., & Santonicola, S. (2020). Occurrence of microplastics in commercial seafood under the perspective

of the human food chain. A review. *Journal of Agricultural and Food Chemistry*, 68(19), 5296–5301. https://doi.org/10.1021/acs.jafc.0c01209

[101] Lehel, J., & Murphy, S. (2021). Microplastics in the food chain: food safety and environmental aspects. *Reviews of Environmental Contamination and Toxicology*, 259, 1–49. https://doi.org/10.1007/398_2021_77

[102] Danopoulos, E., Jenner, L., Twiddy, M., & Rotchell, J. M. (2020). Microplastic contamination of salt intended for human consumption: A systematic review and meta-analysis. *SN Applied Sciences*, 2, 1–18. https://doi.org/10.1007/s42452-020-03749-0

[103] Sharma, S., Sharma, B., & Sadhu, S. D. (2022). Microplastic profusion in food and drinking water: are microplastics becoming a macroproblem? *Environmental Science: Processes & Impacts*, 24(7), 992–1009. https://doi.org/10.1039/D1E M00553G

[104] Chen, C. L. (2015). Regulation and management of marine litter. In *Marine Anthropogenic Litter*, 395–428. https://doi.org/10.1007/978-3-319-16510-3

[105] Onyena, A. P., Aniche, D. C., Ogbolu, B. O., Rakib, M. R. J., Uddin, J., & Walker, T. R. (2021). Governance strategies for mitigating microplastic pollution in the marine environment: a review. *Microplastics*, 1(1), 15–46. https://doi.org/10.3390/microplas tics1010003

[106] Matthews, C., Moran, F., & Jaiswal, A. K. (2021). A review on European Union's strategy for plastics in a circular economy and its impact on food safety. *Journal of cleaner production*, 283, 125263. https://doi.org/10.1016/j.jclepro.2020.125263

[107] Jambeck, J. R., Geyer, R., Wilcox, C., Siegler, T. R., Perryman, M., Andrady, A., ... & Law, K. L. (2015). Plastic waste inputs from land into the ocean. *Science*, 347(6223), 768–771. https://doi.org/10.1126/science.1260352

[108] Convery, F., McDonnell, S., & Ferreira, S. (2007). The most popular tax in Europe? Lessons from the Irish plastic bags levy. *Environmental and Resource Economics*, 38, 1–11. https://doi.org/10.1007/s10640-006-9059-2

[109] Nielsen, T. D., Hasselbalch, J., Holmberg, K., & Stripple, J. (2020). Politics and the plastic crisis: A review throughout the plastic life cycle. *Wiley Interdisciplinary Reviews: Energy and Environment*, 9(1), e360. https://doi.org/10.1002/wene.360

[110] Yuan, Z., Nag, R., & Cummins, E. (2022). Human health concerns regarding microplastics in the aquatic environment-From marine to food systems. *Science of The Total Environment*, 823, 153730. https://doi.org/10.1016/j.scitotenv.2022.153730

4 Thermal Techniques for the Degradation and Remediation of Microplastics

Deepa Sharma, Diwakar Chauhan,
Swapnil L. Sonawane, and Purnima Jain

4.1 INTRODUCTION

Diminutive plastics with dimensions ranging from nanometres to millimeters are known as microplastics [1]. They are classified into two major categories (Figure 4.1): primary and secondary microplastics [2]. The purposefully produced microplastics are coined as primary microplastics while secondary microplastics are obtained due to natural degradation of large plastics or polymeric wastes [3]. Large plastic debris can break down into microplastics either by mechanical abrasion, exposure to sunlight, or microbial activity, and exhibit a remarkable resistance to further degradation leading to their prolonged persistence in the ecosystem [4]. The persistence of microplastics (MPs) in environmental sources is a multifaceted threat due to their accumulation in terrestrial and aquatic environments, infiltrating food chains, which exerts adverse effects on the surroundings. They contribute to toxicity, ingestion harm, bioaccumulation and biomagnification, physical damage, and ecosystem disruption, and could also act as potential pathogen carriers [5]. Moreover, MPs pose significant risks to organisms and ecosystems, and there are many uncertainties about their long-term impacts. Furthermore, they have emerged as a new class of pollutants due to their low rate of degradation, high stability, and high persistence in air, water, and soil. This has led to the need for the development of new and advanced approaches for MP removal, sorting, and upcycling.

Some conventionally used methods include mechanical fragmentation, sorption and filtration, chemical degradation, biodegradation, pyrolysis, and photocatalytic degradation [6]. Sorting and upcycling differ from the degradation of MPs, as the latter involves the chemical bond breaking of plastics and converting them into their corresponding monomers or other smaller molecules [7]. These degradation processes are very time-consuming for the complete mineralization of macromolecules. Thermal degradation has emerged as a promising strategy for addressing the prevalent predicament of MP pollution, providing an efficient and

66   DOI: 10.1201/9781003486947-4

Thermal Techniques for the Degradation and Remediation

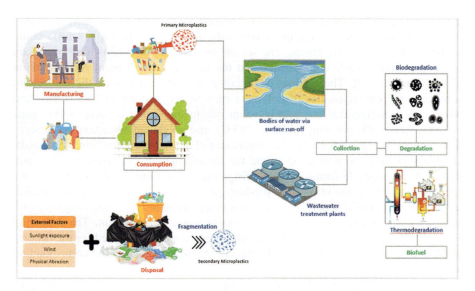

FIGURE 4.1 Illustrating various stages of the microplastic life cycle [11].

versatile sustainable approach to mitigate the impacts of MPs on the environment. Thermal degradation is one of the most traditional and accessible methods of degrading plastics. It offers a promising avenue for remediation as it breaks down these persistent pollutants into smaller and more useful molecules. Thermal processes like pyrolysis or incineration are being used to transform tiny plastics into energy and less toxic byproducts. The thermal degradation temperature and other conditions depend on the nature and type of plastics [8]. For instance, polyethylene (PE) and polypropylene (PP) generally degrade at higher temperatures than polystyrene (PS) and polypropylene terephthalate (PET) [9, 10]. Furthermore, it also demands high energy consumption and emission of harmful byproducts that need to be addressed while considering the use of thermal remediation methods for MP pollution. Therefore, a combination of different remediation techniques tailored to specific environmental contexts and MP characteristics is necessary to mitigate MP pollution effectively.

Thus, this chapter provides an overview of the scientific literature about the thermal degradation and remediation of MPs. Different thermal technologies are presented in this chapter along with a synopsis of the mechanisms involved in their breakdown.

4.2 THERMAL DEGRADATION

The term "thermal degradation" refers to the breakdown of MP polymers at elevated temperatures resulting in their fragmentation and producing small units and other chemical products. It is carried out at temperatures over 100°C causing changes in the polymeric properties of the materials. Thermal degradation of polymers is divided into three major categories. The first involves the disruption and complete degradation of the whole polymer backbone. The second involves the breakdown of side-chain bonds releasing volatile components, and the latter is comprised of a crosslinking mechanism [12, 13]. The thermal degradation of MPs depends on the nature and corresponding characteristics of polymers such as polysulfone, polysiloxanes, and other such materials that are heat resistant due to the strong bonding within their matrix, whereas PP, polyvinyl chloride (PVC), and polybutadiene are easily susceptible to thermal degradation [1]. The importance of thermal techniques in MP remediation lies in their capacity to facilitate the transformation of persistent MP waste into more manageable forms, thereby mitigating its environmental burden [14]. By converting plastic waste into useful products or energy sources, thermal degradation techniques offer a dual benefit of waste reduction and resource recovery, aligning with principles of circular economy and sustainable waste management. The following sections will discuss the various thermal degradation techniques used for the decomposition and remediation of MPs, their salient characteristics, and the corresponding mechanisms.

Thermal degradation techniques encompass a diverse array of methodologies tailored to the unique characteristics of MP waste. Various thermal techniques such as incineration, catalytic cracking, pyrolysis, and thermal depolymerization offer avenues for the management of rapidly multiplying MP waste, each having its unique advantages and disadvantages. When MPs undergo any specific degradation, they typically yield a variety of products and other chemicals depending on various factors like temperature, catalyst, residence time, nature of polymeric material, and other processing conditions.

4.2.1 PYROLYSIS

Pyrolysis is a thermal degradation process carried out in the absence of oxygen or a low-oxygen environment at high temperatures ranging from 300-900°C [15]. It is a thermal depolymerization process that converts MP waste into value-added products such as monomers, fuel, and carbonaceous products. It offers high conversion efficiency due to the effective decomposition of polymers (e.g., PP, PVC, PS, PET) yielding valuable products in the form of liquid or gaseous fuels. This process involves high energy consumption due to the poor conductivity of MPs and the endothermicity of polymer decomposition [16]. For this, catalysts are used to decrease the temperature of the process [17].

Pyrolysis offers many benefits over other traditional plastic waste management methods. For example, recycled plastics often have reduced quality. Their clarity, strength, and flexibility decline with each recycling, a process called downcycling. However, pyrolysis is an upgrading process as it yields value-added products by thermally treating the MPs. Unlike recycling, pyrolysis involves zero cost for sorting, washing, and mixing plastics.

Furthermore, mechanical recycling can only melt and reshape thermoplastic waste into new products whereas in the pyrolysis technique, both thermoplastic and thermoset can be used as a feedstock thereby reducing the persistence of thermoset

Thermal Techniques for the Degradation and Remediation

plastics in the dumpsites [18]. Similarly, nanocomposite materials can also be used as feedstock for this method [19]. Finally, the product and yield composition can vary depending on the composition of the feedstock, process conditions such as heating rate, temperature, and the presence (catalytic pyrolysis), or absence of a catalyst (thermal pyrolysis).

Despite its efficacy in the production of gaseous and liquid fuels, the process of pyrolysis of MPs has some limitations. These include the potential release of hazardous gases and the formation of carbonaceous residues that require appropriate treatment and management strategies before being released into the environment to improve the sustainability of the process.

CLASSIFICATION OF PYROLYSIS BASED ON PROCESSING CONDITIONS

4.2.1.1 Slow Pyrolysis (Non-isothermal)

In this process, the feedstock is heated at a slow rate in the absence of oxygen. The heating rate is usually kept at 10°C/s due to which the volatiles are partially evaporated from the organic material. This results in the formation of char remains as the main product that normally consists of 80% carbon. Due to the formation of solid char during slow pyrolysis, it is also called carbonization.

4.2.1.2 Fast Pyrolysis (Isothermal)

As the name indicates, it involves rapidly heating the feedstock to moderate temperatures ranging from 400–600°C at short residence time and results in yielding a high amount of liquid fuel. The heating rate of fast pyrolysis is kept at 100°C/s and is considered the most common technique both in research and industry for remediation of MPs. The pyrolysis reactor operates isothermally and provides a condition to maximize the production of the liquid. Thus, it is also known as isothermal pyrolysis.

4.2.1.3 Ultra-fast/flash Pyrolysis

Ultra-fast, or flash pyrolysis, is an extremely rapid thermal degradation technique for MPs, with a high heating rate varying between 100–10,000°C/s and short residence time duration. Due to extreme heating conditions, the main products are gases and bio-oil.

4.2.1.4 Thermal Pyrolysis

The depolymerization or cracking of MP materials at extremely high temperatures (350-900°C) in complete absence or low oxygen atmosphere is known as thermal pyrolysis or thermal cracking. During this process, liquid fuel is collected at the condensable portion of the volatile products leaving behind noncondensable gas with high calorific value. Carbonized char or solid residues are among other products that are produced.

It is a thermochemical process that yields a liquid mixture including aromatics, paraffins, isoparaffins, olefins, and naphthenes. The elevated temperature up to 900°C facilitates chain scission, bond breakage, and weakening of intramolecular and inter-molecular forces [20]. Scott *et al.* reported that the majority of the product obtained at temperatures below 700°C was solid, whereas at higher temperatures, gases were the main product [20]. The physicochemical characteristics of the liquid fuel produced by thermal pyrolysis are significantly impacted by the type/nature of MP [21]. A comparative analysis of the chemical structures of low-density polyethylene (LDPE), high-density polyethylene (HDPE), PS, PP, and PET revealed that the $-CH_2-$ and $-CH_3-$ groups are identical in LDPE and HDPE polymer chains, however, aromatic benzene compounds are present in PS and PET in mixed feed [21]. Additionally, an analysis of the chemical characteristics of the consequential liquid fuel revealed that it is a blend of different hydrocarbons.

4.2.1.5 Catalytic Pyrolysis

The heating of plastic waste at an elevated temperature leads to high energy consumption making it a delimiting factor of the cracking process for MPs recycling and upgrading. To overcome this problem, catalysts are used for the degradation process under an inert atmosphere where no oxygen is present. The main purpose of catalysts in MP pyrolysis is to decrease energy consumption, modify the product composition, and reduce the processing time.

A study on waste plastic degradation using catalytic pyrolysis shows the utilization of staged catalysis to produce gasoline-range hydrocarbon oils [22]. In this study, a mesoporous MCM-41 (Mobil Composition of Matter No. 41) catalyst was stacked on top of a microporous ZSM-5 (Zeolite Socony Mobil–5) catalyst to maximize the conversion of plastic waste to gasoline-range hydrocarbons. In addition, various ratios of MCM-41: Zeolite ZSM-5 catalyst were examined and individually compared. It was observed that staged pyrolysis-catalysis yields a high percentage of oil product (83.15 wt%) from HDPE when 1:1 ratio of MCM-41:ZSM-5 of both mesoporous and microporous catalysts was used, producing C_2 (mostly ethene), C_3 (primarily propene), and C_4 (primarily butene and butadiene) as the principal gases. Furthermore, it was noted that the oil products were comprised of 97.72 wt% of gasoline range hydrocarbons and 95.85 wt% of oil was aromatic [22].

Since wax generation is common during the thermal degradation process it has been reduced by incorporating different types of catalysts such as silica, zeolite, or clay. In another study, Jan *et al.* investigated $BaCO_3$ as a non-acidic catalyst for the upconversion of HDPE into fuel oil with low wax production. The study also proves that fuel oil produced via the catalytic process contains relatively low boiling point hydrocarbons in comparison to thermal pyrolysis [23]. Several studies have shown the utilization of silica-alumina as the solid acid catalyst and demonstrated that every type of PE (such as LDPE, HDPE) could be transformed into high-yielding liquid fuel ranging from 77-83 wt% with no generation of wax [24, 25].

Recently Eze *et al.* examined the chemical and physical properties of liquid fuel derived from cracking of mixed waste plastics (MWP) in the presence of clay as a catalyst [18, 26]. Manos et al. carried out PE catalytic pyrolysis using a clay catalyst and compared the results with an ultrastable Y zeolite [27]. It was found that clay catalysts were able to fully degrade HDPE at high processing temperatures, but they were observed to be less active than ultrastable Y zeolite at 600K, although the liquid

Thermal Techniques for the Degradation and Remediation 71

fuel yield was higher in the case of clay catalyst (70%) as compared to Y zeolite catalyst (50%). It was also observed that the liquid products obtained over clay catalysts were heavier, which was attributed to the less acidic properties of clay.

4.2.2 MICROWAVE-ASSISTED DEGRADATION

Microwave-assisted thermal degradation techniques have emerged as an innovative and advanced method of decomposition of MPs that addresses and diminishes the high-temperature and nonselective issues faced by other thermal methods. For this approach, microwave radiation is employed to induce the thermal disintegration of MPs, yielding smaller, less harmful molecules or, at the extreme, valuable resources like fuel and chemicals. This process stands out because it can directly heat the material being treated, which not only enables rapid and even heating but also greatly enhances degrading efficiency relative to other thermal methods [28]. This is the main advantage of microwave-assisted thermal degradation. The use of microwaves can result in concentrating on MP materials with good dielectric properties, which can allow faster heating and a higher rate of degradation [29]. This is a general approach that reduces, in all instances, the amount of energy consumed and hence the carbon footprint associated with the efforts of remediating MPs, making it more sustainable than traditional waste management.

Microplastic-containing waste is put inside the special reactor, which is further subjected to microwave radiation. It allows agitation of the MP and the microwave together to generate heat for the pyrolysis of plastic waste into pyrolysis oil, gas, and char. The exact products of degradation depend on the nature of the plastic, catalysts, and reaction conditions within the reactor [30]. From an environmental point of view, microwave-assisted thermal degradation is a very good solution to reduce the effect of MP pollution. The work turns the waste into useful products and is in line with a circular economy and the recovery of resources [31].

Due to several recognized advantages compared to thermal and catalytic pyrolysis, microwave-induced pyrolysis represents a useful technology for obtaining valuable compounds and fuels [32]. It aids in fast, volumetric, and selective heating of MPs improving the energy recovery process. Microwaves are not able to pass through plastics due to their very low dielectric loss factor. Some typical examples of such plastics are PET, PVC, PS, PP, and PE. These types of plastics are not able to absorb microwave radiation unless it is microwavable-absorbable. Thus, they require supplementary absorbent materials such as iron mesh [33], carbon black [34], silicon carbide, and shredded tires [35], which are high-quality absorbers of high dielectric loss factor and ensure proper heating of the plastic material during pyrolysis.

Despite these, there are certain implementation barriers in the microwave-assisted thermal degradation for MP waste. The proper reactor design should be cost-effective and scalable in development; some other barriers are optimization of process parameters for all sorts of plastics, including their assessment of life cycle environmental impacts [30, 36]. Integration of this technology with current existing waste management systems needs to be evaluated for the future. Applications in the byproducts formed as a result of degradation also need to be evaluated.

4.2.3 INCINERATION

Incineration is the most commonly used method of thermal degradation for the removal of MP contamination. It involves the combustion of MP waste at high temperatures producing gas, heat, and ash as the product of incineration. The primary advantage of this process is its ability to considerably reduce the volume of solid waste by converting complex macromolecules of MP into smaller units. A high amount of heat is generated as a result of combustion, which can be utilized to generate electricity, thereby contributing to both MP degradation and energy recovery and aligning with sustainable waste management principles [37, 38]. Incineration leads to the complete oxidation of plastic waste yielding energy and inert ash residues. It is effective in treating a wide range of MPs, including mixed and contaminated materials.

However, open heating at high temperatures may generate air pollutants and ash containing potentially toxic elements that necessitate proper emission control measures and ash disposal strategies [39, 40]. Moreover, the incomplete combustion of polyolefins, PS, or halogenated plastics can release hazardous byproducts such as volatile organic compounds (VOCs), dioxins, furans, polychlorinated biphenyls, aromatic compounds (such as pyrene and chrysene), and CO and CO_2 depending on the nature of plastic waste [41, 42]. Some released gases are considered greenhouse gases contributing to climate change [43, 44]. Furthermore, the incineration approach requires high installation costs, operation, and maintenance, which restricts its application in MP waste management [45]. Therefore, it is necessary to focus on the short- and long-term effects of incineration on human health and the environment along with technological advancement to improve its efficacy.

4.2.4 GASIFICATION

The gasification method is an advanced thermal treatment that takes place at high temperatures to decompose and convert the polymer of MPs into simpler molecules of smaller size. This approach aims to reduce the negative consequences of MP contamination in the environment. The gasification method is performed in the presence of limited oxygen compared to conventional combustion technology [46]. Moreover, the method also converts the complex polymer chains into simple molecules such as syngas, composed of hydrogen and carbon monoxide, as well as insignificant carbon dioxide and methane. The produced syngas is used as a resource for energy generation, as well as a raw material for chemical reactions, making gasification of MPs an energy-generating technology that addresses MP contamination [17, 47].

The gasification process includes several stages: initial drying, pyrolysis, oxidation, and reduction [48]. In a gasifier, which is typically a high-temperature/pressure vessel, gasification transmutes MPs by removing moisture from the material and thermally decomposes it to produce char, tar, and gases in the absence of oxygen. During subsequent oxidation, limited oxygen is introduced to partially oxidize the char enhancing the production of syngas. Lastly, the final reduction phase triggering a sequence of chemical processes transforms other remaining materials into gas and ash/slag.

Gasification is a thermochemical process that transforms carbonaceous materials into synthetic gas (syngas) made up of carbon monoxide, hydrogen, methane, carbon

Thermal Techniques for the Degradation and Remediation 73

dioxide, and nitrogen [49]. With the use of catalysts such as quartzite sand or Ni-based compounds generation of H_2 can be boosted while reducing methane [50] or even endorse decreased activation energies [51] that facilitate faster reactions and easier conversions. Catalysts such as Ni-based, iron-based, calcined dolomites and magnesites, zeolites, and olivine work in situ to promote chemical processes that affect syngas composition and heating value [52].

Apart from the benefits of plastic gasification, many researchers have shown that mixing MP waste with other residues results in improvement in syngas production, owing to their valuable energetic composition, which contributes to improved process outcomes [53–55]. Ahmed *et al.* (2011) demonstrated that combined samples of woodchips and PE resulted in increased syngas output and composition, as well as higher energy contents and thermal efficiency [53]. The rationale may be based on the hydrogen donor capacity of plastic residues, which stabilizes the radicals created by biomass, as well as the contribution of biomass char to the adsorption of volatiles from PE, which increases hydrogen production.

In conclusion, the gasification of MP minimizes MP waste and creates syngas that can be beneficial to the energy and chemical manufacturing industries, and thus, the benefits may be economic and environmental. Gasification of MP species is environmentally and economically sustainable due to the waste-to-energy conversion, which is beneficial for the environment and economy along with organisms in marine and terrestrial ecosystems.

4.3 UPCYCLING OF MICROPLASTICS

Many methods for recycling plastic waste have been developed to minimize increasing plastic pollution, although each technique possesses some significant challenges in terms of feasibility and cost. Upcycling has been considered an attractive tool for large-scale valorization and remediation of plastic waste. In this process, MP wastes are utilized as a feedstock and transformed into high-value products. Upcycling of plastic pollutants can be carried out either via biodegradation or chemical methods. The biodegradation process facilitates the conversion of organic compounds derived from deteriorated MPs into valuable products by the integration of microorganisms that metabolize these organic compounds via various biochemical processes [56]. However, the chemical method of upcycling uses catalysts to depolymerize the MP waste to form corresponding monomers in the presence of a sophisticated controlled environment. Thus, upcycling expedites the conversion of MPs into high-value and high-performance chemicals, fuels, and materials [7, 57].

4.3.1 Biodegradation

Due to increasing plastic waste, biopolymers have been studied as a tool to reduce plastic pollution and enhance environmental sustainability. Biopolymers are biodegradable materials that are also referred to as thermoplastic polymers (e.g., polylactic acid (PLA), polyvinyl alcohol (PVA), polyhydroxyalkanoate (PHA), and plant-based polymers such as cellulose and starch) [58]. However, thermoplastic polymers derived from polyolefins (e.g., PP, PE) [59] or other functionalized monomers (e.g., PVC, PET) [60] are not biodegradable even though they comprise prooxidant additives that stimulate their photodegradation and/or thermodegradation. Biodegradation of MP simply refers to the breakdown of polymeric materials in the presence of microorganisms that leads to mineralization, carbon recycling, and the generation of biomass. Basically, MP biodegradation is composed of three major phases: biodeterioration, biofragmentation, and assimilation. Since scientists have faced great difficulty in determining the assimilation of oligomers or other monomers by microorganisms irrespective of biopolymers, they have aimed to develop new innovative materials. In recent years, smart plastic polymers have been designed with biodegradation characteristics, facilitating the ease to recycle, reuse, and upcycle plastic waste materials after the end of their life cycle. Irrespective of this, researchers are still working on evolving such new materials and developing other ways to monitor the biodegradability of these materials [61]. The three stages of upcycling of MPs by biodegradation are discussed in the following.

4.3.1.1 Biodeterioration

Biodeterioration is a surface phenomenon that alters the physical, chemical, and mechanical characteristics of a material. It refers to the degradation of a substance in the presence of microorganisms such as bacteria, protozoa, and algae. The microbial activity is governed either by physical and mechanical characteristics of a material or by enzymatic activity, administering the growth of microorganisms on or inside the surface of a suspended material. The surficial microbial development of MP polymers depends on environmental factors like temperature, humidity, and atmospheric pollutants, as well as the constituents of the polymeric materials [61]. It has been observed by researchers that atmospheric pollutants serve as a potential source of nutrients for microbial colonization over the surface of plastics. This was reported by Zanardini *et al.* who showed that sulfuric acid and aromatic and aliphatic hydrocarbons play the role of nutrients and enhance microbial growth over the skin of MPs [62]. In a similar report, Tharanathan and his co-workers showed that organic dyes can also act as potent nutrition for these microorganisms [63]. Efficient biodeterioration is only when there is an intermolecular interaction between the microbes and plastics. This occurs via microbial secretion of sticky matter made up of protein and polysaccharide polymeric matrix that facilitates surficial binding with MPs [64]. Microbial secretion creates a complex interplay of physical, chemical, and biological factors. This microbial sticky matrix is also responsible for attracting atmospheric pollutants, promoting microbial colonization, and protecting microorganisms from unfavorable conditions (such as temperature and humidity). Furthermore, it eases the penetration of microbes from the surface to the interior matrix of polymers, weakening the resistivity of MP to biodegradation [65]. Therefore, the surficial growth of the microbial florae over MPs contributes to the chemical biodeterioration of plastics.

4.3.1.2 Biofragmentation

Fragmentation is a phenomenon of the breakdown of complex macromolecules into small units. This process, which occurs in the presence of a microbial consortium, is known as biofragmentation. Biofragmentation is a vital step preceding assimilation

Thermal Techniques for the Degradation and Remediation 75

where small molecules or monomers are generated by the action of microorganisms under an experimentally controlled environment. As already discussed, biodeterioration leads to microbial colonization on the surface of MPs and enables their penetration into the polymeric matrix. Similarly, biofragmentation is a complementary step of biodeterioration in which breached microorganisms cleave the complex polymeric chains to form oligomers or monomers. Biofragmentation is mainly governed by the enzymes that belong to oxidoreductases and hydrolases (e.g., cellulases, amylases, lipases, esterases) [66]. Despite having prodigious specificity, enzymes are limited by their huge susceptibility to denaturation because of multiple factors such as temperature, radiation, humidity, and pH of the MP matrix [67]. Based on the nature of synthesis, enzymes can be classified as constitutive or inductive. Constitutive enzymes refer to the enzymes that are synthesized throughout the whole life cycle irrespective of the presence of substrate, whereas inductive ones are produced by the cell response caused against any specific substrate. Endopeptidases and endoesterases are constitutive enzymes responsible for catalyzes along the polymer chains while inductive enzymes, also called exoenzymes, as the name suggests, are responsible for marginal catalyses [68]. Enzymes can either remain freely suspended over the MP matrix or can be fixed on particles. However, fixed enzymes have additional benefits in terms of their stability, catalytic activity, and longer durability. Enzymes can follow diverse mechanisms for MP fragmentation such as the secretion of specific enzymes or the generation of free radicals that serves the role of catalysts. The procedure for biofragmentation is easier but it is quite difficult to maintain same reaction conditions all through out the fragmentation process. The biofragmentation process involves the mixing of MPs within a liquid media containing an enzyme of specific activity where by the action of enzymes its depolymerization takes place producing different low molecular weight molecules. The filtration of specific monomers may be possible using Size Exclusion Chromatography [69, 70]. Similarly, to segregate oligomers and monomers in a liquid, Gas Chromatography and High-Performance Liquid Chromatography are employed [71]. The reaction conditions and activity of the enzyme can be modified to obtain a high amount of a specific low-molecular-weight monomer whose purity and other intermediates may be identified using Mass Spectrometry (MS) [72], Nuclear Magnetic Resonance (NMR) [73], and Fourier Transform Infrared Spectroscopy (FTIR) [74, 75].

4.3.1.3 Assimilation

Assimilation is a unique and critical existance of integration between the atoms of fragmented MPs and microbial cells. This event acts as a necessary source of energy, electrons, and other nutrients (i.e., carbon, nitrogen, phosphorus, sulphur) for the formation of the cell structure. Therefore, assimilation facilitates microbial growth and its colonization utilizing nutrient substrates obtained through the fragmentation and biodeterioration of MPs [61]. Monomers neighboring the microbial cells pass through the cellular membranes via particular membrane carriers whereas impermeable molecules can undergo biotransformation reactions to form products that can or cannot be assimilated [76]. The transported molecules are oxidized within the cells through catabolic mechanisms to produce energy in the form of ATP (adenosine triphosphate) and constitutive cell structure molecules. Thus, the catabolic pathways of

TABLE 4.1

Challenges and Shortcomings in the Thermal Degradation of Polymers

Thermal degradation method	Type of microplastic	Challenges	Outcomes	References
Pyrolysis	Polyethylene (PE) Polypropylene (PP) Polyvinyl Chloride (PVC) Polystyrene (PS) Polyethylene Terephthalate (PET) Polycarbonate (PC) Polytetrafluoroethylene (PTFE) Polyamide (Nylon) Polyurethane (PU)	• Products are complicated, • harmful components break down at high temperatures, • creation of coke from liquid materials including much water.	• High efficiency – up to 80% energy recovery – that produces pyrolysis gas, charcoal, and bio-oil from municipal solid waste. • Minimal specifications for the field. • Decreased emissions of SO_2 and NOx. • Great adaptability – the ability to handle a wide range of plastic wastes • It works with nearly all types of plastic waste, whether clean or unclean and unsorted • No need to perform the shredding task. • Its reactor and accessory system house all the processes, from plastic to fuel oil, making it incredibly practical and labour-saving.	[79, 80]
Gasification	Polystyrene (PS) Acrylonitrile Butadiene Styrene (ABS)	• The production of tar. • Better suited for sizable power facilities. • Increased capital and operating expenses. • Metal pipes corrode during a reaction. • Involves a greater use of energy.	• Reducing waste production by 50–90%. • Great adaptability (ability to handle different kinds of waste) • Produces carbon products and H_2-rich synthesis gas from MSW • The calorific value of the synthesis gas can rise if it contains high levels of H_2, CH_4, and CO.	[79, 80]

Incineration	Mixed waste	Recovers energy and eliminates the need to filter environmental pollutants like dioxins and CO_2.	Because it doesn't require a large amount of room and can even recover energy in the form of heat, plastic incineration bypasses some of the restrictions placed on landfills.	[81]
Landfilling	Mixed Waste	It Emits landfill gases, uses space and causes soil and air pollution.	• Lack of use of hazardous substances during the process • Useful for mixed materials	[82, 83]
Combustion	Polyethylene (PE) Polypropylene (PP) Polyvinyl Chloride (PVC) Polystyrene (PS) Polyethylene Terephthalate (PET) Polycarbonate (PC) Polytetrafluoroethylene (PTFE) Polyamide (Nylon) Polyurethane (PU)	• Produces enormous volumes of pollution and greenhouse emissions. • Dioxin and other persistent organic pollutant (POP) production. • Exorbitant startup and ongoing expenses. • Extremely inefficient method for handling garbage that has a lot of moisture	• A frequently used MSW conversion technique. • Utilizing technology and an established industrial infrastructure, MSW is converted into electricity, steam, combined heat and energy, and other energy forms. • A notable 70–80% reduction in the amount of garbage produced. • Minimal running expenses.	[79, 80]

energy production for microbial activity and reproduction can be classified as aerobic, anaerobic, and fermentation.

The aerobic catabolic pathway occurs in the presence of oxygen as the final acceptor, and it is possible with a substrate that can be subjected to oxidation within the cell producing a larger amount of energy than other basic pathways. Similarly, an anaerobic pathway also produces a high amount of energy, but it occurs in the presence of electron acceptors other than oxygen (e.g., NO_3^-, SO_4^{2-}, S, CO_2, Fe^{3+}, fumarate) [68]. The microorganisms lacking any of the above electron transport systems undergo fermentation, which is an incomplete oxidation pathway. In fermentation, cells synthesize some endogenous organic molecules which serve the purpose of electron acceptors. The major products of fermentation comprise mineral or organic molecules excreted into the environment, acting as a carbon source for other organisms. Although the energy released during this process is less than the other two processes due to incomplete oxidation, the ecotoxicity of released mineral molecules in the environment is comparably high in the latter case. For evaluating assimilation efficacy, standardized respirometry methods are used that measure the consumption of oxygen or evolution of CO_2. Since radiolabelled polymers are used to perform $^{14}CO_2$ respirometry, this makes this method very expensive, time-consuming, and hazardous with sophisticated equipment and laboratory requirements [77.

4.3.2 CHEMICAL UPCYCLING

Upcycling refers to the upconversion of plastic waste to produce highly valuable products that are utilized for the synthesis of other materials, thus supporting the circular economy of plastic wastes [57]. Chemical upcycling is not a special subcategory of upcycling process. It differs from biodegradation in terms of the breakdown process of MPs. Biodegradation, as discussed earlier, involves microorganisms for the depolymerization process of plastics while the chemical pathway of upcycling is chemically catalyzed. It converts complex MPs into simple monomer or oligomer units through different chemical reactions including depolymerization, chemical digestion, thermal pyrolysis, incineration, and other catalyzed chemical/thermal processes. The major aim is to valorize massively generated diverse plastic waste as conventional feedstock to produce value-added chemicals, fuels, or materials. Additionally, it could deal with contaminated or complex plastics by adapting pragmatic catalysts and parameters, which is difficult to do with mechanical recycling [78]. Thus, chemical upcycling supports the paradigm of environmental sustainability through low carbon emissions and provides an alternative to utilizing plastic waste for producing high-value chemical products for the circular economy. Hence, it has been discussed separately to provide a better understanding among chemical/thermal techniques of MP degradation and biodegradation processes.

4.4 SHORTCOMINGS AND CHALLENGES

The major shortcomings of thermal degradation and remediation techniques encountered by researchers include energy intensiveness, incomplete degradation,

Thermal Techniques for the Degradation and Remediation 79

release of toxic byproducts, environmental impacts, and difficulty in scale up of these processes as shown in Table 4.1. The thermal degradation methods used for the remediation of MPs are the simplest and most frequently used remediation methods for treating plastic waste. However, use of thermal methods raises concern regarding environmental sustainability and efficacy of upcycling the plastic waste. Therefore, several optimization and modifications are required that address the limitations and challenges that arise during the application of these methods.

Typically, the high-temperature requirement and prolonged heating times during the thermal degradation of MPs make it highly energy-intensive and economically nonfeasible. Additionally, thermal remediations often lead to incomplete digestion or breakdown of MPs causing the production of smaller molecules or even toxic byproducts that further aggravate plastic pollution causing supplementary environmental and health risks. Therefore, the high energy consumption and potential release of byproducts during thermal degradation processes necessitate dynamic monitoring and amendments in the thermal degradation methods to reduce their negative impact on the environment and enhance their viability for scale-up in real time. This highlights the importance of evaluating the overall environmental footprints of thermal remediation techniques for sustainable management practices of MPs.

4.5 CONCLUSION AND FUTURE PROSPECTS

In conclusion, thermal techniques represent a promising route for the degradation and remediation of MPs, offering varied methodologies tailored to the unique challenges posed by plastic waste. Thermal degradation and remediation techniques provide viable solutions for converting persistent MP waste into valuable products or unharmful inert residues. While each thermal technique presents its own set of advantages and disadvantages, their potential holds significant promise for addressing the ubiquitous presence of MPs in the environment. Future research should emphasize optimizing thermal processes, improving energy efficiency, and diminishing environmental impacts. Additionally, interdisciplinary associations between researchers, industries, and policymakers will prove to be essential for rendering scientific advancements into real-world solutions and executing effective management strategies for MPs. By embracing innovation, sustainability, and collaboration, we can carve the way towards a more pristine and healthier environment for present and future generations.

REFERENCES

[1] A. Subair, G. Meera, C. Suchith, KJ. SujithKumar., C. Chingakham, K.L. Priya, M.S. Indu, "Techniques for Removal and Degradation of Microplastics," 2023, pp. 127–153. doi: 10.1007/978-3-031-36351-1_6.

[2] K. Syberg, F. Khan, H. Selck, A. Palmqvist, G. Banta, J. Daley, L. Sano, M. Duhaime, "Microplastics: Addressing ecological risk through lessons learned," *Environ Toxicol Chem*, vol. 34, no. 5, pp. 945–953, May 2015, doi: 10.1002/etc.2914.

[3] M. Lehtiniemi, S. Hartikainen, P. Näkki, J. Engström-Öst, A. Koistinen, and O. Setälä, "Size matters more than shape: Ingestion of primary and secondary microplastics

by small predators," *Food Webs*, vol. 17, p. e00097, Dec. 2018, doi: 10.1016/j.fooweb.2018.e00097.

[4] K. Ziani, C. Ionita-Mindrican, M. Mititelu, S. Neacsu, C. Negrei, E. Morosan, D. Draganescu, O. Preda, "Microplastics: A real global threat for environment and food safety: A state of the art review," *Nutrients*, vol. 15, no. 3, p. 617, Jan. 2023, doi: 10.3390/nu15030617.

[5] Z. Qaiser, M. Aqeel, W. Sarfraz, Z. Fatima Rizvi, A. Noman, S. Naeem, N. Khalid, "Microplastics in wastewaters and their potential effects on aquatic and terrestrial biota," *Case Stud Chem Environ Eng*, vol. 8, p. 100536, Dec. 2023, doi: 10.1016/j.cscee.2023.100536.

[6] S. N. Dimassi, J. N. Hahladakis, M. N. D. Yahia, M. I. Ahmad, S. Sayadi, and M. A. Al-Ghouti, "Degradation-fragmentation of marine plastic waste and their environmental implications: A critical review," *Arab J Chem*, vol. 15, no. 11, p. 104262, Nov. 2022, doi: 10.1016/j.arabjc.2022.104262.

[7] Q. Hou, M. Zhen, H. Qian, Y. Nie, X. Bai, T. Xia, M. Laiq Ur Rehman, Q. Li, M. Ju, "Upcycling and catalytic degradation of plastic wastes," *Cell Rep Phys Sci*, vol. 2, no. 8, p. 100514, Aug. 2021, doi: 10.1016/j.xcrp.2021.100514.

[8] S. L. Madorsky, and S. Straus, "Thermal degradation of polymers at high temperatures," *J Res Natl Bur Stand A Phys Chem*, vol. 63A, no. 3, p. 261, Nov. 1959, doi: 10.6028/jres.063A.020.

[9] E. Esmizadeh, C. Tzoganakis, and T. H. Mekonnen, "Degradation behavior of polypropylene during reprocessing and its biocomposites: Thermal and oxidative degradation kinetics," *Polymers (Basel)*, vol. 12, no. 8, p. 1627, Jul. 2020, doi: 10.3390/polym12081627.

[10] R. Panowicz, M. Konarzewski, T. Durejko, M. Szala, M. Lazinska, M. Czerwinska, P. Prasula, "Properties of Polyethylene Terephthalate (PET) after thermo-oxidative aging," *Materials*, vol. 14, no. 14, p. 3833, Jul. 2021, doi: 10.3390/ma14143833.

[11] A. A. Arpia, W.- H. Chen, A. T. Ubando, S. R. Naqvi, and A. B. Culaba, "Microplastic degradation as a sustainable concurrent approach for producing biofuel and obliterating hazardous environmental effects: A state-of-the-art review," *J Hazard Mater*, vol. 418, p. 126381, Sep. 2021, doi: 10.1016/j.jhazmat.2021.126381.

[12] K. Pielichowski, J. Njuguna, and T. M. Majka, *Thermal degradation of polymeric materials*, 2nd ed. Elsevier, 2022.

[13] J. Izdebska, "Aging and Degradation of Printed Materials," in *Printing on Polymers*, Joanna Izdebska and Sabu Thomas (Eds.) Elsevier, 2016, pp. 353–370. doi: 10.1016/B978-0-323-37468-2.00022-1.

[14] X. Zhao, M. Korey, K. Li, K. Copenhaver, H. Tekinalp, S. Celik, K. Kalaitzidou, R. Ruan, A. Ragauskas, S. Ozcan, "Plastic waste upcycling toward a circular economy," *Chem Eng J*, vol. 428, p. 131928, Jan. 2022, doi: 10.1016/j.cej.2021.131928.

[15] C. Zaman, K. Pal, W. Yehye, S. Sagadevan, S. Shah, G. Adebisi, E. Marliana, R. Rafique, R. Johan, "Pyrolysis: A Sustainable Way to Generate Energy from Waste," in *Pyrolysis*, Mohamed Samer (Ed.), InTech, 2017. doi: 10.5772/intechopen.69036.

[16] C. Vasile, H. Pakdel, B. Mihai, P. Onu, H. Darie, and S. Ciocâlteu, "Thermal and catalytic decomposition of mixed plastics," *J Anal Appl Pyrolysis*, vol. 57, no. 2, pp. 287–303, Feb. 2001, doi: 10.1016/S0165-2370(00)00151-0.

[17] R. Miandad, M. A. Barakat, A. S. Aburiazaiza, M. Rehan, and A. S. Nizami, "Catalytic pyrolysis of plastic waste: A review," *Process Saf Environ Prot*, vol. 102, pp. 822–838, Jul. 2016, doi: 10.1016/j.psep.2016.06.022.

[18] W. U. Eze, I. C. Madufor, G. N. Onyeagoro, and H. C. Obasi, "The effect of Kankara zeolite-Y-based catalyst on some physical properties of liquid fuel from mixed waste

plastics (MWPs) pyrolysis," *Polym Bull*, vol. 77, no. 3, pp. 1399–1415, Mar. 2020, doi: 10.1007/s00289-019-02806-y.

[19] M. Qureshi, A. Oasmaa, H. Pihkola, I. Deviatkin, A. Tenhunen, J. Mannila, H. Minkkinen, M. Pohjakallio, J. Laine-Ylijoki, "Pyrolysis of plastic waste: Opportunities and challenges," *J Anal Appl Pyrolysis*, vol. 152, p. 104804, Nov. 2020, doi: 10.1016/j.jaap.2020.104804.

[20] D. S. Scott, S. R. Czernik, J. Piskorz, and D. S. A. G. Radlein, "Fast pyrolysis of plastic wastes," *Energy Fuels*, vol. 4, no. 4, pp. 407–411, Jul. 1990, doi: 10.1021/ef00022a013.

[21] A. Olufemi, and S. Olagboye, "Thermal conversion of waste plastics into fuel oil," *Int J Petrochem Sci Eng*, vol. 2, no. 8, Nov. 2017, doi: 10.15406/ipcse.2017.02.00064.

[22] D. K. Ratnasari, M. A. Nahil, and P. T. Williams, "Catalytic pyrolysis of waste plastics using staged catalysis for production of gasoline range hydrocarbon oils," *J Anal Appl Pyrolysis*, vol. 124, pp. 631–637, Mar. 2017, doi: 10.1016/j.jaap.2016.12.027.

[23] M. R. Jan, J. Shah, and H. Gulab, "Catalytic degradation of waste high-density polyethylene into fuel products using BaCO3 as a catalyst," *Fuel Process Technol*, vol. 91, no. 11, pp. 1428–1437, Nov. 2010, doi: 10.1016/j.fuproc.2010.05.017.

[24] M. Rahman, B. K. Mondal, N. Ahmed, and M. D. Hossain, "Catalytic pyrolysis of waste high-density (HDPE) and low-density polyethylene (LDPE) to produce liquid hydrocarbon using silica-alumina catalyst," *J Bangladesh Acad Sci*, vol. 47, no. 2, pp. 195–203, Dec. 2023, doi: 10.3329/jbas.v47i2.67950.

[25] A. Marcilla, R. Ruiz-Femenia, J. Hernández, and J. C. García-Quesada, "Thermal and catalytic pyrolysis of crosslinked polyethylene," *J Anal Appl Pyrolysis*, vol. 76, no. 1–2, pp. 254–259, Jun. 2006, doi: 10.1016/j.jaap.2005.12.004.

[26] W. U. Eze, I. C. Madufor, G. N. Onyeagoro, H. C. Obasi, and M. I. Ugbaja, "Study on the effect of Kankara zeolite-Y-based catalyst on the chemical properties of liquid fuel from mixed waste plastics (MWPs) pyrolysis," *Polym Bull*, vol. 78, no. 1, pp. 377–398, Jan. 2021, doi: 10.1007/s00289-020-03116-4.

[27] G. Manos, I. Y. Yusof, N. Papayannakos, and N. H. Gangas, "Catalytic cracking of polyethylene over clay catalysts. Comparison with an ultrastable Y Zeolite," *Ind Eng Chem Res*, vol. 40, no. 10, pp. 2220–2225, May 2001, doi: 10.1021/ie001048o.

[28] P. Patel, and P. Mehta, "Microwave-assisted heating: innovative use in hydrolytic forced degradation of selected drugs," *J Microw Power Electromagn Energy*, vol. 51, no. 3, pp. 205–220, Jul. 2017, doi: 10.1080/08327823.2017.1354746.

[29] C. Yang, H. Shang, J. Li, X. Fan, J. Sun, and A. Duan, "A review on the microwave-assisted pyrolysis of waste plastics," *Processes*, vol. 11, no. 5, p. 1487, May 2023, doi: 10.3390/pr11051487.

[30] P. H. M. Putra, S. Rozali, M. F. A. Patah, and A. Idris, "A review of microwave pyrolysis as a sustainable plastic waste management technique," *J Environ Manage*, vol. 303, p. 114240, Feb. 2022, doi: 10.1016/j.jenvman.2021.114240.

[31] X. Hu, D. Ma, G. Zhang, M. Ling, Q. Hu, K. Liang, J. Lu, Y. Zheng, "Microwave-assisted pyrolysis of waste plastics for their resource reuse: A technical review," *Carbon Resour Convers*, vol. 6, no. 3, pp. 215–228, Sep. 2023, doi: 10.1016/j.crcon.2023.03.002.

[32] K. Ono, and A. Erhard, "Nondestructive Testing, 3. Ultrasonics," in *Ullmann's Encyclopedia of Industrial Chemistry*, Claudia Ley (Ed.) Wiley, 2011. doi: 10.1002/14356007.o17_o02.

[33] Z. Hussain, K. M. Khan, and K. Hussain, "Microwave–metal interaction pyrolysis of polystyrene," *J Anal Appl Pyrolysis*, vol. 89, no. 1, pp. 39–43, Sep. 2010, doi: 10.1016/j.jaap.2010.05.003.

[34] C. Ludlow-Palafox and H. A. Chase, "Microwave-induced pyrolysis of plastic wastes," *Ind Eng Chem Res*, vol. 40, no. 22, pp. 4749–4756, Oct. 2001, doi: 10.1021/ie010202j.

[35] R. J. Meredith, *Engineers' Handbook of Industrial Microwave Heating*, 0-85296th-916th–3rd ed., vol. 25. United Kingdom: The Institution of Electrical Engineers, 2007.

[36] S. Parrilla-Lahoz, S. Mahebadevan, M. Kauta, M. Zambrano, J. Pawlak, R. Venditti, T. Reina, M. Duyar, "Materials challenges and opportunities to address growing micro/nanoplastics pollution: a review of thermochemical upcycling," *Mater Today Sustain*, vol. 20, p. 100200, Dec. 2022, doi: 10.1016/j.mtsust.2022.100200.

[37] T. Astrup, J. Møller, and T. Fruergaard, "Incineration and co-combustion of waste: accounting of greenhouse gases and global warming contributions," *Waste Manage Res*, vol. 27, no. 8, pp. 789–799, Nov. 2009, doi: 10.1177/0734242X09343774.

[38] P. Pandey, M. Dhiman, A. Kansal, and S. P. Subudhi, "Plastic waste management for sustainable environment: techniques and approaches," *Waste Dispos Sustain Energy*, vol. 5, no. 2, pp. 205–222, Jun. 2023, doi: 10.1007/s42768-023-00134-6.

[39] Á. Nagy, and R. Kuti, "The environmental impact of plastic waste incineration," *Acad Appl Res Mil Public Manage Sci*, vol. 15, no. 3, pp. 231–237, Dec. 2016, doi: 10.32565/aarms.2016.3.3.

[40] C.-T. Li, H.-K. Zhuang, L.-T. Hsieh, W.-J. Lee, and M.-C. Tsao, "PAH emission from the incineration of three plastic wastes," *Environ Int*, vol. 27, no. 1, pp. 61–67, Jul. 2001, doi: 10.1016/S0160-4120(01)00056-3.

[41] B. Courtemanche, "A laboratory study on the NO, NO2, SO2, CO and CO2 emissions from the combustion of pulverized coal, municipal waste plastics and tires," *Fuel*, vol. 77, no. 3, pp. 183–196, Feb. 1998, doi: 10.1016/S0016-2361(97)00191-9.

[42] D.-Y. Huang, S.-G. Zhou, W. Hong, W.-F. Feng, and L. Tao, "Pollution characteristics of volatile organic compounds, polycyclic aromatic hydrocarbons and phthalate esters emitted from plastic wastes recycling granulation plants in Xingtan Town, South China," *Atmos Environ*, vol. 71, pp. 327–334, Jun. 2013, doi: 10.1016/j.atmosenv.2013.02.011.

[43] M. Shen, W. Huang, M. Chen, B. Song, G. Zeng, and Y. Zhang, "(Micro)plastic crisis: Un-ignorable contribution to global greenhouse gas emissions and climate change," *J Clean Prod*, vol. 254, p. 120138, May 2020, doi: 10.1016/j.jclepro.2020.120138.

[44] R. R. Bora, R. Wang, and F. You, "Waste polypropylene plastic recycling toward climate change mitigation and circular economy: Energy, environmental, and technoeconomic perspectives," *ACS Sustain Chem Eng*, vol. 8, no. 43, pp. 16350–16363, Nov. 2020, doi: 10.1021/acssuschemeng.0c06311.

[45] Y. Li, R. Zhao, H. Li, W. Song, and H. Chen, "Feasibility analysis of municipal solid waste incineration for harmless treatment of potentially virulent waste," *Sustainability*, vol. 15, no. 21, p. 15379, Oct. 2023, doi: 10.3390/su152115379.

[46] D. Singh, "Advances in Industrial Waste Management," in *Waste Management and Resource Recycling in the Developing World*, Pardeep Singh, Pramit Verma, Rishikesh Singh, Arif Ahamad, and André C. S. Batalhão (Eds.) Elsevier, 2023, pp. 385–416. doi: 10.1016/B978-0-323-90463-6.00027-0.

[47] S. M. Al-Salem, P. Lettieri, and J. Baeyens, "Recycling and recovery routes of plastic solid waste (PSW): A review," *Waste Manage*, vol. 29, no. 10, pp. 2625–2643, Oct. 2009, doi: 10.1016/j.wasman.2009.06.004.

[48] K. Li, R. Zhang, and J. Bi, "Experimental study on syngas production by co-gasification of coal and biomass in a fluidized bed," *Int J Hydrogen Energy*, vol. 35, no. 7, pp. 2722–2726, Apr. 2010, doi: 10.1016/j.ijhydene.2009.04.046.

[49] M. Balat, M. Balat, E. Kırtay, and H. Balat, "Main routes for the thermo-conversion of biomass into fuels and chemicals. Part 2: Gasification systems," *Energy Convers Manag*, vol. 50, no. 12, pp. 3158–3168, Dec. 2009, doi: 10.1016/j.enconman.2009.08.013.

[50] G. Ruoppolo, P. Ammendola, R. Chirone, and F. Miccio, "H2-rich syngas production by fluidized bed gasification of biomass and plastic fuel," *Waste Manage*, vol. 32, no. 4, pp. 724–732, Apr. 2012, doi: 10.1016/j.wasman.2011.12.004.

[51] B. L. F. Chin, S. Yusup, A. Al Shoaibi, P. Kannan, C. Srinivasakannan, and S. A. Sulaiman, "Comparative studies on catalytic and non-catalytic co-gasification of rubber seed shell and high density polyethylene mixtures," *J Clean Prod*, vol. 70, pp. 303–314, May 2014, doi: 10.1016/j.jclepro.2014.02.039.

[52] F. Pinto, R. N. André, C. Carolino, and M. Miranda, "Hot treatment and upgrading of syngas obtained by co-gasification of coal and wastes," *Fuel Processing Technology*, vol. 126, pp. 19–29, Oct. 2014, doi: 10.1016/j.fuproc.2014.04.016.

[53] I. I. Ahmed, N. Nipattummakul, and A. K. Gupta, "Characteristics of syngas from co-gasification of polyethylene and woodchips," *Appl Energy*, vol. 88, no. 1, pp. 165–174, Jan. 2011, doi: 10.1016/j.apenergy.2010.07.007.

[54] J. Alvarez, S. Kumagai, C. Wu, T. Yoshioka, J. Bilbao, M. Olazar, P. Williams, "Hydrogen production from biomass and plastic mixtures by pyrolysis-gasification," *Int J Hydrogen Energy*, vol. 39, no. 21, pp. 10883–10891, Jul. 2014, doi: 10.1016/j.ijhydene.2014.04.189.

[55] R. A. Moghadam, S. Yusup, Y. Uemura, B. L. F. Chin, H. L. Lam, and A. Al Shoaibi, "Syngas production from palm kernel shell and polyethylene waste blend in fluidized bed catalytic steam co-gasification process," *Energy*, vol. 75, pp. 40–44, Oct. 2014, doi: 10.1016/j.energy.2014.04.062.

[56] Y. Zhang, J. N. Pedersen, B. E. Eser, and Z. Guo, "Biodegradation of polyethylene and polystyrene: From microbial deterioration to enzyme discovery," *Biotechnol Adv*, vol. 60, p. 107991, Nov. 2022, doi: 10.1016/j.biotechadv.2022.107991.

[57] X. Chen, Y. Wang, and L. Zhang, "Recent Progress in the Chemical Upcycling of Plastic Wastes," *ChemSusChem*, vol. 14, no. 19, pp. 4137–4151, Oct. 2021, doi: 10.1002/cssc.202100868.

[58] A. Surendren, A. K. Mohanty, Q. Liu, and M. Misra, "A review of biodegradable thermoplastic starches, their blends and composites: recent developments and opportunities for single-use plastic packaging alternatives," *Green Chemistry*, vol. 24, no. 22, pp. 8606–8636, 2022, doi: 10.1039/D2GC02169B.

[59] X. Liu, C. Gao, P. Sangwan, L. Yu, and Z. Tong, "Accelerating the degradation of polyolefins through additives and blending," *J Appl Polym Sci*, vol. 131, no. 18, Sep. 2014, doi: 10.1002/app.40750.

[60] A. Chamas, H. Moon, J. Zheng, Y. Qiu, T. Tabassum, J. Jang, M. Abu-Omar, S. Scott, S. Suh, "Degradation rates of plastics in the environment," *ACS Sustain Chem Eng*, vol. 8, no. 9, pp. 3494–3511, Mar. 2020, doi: 10.1021/acssuschemeng.9b06635.

[61] N. Lucas, C. Bienaime, C. Belloy, M. Queneudec, F. Silvestre, and J.-E. Nava-Saucedo, "Polymer biodegradation: Mechanisms and estimation techniques – A review," *Chemosphere*, vol. 73, no. 4, pp. 429–442, Sep. 2008, doi: 10.1016/j.chemosphere.2008.06.064.

[62] E. Zanardini, P. Abbruscato, N. Ghedini, M. Realini, and C. Sorlini, "Influence of atmospheric pollutants on the biodeterioration of stone," *Int Biodeterior Biodegradation*, vol. 45, no. 1–2, pp. 35–42, Jan. 2000, doi: 10.1016/S0964-8305(00)00043-3.

[63] R. N. Tharanathan, "Biodegradable films and composite coatings: past, present and future," *Trends Food Sci Technol*, vol. 14, no. 3, pp. 71–78, Mar. 2003, doi: 10.1016/S0924-2244(02)00280-7.

[64] F. Cappitelli, P. Principi, and C. Sorlini, "Biodeterioration of modern materials in contemporary collections: can biotechnology help?," *Trends Biotechnol*, vol. 24, no. 8, pp. 350–354, Aug. 2006, doi: 10.1016/j.tibtech.2006.06.001.

[65] S. Bonhomme, A. Cuer, A.-M. Delort, J. Lemaire, M. Sancelme, and G. Scott, "Environmental biodegradation of polyethylene," *Polym Degrad Stab*, vol. 81, no. 3, pp. 441–452, Jan. 2003, doi: 10.1016/S0141-3910(03)00129-0.

[66] J. Wallach, "Biochimie generale (seventh edition)," *Biochem Educ*, vol. 23, no. 1, p. 48, Jan. 1995, doi: 10.1016/0307-4412(95)90207-4.

[67] G. Güner, "HPLC — A practical user's guide," *Biochem Educ*, vol. 23, no. 1, pp. 48–49, Jan. 1995, doi: 10.1016/0307-4412(95)90209-0.

[68] R.-J. Mueller, "Biological degradation of synthetic polyesters—Enzymes as potential catalysts for polyester recycling," *Process Biochem*, vol. 41, no. 10, pp. 2124–2128, Oct. 2006, doi: 10.1016/j.procbio.2006.05.018.

[69] J. A. Ratto, P. J. Stenhouse, M. Auerbach, J. Mitchell, and R. Farrell, "Processing, performance and biodegradability of a thermoplastic aliphatic polyester/starch system," *Polymer (Guildf)*, vol. 40, no. 24, pp. 6777–6788, Nov. 1999, doi: 10.1016/S0032-3861(99)00014-2.

[70] D. N. Bikiaris, G. Z. Papageorgiou, and D. S. Achilias, "Synthesis and comparative biodegradability studies of three poly(alkylene succinate)s," *Polym Degrad Stab*, vol. 91, no. 1, pp. 31–43, Jan. 2006, doi: 10.1016/j.polymdegradstab.2005.04.030.

[71] M. Alberta Araújo, A. M. Cunha, and M. Mota, "Enzymatic degradation of starch-based thermoplastic compounds used in protheses: Identification of the degradation products in solution," *Biomaterials*, vol. 25, no. 13, pp. 2687–2693, Jun. 2004, doi: 10.1016/j.biomaterials.2003.09.093.

[72] U. Witt, T. Einig, M. Yamamoto, I. Kleeberg, W.-D. Deckwer, and R.-J. Müller, "Biodegradation of aliphatic–aromatic copolyesters: evaluation of the final biodegradability and ecotoxicological impact of degradation intermediates," *Chemosphere*, vol. 44, no. 2, pp. 289–299, Jul. 2001, doi: 10.1016/S0045-6535(00)00162-4.

[73] E. Marten, R.-J. Müller, and W.-D. Deckwer, "Studies on the enzymatic hydrolysis of polyesters. II. Aliphatic–aromatic copolyesters," *Polym Degrad Stab*, vol. 88, no. 3, pp. 371–381, Jun. 2005, doi: 10.1016/j.polymdegradstab.2004.12.001.

[74] J.-H. Zhao, X.-Q. Wang, J. Zeng, G. Yang, F.-H. Shi, and Q. Yan, "Biodegradation of poly(butylene succinate-co-butylene adipate) by Aspergillus versicolor," *Polym Degrad Stab*, vol. 90, no. 1, pp. 173–179, Oct. 2005, doi: 10.1016/j.polymdegradstab.2005.03.006.

[75] K. J. Kim, Y. Doi, and H. Abe, "Effects of residual metal compounds and chain-end structure on thermal degradation of poly(3-hydroxybutyric acid)," *Polym Degrad Stab*, vol. 91, no. 4, pp. 769–777, Apr. 2006, doi: 10.1016/j.polymdegradstab.2005.06.004.

[76] A. Sivan, "New perspectives in plastic biodegradation," *Curr Opin Biotechnol*, vol. 22, no. 3, pp. 422–426, Jun. 2011, doi: 10.1016/j.copbio.2011.01.013.

[77] J. Rasmussen, P. H. Jensen, P. E. Holm, and O. S. Jacobsen, "Method for rapid screening of pesticide mineralization in soil," *J Microbiol Methods*, vol. 57, no. 2, pp. 151–156, May 2004, doi: 10.1016/j.mimet.2003.12.004.

[78] M. R. Karimi Estahbanati, X. Y. Kong, A. Eslami, and H. Sen Soo, "Current Developments in the chemical upcycling of waste plastics using alternative energy sources," *ChemSusChem*, vol. 14, no. 19, pp. 4152–4166, Oct. 2021, doi: 10.1002/cssc.202100874.

[79] A. T. Sipra, N. Gao, and H. Sarwar, "Municipal solid waste (MSW) pyrolysis for bio-fuel production: A review of effects of MSW components and catalysts," *Fuel Processing Technology*, vol. 175, pp. 131–147, Jun. 2018, doi: 10.1016/j.fuproc.2018.02.012.

[80] S. Nanda and F. Berruti, "A technical review of bioenergy and resource recovery from municipal solid waste," *J Hazard Mater*, vol. 403, p. 123970, Feb. 2021, doi: 10.1016/j.jhazmat.2020.123970.

[81] F. Zhang, Y. Zhao, D. Wang, M. Yan, J. Zhang, P. Zhang, T. Ding, L. Chen, C. Chen., "Current technologies for plastic waste treatment: A review," *J Clean Prod*, vol. 282, p. 124523, Feb. 2021, doi: 10.1016/j.jclepro.2020.124523.

[82] I. Wojnowska-Baryła, K. Bernat, and M. Zaborowska, "Plastic waste degradation in landfill conditions: The problem with microplastics, and their direct and indirect environmental effects," *Int J Environ Res Public Health*, vol. 19, no. 20, p. 13223, Oct. 2022, doi: 10.3390/ijerph192013223.

[83] K. Korniejenko, B. Kozub, A. Bąk, P. Balamurugan, M. Uthayakumar, and G. Furtos, "Tackling the circular economy challenges—Composites recycling: Used tyres, wind turbine blades, and solar panels," *J Compos Sci*, vol. 5, no. 9, p. 243, Sep. 2021, doi: 10.3390/jcs5090243.

5 Membrane Filtration Technique for Remediation of Microplastics

Anshul Yadav and Kavita Poonia

5.1 INTRODUCTION

Membrane systems' excellent selectivity has led to an increase in their use in separation and purification procedures, ability to operate continuously, and ease of scaling up. Their suitability for wastewater (WW) and drinking water (DW) treatment is particularly notable, as they facilitate water recycling and reuse [1]. Membranes are a type of selective barrier that may discriminate between two phases and limit the passage of substances based on their size, charge, or shape [2, 3]. The three fundamental ideas of sieving, adsorption, and electrostatic phenomena are applied in membrane separation or treatment procedures [4]. The solute (or analyte) and the membrane interact hydrophobically, which causes the adsorption mechanism in the membrane separation process. Rejection is typically made worse by the reduction in membrane pore size brought on by these interactions [5]. The size of molecules and pores in the membrane determine how materials are separated [6]. Hence, a variety of membrane procedures, each with its own unique separation method, have been established.

5.2 MEMBRANE FILTRATION TECHNOLOGIES

Membrane separation technology, as the term suggests, is a cutting-edge separation technique that employs a custom-made film with selective transmission properties. This method is used to separate, purify, and concentrate mixtures, propelled by an outside force [7].

Membrane filtration technology is becoming more widely used in large-scale DW treatment [8] due to its numerous benefits. These include superior effluent quality [9], straightforward process management [10], and rigorous solid-liquid separation with minimal space requirements [11, 12]. Additionally, this method is simple to incorporate into already-existing treatment facilities [13], uses little energy [14], and

86   DOI: 10.1201/9781003486947-5

Membrane Filtration Technique

efficiently gets rid of a wide range of pollutants [15]. Both features of the contamination and the membrane affect how well the pollutant is removed [16]. Despite these advantages, the primary drawback of this technology is the membrane's cost, which can be minimized or even eliminated with proper management of the filtration process.

5.2.1 CHARACTERISTICS OF MEMBRANE SEPARATION TECHNOLOGY

1. The energy consumption is low in the membrane separation process because it does not involve any phase change. This makes reverse osmosis the most energy-efficient method, which is particularly important for addressing the energy crisis in the world [17].
2. The membrane separation process operates at ambient temperature, making it ideal for heat-sensitive substances. This includes the processes of separating, classifying, concentrating, and enriching heavy metals, chemicals, and raw materials in WW. The quality of the raw water has no bearing on the water quality when treating DW using the membrane technique; instead, the water quality is entirely determined by the membrane's properties, such as selectivity and pore size [18].
3. It does not introduce new contaminants or waste materials because it does not require additives and does not change the characteristics of substances throughout the reaction process. It is applicable to several WW treatment procedures [19].
4. The equipment used in the membrane separation method is straightforward, easy to operate, control, and maintain, and offers high separation efficiency. In contrast to conventional water treatment techniques, it has the advantages of occupying less space and providing higher processing efficiency [20].
5. Equipment for membrane separation technology can be standardized and automatically controlled, making it easy to manage and operate. This is also beneficial for promoting industrial development [21, 22].

5.2.2 CLASSIFICATION OF MEMBRANE SEPARATION TECHNOLOGY

Isomerization and anisotropy are two general categories for membranes. Physically and compositionally, isotropic membranes are uniform. They may be thick or nonporous, which restricts their applicability owing to limited permeation fluxes, or microporous, which results in comparatively large permeation fluxes [23].

Depending on the components in their construction, membranes can be categorized as organic or inorganic. Synthetic or naturally occurring organic polymers are used to make organic membranes. Among these polymers are polyethylene (PE), polypropylene, polytetrafluoroethylene (PTFE), and cellulose acetate [24]. Inorganic membranes can be made of metals, silica, zeolites, or ceramic materials [25].

Numerous driving forces affect how media travel across membranes [26]. The membrane techniques will be examined one by one in the following discussion (Figure 5.1).

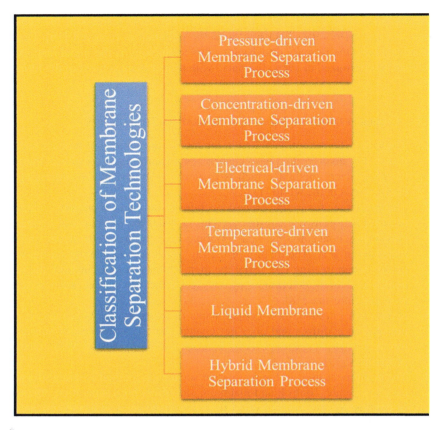

FIGURE 5.1 Classification of membrane separation technologies.

5.2.2.1 Pressure-driven Membrane Separation Process

Pressure-driven membrane procedures are the most widely utilized technology in WW treatment. Using pressure to separate the permeate and retention stages, this method is used to reconcentrate diluted fluids. In comparison to the retention and feed solutions, the solute level in the permeate phase is reduced (Figure 5.2). The pressure used determines the system's overall operating cost [27].

5.2.2.1.1 Microfiltration

Microfiltration (MF) is a physical separation method that uses a porous membrane. It removes germs, turbidity, and dissolved particles via a sieving technique that is dependent on the membrane's pore size. Particles bigger than the pore size (0.1- –0.2μm) of the membrane can be eliminated, while smaller particles can only be eliminated partially. Either organic or inorganic materials can make up the membrane in MF. With MF, you may separate suspensions and emulsions and save around 40% of the organic material. With this approach, the main contaminants removed are

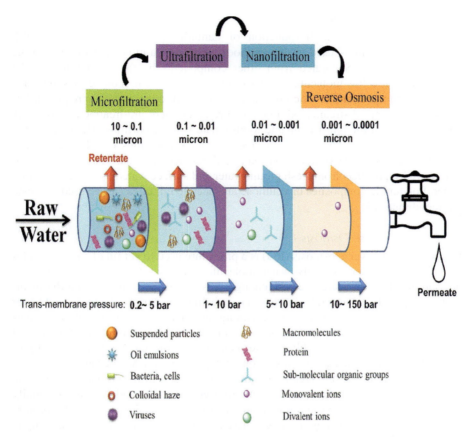

FIGURE 5.2 Pressure-driven membrane separation process for water treatment technologies [27].

silt, algae, protozoa, and bacteria; on the other hand, water (H_2O), minute colloids, viruses, dissolved or organic materials, and monovalent ions like Na^+ and Cl^- can pass through the filter [28].

5.2.2.1.2 Ultrafiltration
Ultrafiltration (UF) membranes are extensively employed as low-energy water filters, helping to eliminate suspended particles, macromolecules, and harmful microbes. The pore diameters of these membranes reach up to around 0.1μm. For particles to separate via these membranes there must be a gradient in either pressure or concentration. Silica, endotoxins, proteins, and viruses can all be retained by UF membranes. Many sectors use this technique, including dairy, food processing, medicines, and drinks. Furthermore, by functioning as a prefilter in reverse osmosis (RO) systems, UF protects the RO membrane [29].

5.2.2.1.3 Nanofiltration

The late 1980s saw the introduction of nanofiltration (NF) membranes [30]. Hydrostatic pressure is used in this method to transport a molecular mixture to the membrane surface. Separated from other components by the membrane are the solvent and a few low molecular weight solutes. Since NF can remove ions that have a significant impact on osmotic pressure, lower working pressures are required. For highly contaminated streams, NF does require an efficient pre-treatment prior to its application, even though it is unable to eliminate soluble portions. The membranes are impacted by free chlorine in the feed water. Applications for NF membranes include desalination, WW treatment, dairy and medicinal fields. They are also used to remove by-products from fresh groundwater and surface water, as well as to soften water. Polyamide composites or cellulose and acetate mixtures can be used to create NF membranes [31].

5.2.2.1.4 Reverse Osmosis

Pressure-driven reverse osmosis (RO) is a procedure used to get rid of small particles and dissolved materials. Only water molecules can pass through it, and osmotic pressure needs to be overcome with enough pressure. Effectiveness of the RO membrane is typically enhanced by increased permeability, superior selectivity, and better fouling resistance. It is thought to be among the most successful membrane separation techniques now in use. In order to eliminate nitrate and minerals, pressurized water flows via a membrane filter. Water is the only molecule that RO does not preserve. The osmotic pressure needed is much higher than for MF because of the pore size. RO is a high-pressure technology that removes salt from salt water. Due to the fact that the flow is against the concentration gradient, RO and NF are essentially different. Water is forced to migrate from the low-pressure side to the high-pressure side by these systems using pressure. There are instances where careful planning is necessary [32].

5.2.2.2 Concentration-driven Membrane Separation Process

Under constant pressure and temperature, a concentration gradient drives the functioning of biological membrane systems. The artificial kidney is a prime example of a synthetic membrane that operates through a concentration-driven process. This group includes dialysis and forward osmosis in which the concentration gradient is the main mechanism for membrane separation.

5.2.2.2.1 Forward Osmosis

The osmotic pressure gradient drives a process called forward osmosis (FO) in which water moves over a semi-permeable barrier between a draw solution and a feed solution. FO's independence from high hydraulic pressure is a definite advantage over conventional pressure-driven membrane technology. Because of its low fouling potential, this method offers possible savings on electricity and membrane maintenance expenses. The efficiency of a FO membrane in the desalination of water has been developed and evaluated. Textile dyeing companies utilize commercially manufactured FO membranes to filter reactive azo dyes from WW.

Membrane Filtration Technique

Polyoxadiazole-cohydrazide (PODH) and polytriazole-co-oxadiazole-cohydrazide are used to polymerize them [33].

5.2.2.2.2 Pervaporation

Pervaporation is a technique used to remove trace amounts of volatile components from liquid mixtures. Vapor pressures are applied through either a porous or nonporous membrane to accomplish this. Membrane permeation and evaporation are combined in this technique to separate the liquid mixture according to their affinities. The petrochemical industries frequently employ this technique to separate volatile organic molecules from hydrocarbons. The difference in concentration is what propels this strategy. It works through a technique called solution-diffusion, which causes vapor to develop as it penetrates. Afterwards, the process's generated vapor is eliminated by either adding an inert medium later on in the process or applying low pressure. The separation of an ethanol-water mixture is a useful application for this technique [2].

5.2.2.2.3 Gas Separation

The pervaporation process and the Gas Separation (GS) process work from the same premise. The feed is first taken up by the membrane, then diffuse through it, and lastly the permeate is desorbed on the side that is under low pressure. One essential component of the GS process is selectivity. The solution diffusion process facilitates the passage of gaseous molecules over the membrane. When employing asymmetric, homogenous, or polymeric membranes to separate gaseous mixtures and polar vapors, this method works especially well. Polymeric membranes with hollow fiber configurations are frequently employed in GS [34].

5.2.2.3 Electrical-driven Membrane Separation Process

The technique, known as electrodialysis, uses ion exchange membranes to apply an electric potential to a water-based solution in order to selectively extract ionic components. These dense IEMs are made of polymers with fixed ionic charge groups inserted in the polymer matrix. There are two distinct types of IEMs: Anion Exchange Membranes (AEMs) and Cation Exchange Membranes (CEMs). In their polymer matrix, CEMs have negatively charged groups while AEMs have positively charged groups. Uses in the pharmaceutical sector, saltwater desalination, and food organic acid extraction are the main applications for this method [31].

5.2.2.3.1 Ion Exchange Membrane

A kind of semipermeable membrane in which ionic groups are attached to a polymeric structure is known as an Ion Exchange Membrane (IEM). In order to do this, magnetic and fluidized ion exchange techniques are combined. Ionic groups and concentration have proven useful in a variety of applications. A nonporous anion exchanger is effectively used to remove nominal organic matter. IEMs can be categorized based on the functionality of the ions and the structure of the polymer backbone [2].

5.2.2.4 Temperature-driven Membrane Separation Process

5.2.2.4.1 Membrane Distillation

Membrane distillation has long been seen as a viable method for treating wastewater and desalinating saltwater. In contrast to hydrophilic membranes, hydrophobic membranes are capable of removing nearly all colloids, macromolecules, volatile and non-volatile substances, and salts. By promoting greater recalcitrant biodegradation, this membrane filtration system lowers the amount of sludge produced and the environmental impact of the process, while still delivering superior effluent quality. Due to its exceptional stability, it is more cost-effective than a Reverse Osmosis Membrane Bioreactor (RO-MBR). However, its potential for Chemical Oxygen Demand (COD) removal from feed water is limited [35].

5.2.2.5 Liquid Membrane

In this procedure, a thin layer of organic liquid that functions as a semipermeable barrier separates two aqueous phases with distinct compositions. Liquid Membranes (LMs) do not require the usage of solid membranes, in contrast to conventional membrane processes. In LM, one-stage extraction, high selectivity, and stripping are all features of non-equilibrium mass transfer. Three types of LM are supported: LM, emulsion LM, and bulk LM. Supported LMs consist of an organic phase-immobilizing inert microporous support. An immiscible liquid layer sits between two miscible liquids in an emulsion liquid. A restricted diffusion path that is away from the border layer is used by bulk LM [36]. Diffusion is the main way that mass moves across the membrane. Nevertheless, the process of separation can also be explained step-by-step by different methods. First, sorption takes place at the feed-membrane interface after diffusion takes place in the feed solution across the boundary layer. Diffusion on the receiving side across the boundary layer occurs after convective transport occurs in the membrane [37].

5.2.2.6 Integrated/hybrid Membrane Separation Processes

Membrane fouling is the main drawback of the membrane approach. In response, hybrid techniques that lower operating costs while enhancing water quality have been created. A membrane process and conventional process are combined to create a membrane hybrid process [38]. Hybrid processes fall into one of two categories: mixtures of one membrane process with another or combinations of two or more separate membrane processes [39]. A combination of multiple treatment approaches is usually utilized to address the problem because no one treatment strategy can accomplish all therapy objectives. Some of the most recent integrated or hybrid membrane technologies are as follows.

5.2.2.6.1 FO-MBR

This method is more energy-saving compared to traditional techniques. It allows for the extraction of phosphorus from the input and generates good-quality WW. It is also effective in eliminating trace organic pollutants from WW with high total suspended solids (TSS), performing better than RO-MBR. This method has less fouling than RO-MBR and is mostly reversible. Nevertheless, this method's

Membrane Filtration Technique

membrane stability is questionable, and a rise in salinity may reduce water flow and microbial kinetics [40].

5.2.2.6.2 RO-MBR

This method is a more cost-effective substitute for Forward Osmosis Membrane Bioreactor (FO-MBR) as it uses less energy compared to the standard MBR. However, its efficiency in treating high saline WW is lower than the FO-MBR process. Despite this, the treatment process yields stable and superior-quality product water [41].

5.2.2.6.3 Advanced Oxidation Processes/electrocoagulation MBR

Colors and tenacious pollutants, such as medications, may be removed with this simple-to-operate technology. It generates less sludge and has a decreased chance of fouling while operating. However, its main drawback is its ineffectiveness in treating WW contaminated with high TSS. Its high operational cost also restricts its usage [42].

5.2.2.6.4 Granular MBR

This approach is more shock-resistant and shows a greater rate of nitrification and denitrification. It has a lower potential for fouling and leaves a smaller footprint during operation. However, fouling can become a significant issue in the later stages of operation, and it takes an extended period to form granules during start-up [43].

5.2.2.6.5 Biofilm/bio-entrapped MBR

This process can minimize the amount of suspended solids (SS), shows a reduced propensity for fouling, and has a noticeably high rate of nitrification and denitrification. Significant fouling, however, may occur over time [44].

5.2.2.6.6 Coagulation-membrane Process

Coagulation with membrane filtration reduces membrane fouling and improves the removal of pollutants. This combination has been used in numerous research to treat surface water using coagulants such as $Fe_2(SO_4)_3$, $Al_2(SO_4)_3$, $AlCl_2(OH)$, $FeCl_3$, and $C_{12}H_{24}N_2O_9$. Permeate quality increased and membrane fouling decreased during these experiments. Moreover, heavy metal ions including antimony (Sb) and arsenic (As) could be effectively eliminated by combining coagulation with ultrafiltration (UF) membrane [45].

5.2.2.6.7 Adsorption-membrane Process

Adsorption technology is primarily employed in the treatment of water. Powdered activated carbon (PAC) is capable of eliminating organic compounds. The method of the hybrid adsorption membrane lowers the rate of membrane fouling [46].

5.2.2.6.8 Prefiltration-membrane Process

This technique utilizes sand and packed bed materials as initial barriers to eliminate coarse substances and microorganisms. The application of granular media filters can mitigate both fouling on the membrane surface and pore blockage [47].

5.3 APPLICATIONS OF MEMBRANE TECHNOLOGY

5.3.1 INDUSTRIAL WW TREATMENT

The characteristics of industrial WW may be demonstrated by a variety of indicators, including as pH value, color, turbidity, biological variables, ammonium nitrogen, suspended solids (SS), COD, biological oxygen demand (BOD), heavy metals, and pH value. Membrane techniques are frequently employed in the management of urban WW, which results in increased expenses for both the purification of water and the discharge of WW. This method aids in the direct retrieval of recycled substances, by-products, and solvents. It also contributes to the avoidance of large-scale, highly polluted WW discharges [36].

5.3.2 FOOD INDUSTRY

The production of fish, dairy products, animals, vegetables, and drinks is only one of the many subsectors that make up the food industry. As a result, the quality of WW differs across each sector, and is often characterized by high levels of organic matter, but it is possible to separate and use the high-value components found in these waste liquids, including phenols, carotenoids, pectin, lactose, and proteins [48].

5.3.3 PULP AND PAPER INDUSTRY

Because the pulp and paper industries depend so significantly on water for their operations, a sizable amount of effluent is produced. The pulp and paper industry's present WW treatment techniques can be made more effective by using membrane filtration. MBR systems can typically remove 82% to 99% of COD and almost all of SS in 0.12 to 2.5 days of hydraulic retention time (HRT). The effluent's COD and color intensity may be reduced by around 90% using the NF treatment procedure.

5.3.4 TEXTILE INDUSTRY

The textile processing industry (TPI) uses water as the principal medium for removing impurities and applying dyes and finishing agents. Since energy recovery and reuse are the current trends in industrial WW treatment, treating TPI WW using a combination of aerobic MBR and *anaerobic* MBR technologies is a viable option. An MBR is the energy recovery technique used; color reduction is achieved by applying aerobic MBR thereafter, resulting in reusable effluent [41].

Membrane Filtration Technique

5.3.5 Tannery Industry

Since tanneries use a lot of water in their operations, managing their effluent is a big problem. It has been discovered that chromium may be successfully removed from a hybrid system using inexpensive MBR minerals, while additional minerals aid in reducing fouling. Although aerobic MBR is a promising technique, there is minimal use of it in pilot or large-scale applications for treating tannery effluent [42].

5.3.6 Landfill Leachate

Rainwater seeping through and moisture released from garbage in landfills creates leachate, a WW with rich organic matter and nitrogen from ammonia. The age and maturity of the landfill site have an impact on this leachate's chemical makeup. When comparing younger leachate to older or more mature leachate, the former often has a larger concentration of organic components. MBR, RO treatment, and stripping coupled with flocculation are effective ways to reduce pollutants in leachate [43].

5.3.7 Pharmaceutical WWs

The waste disposal from the pharmaceutical industry encompasses a diverse array of compounds, each with distinct structures, functions, actions, and operations. When pharmaceutical WW containing Cephalosporin is treated with an MBR enhanced degradation through bioaugmentation results. MBRs that utilize specific microorganisms could potentially rival existing treatment processes for pharmaceutical WW [35].

5.3.8 Oily and Petrochemical WWs

Because of its poisonous and enduring qualities, WW from the petrochemical and oil sectors is a major source of pollution. These pollutants come from a variety of sources, including as the extraction and refining of crude oil, the manufacturing of petrochemicals, the fabrication of metal, the creation of lubricants and coolants, and the washing of cars. In one study, oil-contaminated WW was treated in a modified full-scale facility that went from chemical demulsification to an UF procedure and then a MBR approach. This method worked well in getting rid of all phenolic chemicals, oil, and tar while also reducing 90% of the COD [2].

5.3.9 Municipal WW Treatment

Different countries may have different WW volumes and types of contaminants from municipal sources due to a variety of reasons. Municipal WW is usually treated to get rid of unwanted substances. In an oxygen-rich atmosphere, organic matter is biodegraded by microorganisms in to smaller molecules (such as CO_2, NH_3, PO_4, etc.) [36].

5.4 ADVANTAGES OF MST

Membrane Separation Technology (MST) has numerous benefits including the following [49]:

- Techniques for membrane separation are applicable to a broad spectrum of separations since they may be applied at larger sizes as well as at the molecular level. The process does not require a phase change, which means it generally requires less energy. The only situation in which more energy could be required is when pushing the permeate stream over the membrane requires raising the feed stream's pressure [50].
- Membrane methods are cost-effective and eco-friendly due to their simplicity, efficiency, and use of non-detrimental materials. They are commonly employed for water softening. A key advantage of membrane methods is their ability to carry out gentle molecular separation, a feature often absent in other separation processes like centrifugation. Another significant advantage of membrane techniques is their capacity to handle large volumes and continuously generate product streams [51].
- Membrane methods provide a straightforward, cost-effective, and user-friendly solution for removing undesired elements from WW. They do not require intricate control systems. Membranes are highly selectively engineered to isolate only the particular components that require isolation. Generally, the selectivity rates for membrane separation surpass those of volatility in distillation procedures [52].
- The process conveniently facilitates the elimination of bacteria and particles. Its simplicity and automated operation require minimal supervision, making it ideal for applications in smaller systems [50].
- This method effectively eliminates almost all ionized impurities and the majority of non-ionized substances. It is ideal for smaller setups that experience significant seasonal variations in water usage. The process is unaffected by changes in flow or the amount of total dissolved solids, and it can be put into operation instantly, without requiring any initial settling period [50].
- A wide variety of polymers and inorganic substances can be utilized to create membranes, which increases the potential for controlling separation selectivity. Furthermore, membrane methods can retrieve minor elements from the input stream without increasing energy expenditure [53].

5.5 DISADVANTAGES OF MST

Two significant issues associated with membrane separation methods are membrane fouling and membrane modules.

Seawater and brackish water are now treated using membrane processes such as MF, NF, UF, and RO in addition to repeating the process of reusing water [54]. Polymeric membranes are the most widely used kind of membrane, mostly composed of polysulfone and polyethersulfone. Nevertheless, fouling can occur in these membranes because of their hydrophobic nature [55, 56]. Membrane performance

Membrane Filtration Technique

is lowered as a result of the blockage of membrane pores. Because more cleaning procedures are required, it also increases operating expenses [57]. The chemistry of the feed, microbes, solute absorption causing pore blockage, and the deposition of inorganic elements on the membrane surface are some of the factors that cause membrane fouling [58]. The membrane can become contaminated either irreversibly or reversibly [59]. Particles that stick to the membrane surface cause reversible fouling, but particles that attach so firmly that physical cleaning is unable to remove them cause irreversible fouling [60].

5.5.1 MEMBRANE FOULING

5.5.1.1 Factors Influencing Membrane Fouling

Membrane fouling is contingent upon a number of systemic factors, such as feed qualities (pH and ion strength), membrane features (roughness and hydrophobicity), and operating parameters (temperature, transmembrane pressure, and cross-flow rate) [61]. Numerous elements interact in various ways to intensify the process of membrane fouling. The following is an overview of the elements that lead to fouling.

5.5.12 Membrane Characteristics

Fouling is less common in hydrophilic membranes, such as ceramic membranes, and more common in hydrophobic membranes, such as polymeric membranes. During operation, the roughness of the surface creates a niche for colloidal particles to gather on the membrane surface, and fouling gets worse as the surface roughness gets rougher. Fouling risk increases with membrane pore size because bigger hole sizes increase the possibility of contaminant occlusion [62]. A greater tendency for membrane fouling is associated with an increase in hydrophobicity, whereas a reduction in membrane fouling is connected to an increase in hydrophilicity. Colloidal particle deposition has the potential to cause membranes to become negatively charged. This can then draw positively charged ions from mixed liquid suspended solids (MLSS) and cause inorganic membrane fouling [63].

5.5.1.3 Operating Conditions

By using a cross-flow filtering mode, the risk of membrane fouling is decreased since fewer cake layers are formed on the membrane. Membrane fouling rates decrease with increasing aeration rates. Because of the release of more bacterial extracellular polymeric substances (EPS) and a larger concentration of filamentous bacteria, lower temperatures increase the danger of membrane fouling. Increased membrane efficiency and longer operating times result from a lower rate of membrane fouling caused by a greater COD/N ratio in the feed [64]. In contrast, studies indicate that a lower COD/N ratio is associated with reduced fouling. As the HRT decreases, fouling tends to increase. However, an overly long HRT can lead to the accumulation of fouling agents. Operating at a high solids retention time (SRT) that results in low EPS production can limit fouling. However, because of the addition of MLSS and higher sludge viscosity, fouling increases at very high SRT. Moreover, fouling gets worse when the organic loading rate increases. When EPS synthesis is enhanced by elevating the food-to-microbe ratio by high biomass input, fouling is found to significantly rise [65].

5.5.1.4 Feed/biomass Properties

When the floc size drops, the membrane fouling rises. More membrane fouling is caused by the release of bound EPS as a result of elevated salinity. Membrane fouling rates increase in response to a pH reduction.

Increased fouling may be the result of higher levels of MLSS. However, studies seem to indicate that it has little to no effect on fouling. Fouling is more likely when EPS are abundant in the feed. Moreover, greater membrane fouling may result from higher viscosity [66].

5.5.1.5 Control of Membrane Fouling

The different strategies used to address the problems caused by fouling are outlined as follows [67].

5.5.1.5.1 Air Spargin

This reduces the amount of polarization and fouling. Shear stress is applied to the membrane's surface to lessen turbulence variations. A higher rate of aeration may make the membrane fouling worse.

5.5.1.5.2 Mechanical Cleaning

Mechnical cleaning is accomplished by applying shear stress to the membrane's surface.

5.5.1.5.3 Ultrasonic Mitigation

This technique removes both soluble and insoluble particles by using water-based medium and ultrasonic help. It removes the biofilm layer from the membrane's surface and significantly lowers concentration polarization [68].

5.5.1.5.4 Chemical Cleaning

This calls for the use of oxidizing agents, surfactants, chelates, acids, and bases in addition to the recently developed nitrite and rhamnolipids acids. By solubilizing and neutralizing the bases that cause the foulant's hydrolysis, solubilization, and saponification, these components help to eliminate fouling [69].

5.5.1.5.5 Fouling Release Surfaces and Nanomaterials

Membrane fouling may be controlled by creating membranes with antifouling surfaces that have unique chemical and physical properties. Hydrophilic surfaces have demonstrated significant promise in limiting non-specific interactions and managing a range of foulants. The fouling can also be reduced by post-modifying the membranes using polymeric antifouling substances or inorganic nanomaterials [70].

5.5.1.5.6 Cell Entrapment

Cells are bound to a solid substrate or enclosed within another cell type to inhibit their unconstrained motility in cell immobilization techniques such as passive immobilization and Canthi's method, which creates a porous polymer matrix in which cells are artificially trapped. While this method may not be entirely dependable for eliminating pathogens and large particles, it serves as a viable substitute for traditional biological treatment systems [42].

5.5.1.5.7 Biological Mitigation

This is a novel method with significant potential in managing biofouling. The formation of microbial attachment or biofilm is hindered by obstructing the synthesis of adenosine triphosphate. The prevention of early microbial adhesion is more successful than the disruption of existing biofilm by enzymes that target EPSs, such as proteinase K, subtilisin, trypsin, and others. Despite certain limitations like instability, temperature, and pH, protease is superior to conventional chemicals for controlling irreversible membranes [71].

5.5.1.5.8 Electrically based Mitigation

Membrane fouling may be efficiently avoided by using electrical techniques such as electric fields' effects on charged particles, electrophoresis (EP), and electrostatic repulsion. These methods, which generally consist of external methods like electro-coagulation and EP or internal methods like microbial fuel cells, are used to reduce fouling in MBRs.

5.5.2 MEMBRANE MODULES

Thousands or even hundreds of square meters of membrane are needed to get the necessary separation in industrial membrane plants. There exist numerous cost-effective membrane packaging methods that offer a large surface area for efficient and effective separation [72]. Typically, membrane module designs are utilized to prevent membrane fouling as shown in Figure 5.3.

5.5.2.1 Plate-and-Frame Modules

Initially, membrane systems were offered as plate-and-frame modules. However, because hollow and spiral-wound fiber modules are less expensive, they have essentially taken their place. These days, plate-and-frame modules are rarely employed in UF and RO operations, especially when fouling conditions are significant [73].

5.5.2.2 Tubular Modules

When a strong resistance to membrane fouling is needed, as is often the case in UF applications, tubular modules are very useful. These membranes are made up of tiny, 0.5–1 cm diameter tubes contained within bigger tubes. Within the tubular membrane system, many of these tubes are arranged in series [74].

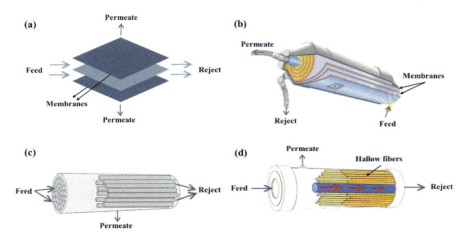

FIGURE 5.3 Types of membrane modules configurations: (a) plate and frame, (b) spiral wound, (c) tubular, and (d) capillary fiber [72].

5.5.2.3 Spiral-wound Modules

Modules at the commercial scale are made out of a central collecting pipe encircling a number of membrane envelopes, each measuring 10–20 feet square. A typical commercial spiral-wrapped module measures 3.33 feet in length and 0.66 feet in diameter. Designs with several envelopes, where the permeate passes through the center pipe, decrease the pressure drop [75].

5.5.2.4 Hollow-fiber modules

Typically, hollow fiber modules are between 3 and 5 feet tall and have a diameter of 10 to 20 cm. Usually, they are employed with the feed stream external to the fiber. Within the membrane, water enters the fiber's lumen. A sizable quantity of fibers is bundled together, "potted" at both ends in epoxy resin, and contained in an outer shell [76].

5.6 CONCLUSION

Research has indicated that traditional drinking water treatment plants (DWTPs) and wastewater treatment plants (WWTPs) are not fully capable of getting rid of MPs from water. Consequently, these MPs can be found in DWTP and WWTP effluents. One cutting-edge treatment technique that is thought to be very successful and promising for eliminating MPs from water is membrane technology. As a novel separation technique, membrane separation has found extensive applications across various sectors, as discussed in this chapter. It is seen as a future direction for water treatment. When used in wastewater treatment, it not only delivers excellent results but also allows for recycling, thereby yielding significant economic, social, and environmental advantages. Nevertheless, membrane fouling, the high cost of membrane materials, and a comparatively limited operating lifespan are the present problems

Membrane Filtration Technique

with membrane separation technology. These issues have somewhat restricted its widespread application. As novel membrane materials appear and membrane separation technology continues to progress, these problems are expected to be resolved. This will improve the water treatment industry's use of membrane separation technologies. In conclusion, advanced membrane technology is poised to address the challenges in the wastewater treatment process and ensure long-term performance.

REFERENCES

1. Sonune, Amit, and Rupali Ghate. "Developments in wastewater treatment methods." *Desalination* 167 (2004): 55–63. https://doi.org/10.1016/j.desal.2004.06.113
2. Ravanchi, Maryam Takht, Tahereh Kaghazchi, and Ali Kargari. "Application of membrane separation processes in petrochemical industry: A review." *Desalination* 235, no. 1-3 (2009): 199–244. https://doi.org/10.1016/j.desal.2007.10.042
3. Van der Bruggen, Bart, Carlo Vandecasteele, Tim Van Gestel, Wim Doyen, and Roger Leysen. "A review of pressure-driven membrane processes in WW treatment and DW production." *Environmental Progress* 22, no. 1 (2003): 46–56. https://doi.org/10.1002/ep.670220116
4. Padaki, Mahesh, R. Surya Murali, Ms S. Abdullah, Nurasyikin Misdan, A. Moslehyani, M. A. Kassim, Nidal Hilal, and A. F. Ismail. "Membrane technology enhancement in oil–water separation. A review." *Desalination* 357 (2015): 197–207. https://doi.org/10.1002/ep.670220116
5. Li, Kai, Tinglin Huang, Fangshu Qu, Xing Du, An Ding, Guibai Li, and Heng Liang. "Performance of adsorption pre-treatment in mitigating humic acid fouling of ultrafiltration membrane under environmentally relevant ionic conditions." *Desalination* 377 (2016): 91–98. https://doi.org/10.1016/j.desal.2015.09.016
6. Zhao, Dandan, Yang Yu, and J. Paul Chen. "Treatment of lead contaminated water by a PVDF membrane that is modified by zirconium, phosphate and PVA." *Water Research* 101 (2016): 564–573. https://doi.org/10.1016/j.watres.2016.04.078
7. Wenten, I. Gede. "Reverse osmosis applications: Prospect and challenges." *Desalination* 391 (2016): 112–125. https://doi.org/10.1016/j.desal.2015.12.011
8. Chew, Chun Ming, Mohamed Kheireddine Aroua, Mohd Azlan Hussain, and W. M. Z. Wan Ismail. "Practical performance analysis of an industrial-scale ultrafiltration membrane water treatment plant." *Journal of the Taiwan Institute of Chemical Engineers* 46 (2015): 132–139. https://doi.org/10.1016/j.jtice.2014.09.013
9. Song, Jia, Zhenghua Zhang, and Xihui Zhang. "A comparative study of pre-ozonation and in-situ ozonation on mitigation of ceramic UF membrane fouling caused by alginate." *Journal of Membrane Science* 538 (2017): 50–57. https://doi.org/10.1016/j.memsci.2017.05.059
10. Zhang, Xiwang, David K. Wang, Diego Ruben Schmeda Lopez, and João C. Diniz da Costa. "Fabrication of nanostructured TiO_2 hollow fiber photocatalytic membrane and application for WW treatment." *Chemical Engineering Journal* 236 (2014): 314–322. https://doi.org/10.1016/j.cej.2013.09.059
11. Kimura, Katsuki, Yasushi Hane, Yoshimasa Watanabe, Gary Amy, and Naoki Ohkuma. "Irreversible membrane fouling during ultrafiltration of surface water." *Water research* 38, no. 14–15 (2004): 3431–3441. https://doi.org/10.1016/j.watres.2004.05.007
12. Kingsbury, Benjamin F. K., and K. Li. "A morphological study of ceramic hollow fibre membranes." *Journal of Membrane Science* 328, no. 1-2 (2009): 134–140. https://doi.org/10.1016/j.memsci.2008.11.050

13. Le, Ngoc Lieu, and Suzana P. Nunes. "Materials and membrane technologies for water and energy sustainability." *Sustainable Materials and Technologies* 7 (2016): 1–28. https://doi.org/10.1016/j.susmat.2016.02.001

14. Teodosiu, Carmen, Andreea-Florina Gilca, George Barjoveanu, and Silvia Fiore. "Emerging pollutants removal through advanced DW treatment: A review on processes and environmental performances assessment." *Journal of Cleaner Production* 197 (2018): 1210–1221. https://doi.org/10.1016/j.susmat.2016.02.001

15. Im, Dongbum, Norihide Nakada, Yasuyuki Fukuma, and Hiroaki Tanaka. "Effects of the inclusion of biological activated carbon on membrane fouling in combined process of ozonation, coagulation and ceramic membrane filtration for water reclamation." *Chemosphere* 220 (2019): 20–27. https://doi.org/10.1016/j.susmat.2016.02.001

16. Snyder, Shane A., Samer Adham, Adam M. Redding, Fred S. Cannon, James De Carolis, Joan Oppenheimer, Eric C. Wert, and Yeomin Yoon. "Role of membranes and activated carbon in the removal of endocrine disruptors and pharmaceuticals." *Desalination* 202, no. 1–3 (2007): 156–181. https://doi.org/10.1016/j.desal.2005.12.052

17. Belaissaoui, Bouchra, Joan Claveria-Baro, Ana Lorenzo-Hernando, David Albarracin Zaidiza, Elodie Chabanon, Christophe Castel, Sabine Rode, Denis Roizard, and Eric Favre. "Potentialities of a dense skin hollow fiber membrane contactor for biogas purification by pressurized water absorption." *Journal of Membrane Science* 513 (2016): 236–249. https://doi.org/10.1016/j.memsci.2016.04.037

18. Thakur, Vijay Kumar, and Stefan Ioan Voicu. "Recent advances in cellulose and chitosan based membranes for water purification: A concise review." *Carbohydrate Polymers* 146 (2016): 148–165. https://doi.org/10.1016/j.carbpol.2016.03.030

19. Gao, Jianwen. "Membrane separation technology for WW treatment and its study progress and development trend." In 2016 4th international conference on mechanical materials and manufacturing engineering. Atlantis Press, 2016. https://doi.org/10.2991/mmme-16.2016.202

20. Zhang, Feng, Shoujian Gao, Yuzhang Zhu, and Jian Jin. "Alkaline-induced super hydrophilic/underwater super oleophobic polyacrylonitrile membranes with ultralow oil-adhesion for high-efficient oil/water separation." *Journal of Membrane Science* 513 (2016): 67–73. https://doi.org/10.1016/j.memsci.2016.04.020

21. Liu, Jie, Junsheng Yuan, Zhiyong Ji, Bingjun Wang, Yachao Hao, and Xiaofu Guo. "Concentrating brine from seawater desalination process by nanofiltration–electro dialysis integrated membrane technology." *Desalination* 390 (2016): 53–61. https://doi.org/10.1016/j.desal.2016.03.012

22. Yan, Fang, Hao Chen, Yang Lü, Zhenhua Lü, Sanchuan Yu, Meihong Liu, and Congjie Gao. "Improving the water permeability and antifouling property of thin-film composite polyamide nanofiltration membrane by modifying the active layer with triethanolamine." *Journal of Membrane Science* 513 (2016): 108–116. https://doi.org/10.1016/j.memsci.2016.04.049

23. Sagle, Alyson, and Benny Freeman. "Fundamentals of membranes for water treatment." *The Future of Desalination in Texas* 2, no. 363 (2004): 137.

24. Aliyu, U. M., S. Rathilal, and Y. M. Isa. "Membrane desalination technologies in water treatment: A review." *Water Practice & Technology* 13, no. 4 (2018) 738–752. https://doi.org/10.2166/wpt.2018.084

25. Mallada, Reyes, and Miguel Menéndez, eds. *Inorganic membranes: Synthesis, characterization and applications.* Elsevier, 2008.

26. Jhaveri, Jainesh H., and Z. V. P. Murthy. "A comprehensive review on anti-fouling nanocomposite membranes for pressure driven membrane separation processes." *Desalination* 379 (2016): 137–154. https://doi.org/10.1016/j.desal.2015.11.009

27. Liao, Yuan, Chun-Heng Loh, Miao Tian, Rong Wang, and Anthony G. Fane. "Progress in electrospun polymeric nanofibrous membranes for water treatment: Fabrication, modification and applications." *Progress in Polymer Science* 77 (2018): 69–94. https://doi.org/10.1016/j.progpolymsci.2017.10.003

28. Bharagava, Ram Naresh, and Pankaj Chowdhary, eds. *Emerging and eco-friendly approaches for waste management.* Berlin, Germany: Springer, 2019. https://doi.org/10.1186/s12302-020-00383-w

29. Gitis, Vitaly, and Nicholas Hankins. "Water treatment chemicals: Trends and challenges." *Journal of Water Process Engineering* 25 (2018): 34–38. https://doi.org/10.1016/j.jwpe.2018.06.003

30. Schäfer, Andrea, Anthony G. Fane, and T. David Waite, eds. *Nanofiltration: Principles and applications.* Elsevier, 2005. https://lccn.loc.gov/2002043919

31. Bolong, Nurmin, A. F. Ismail, Mohd Razman Salim, and T. Matsuura. "A review of the effects of emerging contaminants in WW and options for their removal." *Desalination* 239, no. 1–3 (2009): 229–246 https://doi.org/10.1016/j.desal.2008.03.020.

32. Divya, M., S. Aanand, A. Srinivasan, and B. Ahilan. "Bioremediation–an eco-friendly tool for effluent treatment: a review." *International Journal of Applied Research* 1, no. 12 (2015): 530–537.

33. Ismail, Fauzi, Kailash Chandra Khulbe, and Takeshi Matsuura. *Reverse osmosis.* Elsevier, 2018.

34. Dewi, Rosmaya, Norazanita Shamsuddin, Muhammad S. Abu Bakar, Jose H. Santos, Muhammad Roil Bilad, and Lee Hoon Lim. "Progress in emerging contaminants removal by adsorption/membrane filtration-based technologies: a review." *Indonesian Journal of Science and Technology* 6, no. 3 (2021): 577–618. https://doi.org/10.17509/ijost.v6i3

35. Peters, Th. "Membrane technology for water treatment." *Chemical Engineering & Technology* 33, no. 8 (2010): 1233–1240. https://doi.org/10.1002/ceat.201000139

36. Jyoti, Jain, Dubey Alka, and Singh Jitendra Kumar. "Application of membrane-bioreactor in waste-water treatment: A review." *International Journal of Chemistry and Chemical Engineering* 3, no. 2 (2013): 115–122.

37. Dawood, Sara, and Tushar Sen. "Review on dye removal from its aqueous solution into alternative cost effective and non-conventional adsorbents." *Journal of Chemical and Process Engineering* 1, no. 104 (2014): 1–11. http://hdl.handle.net/20.500.11937/48131

38. Magara, Yasumoto, Shoichi Kunikane, and Masaki Itoh. "Advanced membrane technology for application to water treatment." *Water Science and Technology* 37, no. 10 (1998): 91–99. https://doi.org/10.1016/S0273-1223(98)00307-2

39. Naga Babu, Andraju. "Development and Characterization of Non-conventional Low Cost Adsorbents for the Removal of Diverse Pollutants from WW." (2019). Ph.D. Thesis, http://hdl.handle.net/10603/286784

40. Mulligan, Catherine N., R. N. Yong, and B. F. Gibbs. "Surfactant-enhanced remediation of contaminated soil: A review." *Engineering Geology* 60, no. 1–4 (2001): 371–380. https://doi.org/10.1016/S0013-7952(00)00117-4

41. Collivignarelli, Maria Cristina, Alessandro Abbà, Marco Carnevale Miino, and Silvestro Damiani. "Treatments for colour removal from WW: State of the art." *Journal of Environmental Management* 236 (2019): 727–745. https://doi.org/10.1016/j.jenvman.2018.11.094

42. Ahmad, Akil, Siti Hamidah Mohd-Setapar, Chuo Sing Chuong, Asma Khatoon, Waseem A. Wani, Rajeev Kumar, and Mohd Rafatullah. "Recent advances in new generation dye removal technologies: Novel search for approaches to reprocess WW." *RSC Advances* 5, no. 39 (2015): 30801–30818. https://doi.org/10.1039/C4RA16959J

43. Koc-Jurczyk, Justyna. "Removal of refractory pollutants from landfill leachate using two-phase system." *Water Environment Research* 86, no. 1 (2014): 74–80. https://doi.org/10.2175/106143013X13807328848810

44. Sonune, Amit, and Rupali Ghate. "Developments in WW treatment methods." *Desalination* 167 (2004): 55–63. https://doi.org/10.1016/j.desal.2004.06.113

45. Pavithra, K. Grace, and V. J. Jaikumar. "Removal of colorants from WW: A review on sources and treatment strategies." *Journal of Industrial and Engineering Chemistry* 75 (2019): 1–19. https://doi.org/10.1016/j.jiec.2019.02.011

46. Zoubeik, Mohamed, Mohamed Ismail, Amgad Salama, and Amr Henni. "New developments in membrane technologies used in the treatment of produced water: A review." *Arabian Journal for Science and Engineering* 43 (2018): 2093–2118. https://doi.org/10.1007/s13369-017-2690-0

47. Obotey Ezugbe, Elorm, and Sudesh Rathilal. "Membrane technologies in WW treatment: A review." *Membranes* 10, no. 5 (2020): 89. https://doi.org/10.3390/membranes10050089

48. Barceló, Damià, and Mira Petrovic, eds. *Emerging contaminants from industrial and municipal waste: Removal technologies.* Vol. 5. Springer Science & Business Media, 2008.

49. Gao, Wei, Heng Liang, Jun Ma, Mei Han, Zhong-lin Chen, Zheng-shuang Han, and Gui-bai Li. "Membrane fouling control in ultrafiltration technology for DW production: A review." *Desalination* 272, no. 1–3 (2011): 1–8. https://doi.org/10.1016/j.desal.2011.01.051

50. Bera, Sweta Parimita, Manoj Godhaniya, and Charmy Kothari. "Emerging and advanced membrane technology for WW treatment: A review." *Journal of Basic Microbiology* 62, no. 3–4 (2022): 245–259. https://doi.org/10.1002/jobm.202100259

51. Qu, Fangshu, Heng Liang, Jian Zhou, Jun Nan, Senlin Shao, Jianqiao Zhang, and Guibai Li. "Ultrafiltration membrane fouling caused by extracellular organic matter (EOM) from Microcystisaeruginosa: Effects of membrane pore size and surface hydrophobicity." *Journal of Membrane Science* 449 (2014): 58–66. https://doi.org/10.1016/j.memsci.2013.07.070

52. Wang, Nan, Xing Li, Yanling Yang, Zhiwei Zhou, Yi Shang, and Xiaoxuan Zhuang. "Photocatalysis-coagulation to control ultrafiltration membrane fouling caused by natural organic matter." *Journal of Cleaner Production* 265 (2020): 121790. https://doi.org/10.1016/j.jclepro.2020.121790

53. Wang, Hui, Minkyu Park, Heng Liang, Shimin Wu, Israel J. Lopez, Weikang Ji, Guibai Li, and Shane A. Snyder. "Reducing ultrafiltration membrane fouling during potable water reuse using pre-ozonation." *Water Research* 125 (2017): 42–51. https://doi.org/10.1016/j.watres.2017.08.030

54. Erkanlı, Mert, Levent Yilmaz, P. Zeynep Çulfaz-Emecen, and Ulku Yetis. "Brackish water recovery from reactive dyeing WW via ultrafiltration." *Journal of Cleaner Production* 165 (2017): 1204–1214. https://doi.org/10.1016/j.jclepro.2017.07.195

55. Marino, Tiziana, Enrico Blasi, Sergio Tornaghi, Emanuele Di Nicolò, and Alberto Figoli. "Polyethersulfone membranes prepared with Rhodiasolv ŪPolarclean as water soluble green solvent." *Journal of Membrane Science* 549 (2018): 192–204. https://doi.org/10.1016/j.memsci.2017.12.007

56. Ahmed, Farah, Boor Singh Lalia, Victor Kochkodan, Nidal Hilal, and Raed Hashaikeh. "Electrically conductive polymeric membranes for fouling prevention and detection: A review." *Desalination* 391 (2016): 1–15. https://doi.org/10.1016/j.desal.2016.01.030

57. Laohaprapanon, Sawanya, Angelita D. Vanderlipe, Bonifacio T. Doma Jr, and Sheng-Jie You. "Self-cleaning and antifouling properties of plasma-grafted poly

(vinylidenefluoride) membrane coated with ZnO for water treatment." *Journal of the Taiwan Institute of Chemical Engineers* 70 (2017): 15–22. https://doi.org/10.1016/j.jtice.2016.10.019

58. Zinadini, Sirus, and Foad Gholami. "Preparation and characterization of high flux PES nanofiltration membrane using hydrophilic nanoparticles by phase inversion method for application in advanced WW treatment." *Journal of Applied Research in Water and WW* 3, no. 1 (2016): 232–235.

59. Ding, Q., H. Yamamura, N. Murata, N. Aoki, H. Yonekawa, A. Hafuka, and Y. Watanabe. "Characteristics of meso-particles formed in coagulation process causing irreversible membrane fouling in the coagulation-microfiltration water treatment." *Water Research* 101 (2016): 127–136. https://doi.org/10.1016/j.watres.2016.05.076

60. Zhao, Fangchao, Huaqiang Chu, Yalei Zhang, Shuhong Jiang, Zhenjiang Yu, Xuefei Zhou, and Jianfu Zhao. "Increasing the vibration frequency to mitigate reversible and irreversible membrane fouling using an axial vibration membrane in microalgae harvesting." *Journal of Membrane Science* 529 (2017): 215–223.] https://doi.org/10.1016/j.memsci.2017.01.039

61. Mi, Baoxia, and Menachem Elimelech. "Organic fouling of forward osmosis membranes: Fouling reversibility and cleaning without chemical reagents." *Journal of Membrane Science* 348, no. 1–2 (2010): 337–345. https://doi.org/10.1016/j.memsci.2009.11.021

62. Liao, Yichen, Alnour Bokhary, Esmat Maleki, and Baoqiang Liao. "A review of membrane fouling and its control in algal-related membrane processes." *Bioresource Technology* 264 (2018): 343–358. https://doi.org/10.1016/j.biortech.2018.06.102

63. Drews, Anja. "Membrane fouling in membrane bioreactors—Characterisation, contradictions, cause and cures." *Journal of Membrane Science* 363, no. 1–2 (2010): 1–28. https://doi.org/10.1016/j.memsci.2010.06.046

64. Hilal, Nidal, Oluwaseun O. Ogunbiyi, Nick J. Miles, and Rinat Nigmatullin. "Methods employed for control of fouling in MF and UF membranes: a comprehensive review." *Separation Science and Technology* 40, no. 10 (2005): 1957–2005. https://doi.org/10.1081/SS-200068409

65. Vrouwenvelder, J. S., J. A. M. Van Paassen, L. P. Wessels, A. F. Van Dam, and S. M. Bakker. "The membrane fouling simulator: A practical tool for fouling prediction and control." *Journal of Membrane Science* 281, no. 1–2 (2006): 316–324. https://doi.org/10.1016/j.memsci.2006.03.046

66. Chang, In-Soung, Pierre Le Clech, Bruce Jefferson, and Simon Judd. "Membrane fouling in membrane bioreactors for WW treatment." *Journal of Environmental Engineering* 128, no. 11 (2002): 1018–1029. https://doi.org/10.1061/(ASCE)0733-9372(2002)128:11(1018)

67. Iorhemen, Oliver Terna, Rania Ahmed Hamza, and Joo Hwa Tay. "Membrane fouling control in membrane bioreactors (MBRs) using granular materials." *Bioresource Technology* 240 (2017): 9–24. https://doi.org/10.1016/j.biortech.2017.03.005

68. Peng, Na, Natalia Widjojo, Panu Sukitpaneenit, May May Teoh, G. Glenn Lipscomb, Tai-Shung Chung, and Juin-Yih Lai. "Evolution of polymeric hollow fibres as sustainable technologies: Past, present, and future." *Progress in Polymer Science* 37, no. 10 (2012): 1401–1424. https://doi.org/10.1016/j.progpolymsci.2012.01.001

69. Mishima, I., and J. Nakajima. "Control of membrane fouling in membrane bioreactor process by coagulant addition." *Water Science and Technology* 59, no. 7 (2009): 1255–1262. https://doi.org/10.2166/wst.2009.090

70. Bagheri, Majid, Ali Akbari, and Sayed Ahmad Mirbagheri. "Advanced control of membrane fouling in filtration systems using artificial intelligence and machine

learning techniques: A critical review." *Process Safety and Environmental Protection* 123 (2019): 229–252. https://doi.org/10.1016/j.psep.2019.01.013

71. Kimura, Katsuki, and Yasumitsu Oki. "Efficient control of membrane fouling in MF by removal of biopolymers: Comparison of various pretreatments." *Water Research* 115 (2017): 172–179. https://doi.org/10.1016/j.watres.2017.02.033

72. Abushawish, Alaa, Ines Bouaziz, Ismail W. Almanassra, Maha Mohammad AL-Rajabi, Lubna Jaber, Abdelrahman K. A. Khalil, Mohd Sobri Takriff, Tahar Laoui, Abdallah Shanableh, Muataz Ali Atieh, and et al. "Desalination pretreatment technologies: Current status and future developments" *Water* 15, no. 8 (2023): 1572. https://doi.org/10.3390/w15081572

73. Lee, S. "Performance comparison of spiral-wound and plate-and-frame forward osmosis membrane module." Membranes 10, no. 11(2020): 318.

74. Valladares Linares, Rodrigo, Luca Fortunato, N. M. Farhat, S. S. Bucs, M. Staal, E. O. Fridjonsson, M. L. Johns, Johannes S. Vrouwenvelder, and T. Leiknes. "Mini-review: Novel non-destructive in situ biofilm characterization techniques in membrane systems." *Desalination and Water Treatment* 57, no. 48–49 (2016): 22894–22901. https://doi.org/10.1080/19443994.2016.1180483

75. Fritzmann, Clemens, Jonas Löwenberg, Thomas Wintgens, and Thomas Melin. "State-of-the-art of reverse osmosis desalination." *Desalination* 216, no. 1–3 (2007): 1–76. https://doi.org/10.1016/j.desal.2006.12.009

76. Togo, Norihiro, Keizo Nakagawa, Takuji Shintani, Tomohisa Yoshioka, Tomoki Takahashi, Eiji Kamio, and Hideto Matsuyama. "Osmotically assisted reverse osmosis utilizing hollow fiber membrane module for concentration process." *Industrial & Engineering Chemistry Research* 58, no. 16 (2019): 6721–6729. https://doi.org/10.1021/acs.iecr.9b00630

6 Rapid Sand Filtration Technique for Remediation of Microplastics

Aditi Pandey, Asha Bhausaheb Kadam, and Achal Mukhija

6.1 INTRODUCTION

Microplastics (MPs), plastics smaller than 5 mm in size, have been recognized as imminent pollutants that are hazardous to both human health and aquatic ecosystems [1–3]. With growing urbanization and industrialization plastic pollution is becoming more common in the aquatic environment, which includes water bodies and oceans. MPs have been discovered in several locations, including Asia, Europe, North America, and Antarctica [4–9]. Their ubiquity in drinking water and possible negative impact on human health has raised serious concerns. MPs absorb organic pollutants like hexachlorocyclohexane (HCH), dichloro-diphenyl-trichloro ethane (DDT), and polychlorinated biphenyls (PCB), and they pose a major risk, particularly to humans. When ingested or inhaled they often release absorbed contaminants and additives into the body of an individual and have carcinogenic effects. MPs enter aquatic bodies through upstream agricultural, industrial, and urban regions or by way of diversions from runoff water from highways. They provide a habitat for harmful microorganisms and are equally toxic to marine organisms. These can even be transported by wind or water, allowing them to be detected in mussels, oysters, seabirds, and wild fish. When ingested they cause genetic mutation and thereby influence ecosystems. Thus, water remediation is imperative for several reasons related to ecology, environmental health, and ensuring human health.

Water remediation refers to the process of treating and purifying water contaminated by pollutants, such as chemicals and industrial waste. Following a contamination incident, the drinking water or wastewater utility remediation process entails prompt action, careful evaluation, efficient remediation planning and execution, continual surveillance, and eventually the restoration of regular service. Successful remediation requires several essential elements, including ensuring public

DOI: 10.1201/9781003486947-6

safety, efficient communication, regulatory compliance, and system improvements. There are different techniques for water pollution remediation including physical remediation and chemical remediation.

Physical remediation is the process of removing pollutants from water by mechanical or physical means. Particulate matter, sediments, and other solid contaminants are removed through filtration, sedimentation, flotation, skimming, adsorption, and screening. For example, the use of settling ponds, filtration systems, and activated carbon filters. Chemical remediation is the process of eliminating, neutralizing or transforming toxins in water by chemical processes. Both organic and inorganic chemicals are targeted by these approaches, and these processes aim to alter or neutralize contaminants in water chemically.

In 2017, MPs were first reported to be found in drinking water by Orb Media [10]. The first attempt for the determination of the presence of MPs in drinking water conducted in Germany in 2019, revealed that MPs were present in a DWTP (drinking water treatment plant) [11]. MPs have also been reportedly found in water treatment facilities, tap water, and bottled water [11–14]. The design of drinking water treatment units consists of several processes, such as pre-sedimentation, coagulation and flocculation, sedimentation, filtration, and advanced stages like chlorination. Reported studies mostly concentrate on the existence and frequency of MPs in DWTPs [11–14] whereas the effectiveness of MP removal from each DWTP method is described in a limited number of experiments.

One of the primary routes in which MPs reach the environment is through wastewater treatment plants (WWTPs). According to studies, sand filters may cut MPs in WWTPs by 73.8% to 99.2%. Sand filters reduce solid content in wastewater and increase MP removal rates through an assortment of processes. Mechanisms of sand filtration involve physical straining, sedimentation, adsorption, and biological action in slow sand filters. Different aspects related to the design, function, and efficiency of rapid sand filters (RSFs) in water treatment make them indispensable constituents of wastewater treatment systems.

Through rapid sand filtration technique, relatively large suspended particles can be removed. Water passing between the layers of coarse sand and gravel, which act as a natural filter, physically traps particles larger than the pore spacing. After that, as water flows through the filter bed, the sedimentation process takes place in which larger particles settle down in the sand layers due to gravity forces. Both van der Waals interactions and electrostatic forces are involved in the process due to which MPs get stuck to the surface of sand grains by adsorption. Subsequently, the shear forces produced by the water flow in the sand bed assist in the detection and removal of MPs.

RSFs are often classified into two types: fast gravity sand filters and rapid pressure sand filters. Rapid-gravity sand filters work and rely on gravity parameters resulting in lower energy costs. They are commonly employed in municipal drinking water treatment plants and are effective in eliminating bigger and suspended solid particles, including larger MPs. Rapid-pressure sand filters on the other hand employ high pressure usually between 3 and 7 bars (45 and 100 psi), and water is forced through the sand bed in the filtration process. Faster filtering rates and increased throughput are made possible by the pressurized system which makes them useful in a variety of settings, including industrial applications. Compared to rapid-gravity sand filters, rapid-pressure sand filters are smaller in size, which makes them appropriate for installations with limited space. These techniques need sufficient pre-treatment (typically flocculation-coagulation) and post-treatment (usually chlorine disinfection) to provide safe drinking water.

Rapid sand filtration is frequently used in urban and industrial settings to filter vast amounts of water where land is a major constraint and resources such as materials, skilled personnel, and a steady supply of electricity are available.

6.2 RAPID SAND FILTRATION

Rapid sand filtration also known as mechanical filtration process is a purely physical water purification method that provides rapid and efficient removal of comparatively large suspended particulates. It is a mainstay in many water treatment plants due to its efficiency in eliminating particles and simplicity of use. It is particularly useful in industrialized countries to effectively remove MPs and other suspended particles from water. RSFs first appeared in the United States at the end of the 1800s, and they rapidly gained popularity. In contrast, slow sand filters required less land area for necessary equipment, which is why they became the predominant form of water filtration by the 1920s. Most often, rapid sand filtration is utilized in conjunction with other techniques for purifying water. Presently, the most extensively employed water treatment technologies for treating large amounts of drinking water in industrialized countries combine coagulation, flocculation, sedimentation, filtration, and disinfection (e.g., chlorination, ozonation) [15]. Rapid sand filtration method is also frequently employed for agricultural water treatment, irrigation systems, and to preserve aquaculture's water quality. (Figure 6.1). It prevents clogging and removes particulates, organic matter, and unwanted nutrients from the water.

In a rapid sand filter system, rapid filters pass quickly through the filter beds. Particles of greater size remain trapped in the filter, whereas tiny materials can flow through because they fit into the pores between the sand grains. Particles are removed

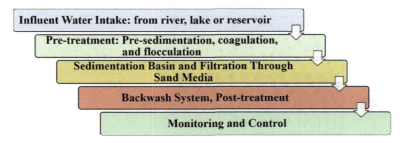

FIGURE 6.1 Flow chart of rapid sand filtration technique.

from the sand substrate at a significant depth by rapid sand filtering. Physical straining is the most significant mechanism associated with fast filters.

However, for the supply of safe drinking water, essential preparatory and final disinfection steps must be followed that include pre-sedimentation, flocculation, coagulation, and chlorination processes. Furthermore, RSFs necessitate regular backwashing and maintenance to avoid clogging and preserve filter efficiency.

6.2.1 Wastewater Treatment Process and Basic Design Principles of Rapid Sand Filtration Technique

The rapid sand filtration technique is an essential element of many water treatment facilities and works best when it is precisely engineered and operated. The key components in the design of the rapid sand filtration system are as follows (Figure 6.2):

1. Chamber: filter tank or filter box
2. Media of the filter system (sand)
3. Gravel support
4. Underdrain system
5. Wash water troughs

Typically constructed of reinforced concrete, the filter chamber is 1.5–2 meters high and filled with sand and gravel with different sizes of grains. The finest grains are usually found in the top layer, with larger grains found in the lower levels. Particles of various sizes can be captured with the use of this arrangement. A gravel support layer

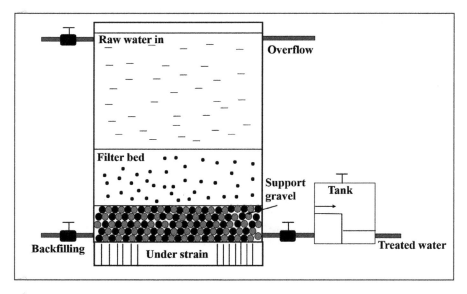

FIGURE 6.2 Basic design of rapid sand filter system.

Rapid Sand Filtration Technique

with coarser gravel is present beneath the sand layers. This layer serves to hold the sand in place and prevent it from washing into the underdrain system. Additionally, it helps to distribute incoming water evenly throughout the filter bed. After passing through the layers of gravel and graded sand, the water reaches the top of the sand bed where it is filtered. The chamber is drained by an underdrain system consisting of a network of perforated pipes or a series of channels located beneath the gravel layer [16].

The filtered water is gathered by this system and sent to the output. Additionally, backwashing is an essential step since it properly distributes the backwash water across the filter bed. This procedure aids in filter cleaning by eliminating gathered sand particles, which clog filters and reduce their efficiency. Cleaning the filter bed also restores the filter's initial functionality. In certain systems, backwashing and air scouring are combined to improve cleaning by agitating the filter media significantly.

The fundamental physical principles governing the filtering mechanism are mechanical straining and adsorption. These principles are essential for successfully filtering out suspended particles and other contaminants from water. First, mechanical straining is the primary mechanism of removing large suspended particles from water as they travel through the filter medium. Second, adsorption is a secondary mechanism involving van der Waals forces that lead smaller particles to physically adsorb and adhere to the sand grain surface and enhance the filtration process by removing dissolved substances and very fine particles that are not effectively captured by mechanical straining [17].

6.2.2 Health Aspects of Rapid Sand Filtration Technique

Rapid sand filtration is a very effective way to remove many pathogenic bacteria and protozoa (e.g., Giardia and Cryptosporidium), which are common causes of waterborne diseases. Additionally, it mitigates viral contaminants, especially when combined with disinfectant and pre-treatment. Furthermore, eliminating suspended particles from water decreases its turbidity, which is beneficial for aesthetic reasons as well as the fact that excessive turbidity protects pathogens from disinfection procedures [18–20]. The filter reduces the amount of organic and inorganic impurities that are harmful to human health or degrade the quality of the water by eliminating suspended particles. Likewise, the coagulation-flocculation process also reduces the turbidity to less than 0.1 Nephelometric Turbidity Units (NTU), which is considered a high standard for drinking water quality [16]. Thus, rapid sand filtration systems can be used to supply safe and clean drinking water that meets safety standards when used in conjugation with other specific pre- and post-treatments.

6.2.3 Construction, Operation, Function, and Maintenance of Rapid Sand Filters

For the removal of water contaminants, rapid sand filtration systems are constructed, operated, and maintained with careful planning, consistent monitoring, and regular maintenance.

Construction

Rapid sand filtration systems are constructed by keeping in mind the volume of water that has to be treated, which further determines the size and number of filter tanks required. The filter system is constructed by using concrete or steel and includes sections for the filter bed and backwash. This requires the supervision of highly skilled and experienced personnel. In addition, a backwash water distribution system called an underdrain system is built at the bottom of the filter bed. It is made up of specifically engineered plates or perforated pipes that collect filtered water. Effective water treatment can be made possible by the very precise implementation of a rapid sand filtration system.

Operation, Function, and Maintenance

A rapid sand filtration system functions in two ways: filtration and backwashing. In the filtration process suspended particles move through the filter medium and attach to the surfaces of the sand grains and get removed from the turbid water. In the backwashing process, water flows through the filter in the opposite direction at a speed that fluidizes the filter bed material and removes the particles that have been trapped in the filter. Depending on the quality of the input water, backwash intervals usually last between 24 and 72 hours [16]. Every filter bed cleaning must be followed by a 5- to 10-minute purification process. It takes multiple parallel filtration units to ensure a steady flow of water. Maintaining a steady flow rate through the filter optimizes filtration efficiency and prevents media disturbance. It is also necessary to monitor the backwash process; in particular, the flow rate needs to be regulated to prevent the filter media from eroding. Maintaining effective performance also requires periodic reprocessing of the filter bed [16]. Thus, operation and maintenance demand highly competent and experienced personnel.

6.3 TECHNIQUE OF RAPID SAND FILTRATION FOR REMEDIATION OF MICROPLASTICS

Following primary and secondary treatment, the samples are subjected to four further treatments. Pre-sedimentation, coagulation flocculation, and sedimentation in the operating unit, to examine the effectiveness of MP removal from wastewater treatment facilities [19]. A single-media RSF is regarded as a useful, economical, and acceptable technology for treating both wastewater and drinking water.

6.3.1 RAPID SAND FILTER REACTOR AND FILTER MEDIA

Effective water treatment relies on a combination of a rapid sand filter reactor design and well-selected filter media. A rapid sand filter reactor is a water treatment unit designed to utilize the filter media for effective water purification. The reactor is made up of many parts and uses certain procedures to achieve high filtering efficiency. In a schematic layout of a rapid sand filter reactor, the following components are present (i.e., inlet and outlet, filter media bed, pump and flow meter, underdrain system, and backwash system). An inlet distribution system is made up of a system of pipes

Rapid Sand Filtration Technique

or channels through which influent water is distributed equally across the filter bed surface. The filter bed consists of multiple layers of sand/gravel layer and under-strain systems, each serving a specific purpose in the filtration process. A submersible pump is provided into the reactor for the continuous flow and uniform distribution of water throughout the filter bed, both of which are necessary for efficient filtering performance. Furthermore, the flow meter provided in the filter system measures and manages the water flow rate that enters the reactor [15].

The treatment effectiveness of the sand filter greatly depends on the selection and preparation of the sand and gravel that are needed for it. The filter media of RSFs absorb and remove suspended particles from water. Silica sand is frequently used as a filter medium in rapid sand filtration systems due to its chemical and physical characteristics, which promote prolonged filtration and efficiency. The effective size (ES) and uniformity coefficient (UC) of the media used in filtration systems determine their effectiveness. These parameters provide important details regarding hydraulic properties and the filtering performance of the media used. The ES value of the media used impacts the hydraulic characteristics of the filter and its ability to retain tiny particles. Filter media with low ES value shows improved removal of tiny particles. The UC value of filter media on the other hand indicates the uniformity of the sand's grading. Filter media with low UC value (near 1) indicate the uniform range of particle sizes.

A sequence of sieves with different mesh sizes is used to filter silica sand. For instance, sand may be screened to 40–70 mesh and 20–40 mesh sizes, where the numbers denote the number of sieve openings per inch. Sand is sorted using this procedure into size fractions appropriate for various filtering uses. In sieve analysis, an increasing number of sieves with decreasing openings are stacked on top of a sample of sand. Particles can flow through the sieves to the finest sieve that can hold them because they are mechanically shaken. For every fraction, the weight of sand retained on each sieve is measured, and the percentage of the total sample weight is computed. The sand sample's particle size distribution is graphically represented by the distribution accumulation curve. It displays the weight of the sample that is cumulatively finer than a specified particle size as a percentage. A graph is plotted using the percentage of sample weight that passes through each sieve size as a result of the sieve analysis. The cumulative percentage of the sample weight finer than the specified size is shown on the y-axis, while the x-axis shows the particle size (often on a logarithmic scale). The particle sizes that correspond to 10% and 60% on the cumulative percentage axis are identified to get the D10 and D60 values from the curve. These details are essential for figuring out the UC as well as the filter media's grade and possible performance.

These systems are an essential part of drinking water and wastewater treatment operations because of their durability, affordability, and ability to manage a range of water quality and flow rates.

6.3.2 Microplastics Artificial Sample

Artificial MP samples, often referred to as synthetic MPs, are developed for several scientific, regulatory, and environmental monitoring applications. Artificial samples are synthesized for standardization and control, which allows for consistency in experiments and comparability across different studies. To identify and validate analytical techniques for the detection and quantification of MPs in environmental samples (such as water, soil, and air), artificial MPs are employed as standards. For the preparation of an artificial MP sample appropriate polymer type has to be selected and it should disperse uniformly across the medium. Artificial MP samples are prepared by using a wide variety of synthetic polymers such as polyethylene, polypropylene, and polyvinyl chloride.

Several techniques are utilized for producing artificial MP particles with the desired size, shape, and surface characteristics. It includes mechanical fragmentation, extrusion and cutting, and microsphere manufacturing. The artificial MP particle samples can be filtered or processed to achieve the desired size ranges. Thereafter, the synthesized MP particles are mixed with filter media to form a stock solution, which is then sonicated and continuously stirred to ensure uniform dispersion of MPs in the stock solution. These types of artificially synthesized MPs are used for studying the behavior, action, and effects of MPs in controlled laboratory environments. This ultimately contributes to the development of mitigation and environmental monitoring programs.

6.3.3 MICROPLASTIC IDENTIFICATION

Accurate identification of MPs is important for understanding their distribution, origins, and possible effects on both human and environmental health. For the identification of MPs in environmental settings, multiple procedures are followed including collection, processing, and analysis. Samples can be collected from water, sediment, and biota and prepared via sieving, density separation, chemical digestion, and filtration. Thereafter, the identification of samples can be performed visually by stereo microscopy, light microscopy, and polarized light microscopy and spectroscopically by Fourier Transform Infrared (FTIR) spectroscopy or Raman spectroscopy. Thermal analysis techniques (i.e., pyrolysis-gas chromatography/mass spectrometry and differential scanning calorimetry) are employed to identify the polymer type. Furthermore, information regarding surface morphology, texture, elemental composition, and chemical state can be obtained by using other advanced techniques like Scanning Electron Microscopy (SEM) and X-ray Photoelectron Spectroscopy (XPS) [16]. Thus, by the use of these spectroscopic and microscopic techniques, MPs can be distinguished from other materials. In addition to this information regarding polymer type, their sources, and their effects on the natural environmental condition as well as on human health can be comprehended.

6.3.4 PRE-TREATMENT

Pre-treatment in the water treatment process is a vital step that sets the water up for further filtering and purification procedures. Eliminating organic debris, big particles, and other impurities that might harm equipment or impede the filtering process is the main objective of pre-treatment. Pre-treatment aims to mimic a series of water treatment units before the water enters the filter unit. Examples of pre-treatment processes include pre-sedimentation, coagulation, flocculation, and sedimentation.

Rapid Sand Filtration Technique

6.3.4.1 Pre-sedimentation

Pre-sedimentation is a crucial step in the pre-treatment of water, especially for raw water sources like rivers and lakes that have significant turbidity or sediment content. It is mostly used to get rid of big, heavy particles from the water before it gets additional treatment. Sand, silt, and other suspended particles fall under this category. This step enhances overall efficacy by reducing the load on subsequent treatment phases. The pre-sedimentation stage of MP removal is equivalent to the settling of particles in the gravity-based processing of wastewater. In the pre-sedimentation unit, discrete particle deposition takes place [20, 21]. Particle deposition is influenced by several factors, including temperature, particle concentration in the liquid, shape (flat, round, or irregular), density, liquid-specific gravity, viscosity, and suspension properties [22].

It lowers the turbidity of the water by eliminating bigger particles, which improves the effectiveness of following filtering and treatment procedures. Pre-sedimentation takes place through the following steps: At first, water enters the pre-sedimentation basin or tank, and water is stored in a clarifier or sedimentation basin, where the flow velocity is greatly decreased. Due to gravity action on the suspended particles brought on by this interval of time, the particles settle to the bottom of the basin. When particles settle, sludge accumulates at the basin's bottom, which must be removed periodically. The goal of routinely clearing this sludge is to avoid buildup, which can reduce the basin's efficiency and capacity. Sludge can be mechanically removed with the use of pumps or scrapers. The water is collected from the surface of the basin after a substantial drop of suspended particles, and it is then delivered to the next stage of treatment, which is coagulation and flocculation, disinfection, or filtering.

6.3.4.2 Coagulation and Flocculation

The coagulation process in the pre-treatment of water makes it easier to remove tiny particles. In this process, destabilizing/aggregating agents like alum (coagulant) are used for aggregating the tiny particles. Coagulants used for water treatment are commonly categorized as Al- and Fe-based coagulants. Aluminum sulfate ("alum"), ferric chloride, and ferric sulfate are the most widely used coagulants for water treatment. With the addition of coagulant chemicals of charges opposite to those of the suspended solids, aggregate formation takes place due to the chemical interaction between coagulants and MP particles [23]. The aggregates formed in this step are usually referred to as micro flocs.

Furthermore, the flocculation process is a slow stirring process that is performed to enhance the elimination of MP particles from the water. After the aggregation of MP particles in the coagulation step, further aggregation of tiny unstable particles (called flocs) into bigger floc aggregates takes place in a specifically built basin or tank known as a flocculation. Since larger aggregates are formed in the flocculation process and hence are easier to remove by filtration or sedimentation, by aggregating small particles into larger flocs, the effectiveness of the sedimentation and filtering processes is increased. The size and strength of flocs formed in these steps depend on the stirring speed of the flocculator, the chemistry involved during the coagulation step, and also on the properties of the water used.

It is necessary to remove the larger aggregates that result from the coagulation and flocculation process before proceeding with additional treatment procedures. The coagulation-flocculation process lowers the turbidity of the water sample and increases sedimentation and filtering efficiency. The coagulation-flocculation process is affected by variables like temperature, type of coagulant used, and chemical-physical forces involved in the process [24]. The efficient removal of suspended MP particles and pollutants in this step increases the quality of the treated water.

6.3.4.3 Sedimentation

After coagulation and flocculation, the sedimentation process is involved in the water treatment process. It is one of the most important steps in the pre-treatment process of water, which involves the removal of suspended materials, including the flocs created during the coagulation-flocculation process. During this process, flocs formed in the previous step get settled under the force of gravity, resulting in the production of water with a high degree of purity. The sediment formed at the bottom due to the settling of particles is known as sludge. This process removes a significant quantity of suspended materials from the water, which is required for the effectiveness of subsequent treatment operations such as filtration and disinfection.

6.3.5 FILTRATION TESTS

Filtration tests are essential to evaluate the efficiency of the filter media used in rapid sand filtration systems. These tests are performed to study how well the system removes impurities from water and determine the filter's capacity and characteristics of operation under various circumstances. It involves the study of influent and effluent water quality, along with water flow rate, pressure drop, and backwash rate and frequency. The assessment of water quality is conducted both before and after filtration, with a specific focus on suspended solids, turbidity, and the existence of specific contaminants such as MPs. It involves the assessment of parameters like the speed at which water is filtered and the difference in pressure across the filter medium. These parameters are equally essential to comprehend the filter's capacity and effectiveness and also show an indication of fouling or clogging [25–30]. Thus, through a methodical assessment of parameters, these tests provide details for refining filter configuration and functioning.

6.3.6 FILTER MEDIA ANALYSIS

Analysis of filter media is important since it directly affects the filter's capacity to remove MP particles and pathogens. It involves the assessment of filter media to maximize the effectiveness of the rapid sand filtering process. For filter media analysis, samples are collected and prepared from different layers and sections of the filter bed. At first, filter media undergoes a drying cleaning process to remove any adhered particles or contaminants that could affect the analysis.

Thereafter, for the determination of particle size distribution, sieve analysis is performed. Key factors that affect the filtering system's performance and efficiency include its effective size, uniformity coefficient, porosity, specific surface area, and durability [31–36]. Water treatment facilities increase water quality and treatment efficacy by carrying out thorough media analyses to make sure the chosen media fulfills the requirements for pollutant removal, flow rates, and operational stability.

6.4 MECHANISM OF MICROPLASTIC REMOVAL PROCESS USING A RAPID SAND FILTER

Using RSFs to remove MPs includes several physical and chemical processes that work together to improve filtration effectiveness. Particulate removal using granulated media involves three processes: transportation (which includes Brownian motion); sedimentation and particle attraction; sticking ability, which includes mechanical straining, adsorption (physical-chemical), and biological processes; and resistance (which includes particle collisions and repelling forces) [25, 37, 38]. First, when passing through filter media, relatively large suspended particles are caught between sand grains (mechanical straining). Second, Van der Waals forces (physical adsorption) cause smaller particles to adhere to the surface of the sand grains. Chemical filters such as coagulants or flocculants can be used to increase the adhesion force [17, 39].

The size of MPs has a major impact on the mechanical straining process during their removal using the rapid sand filtration process. Larger MPs are more easily removed by straining because they are more likely to be found on the surface or in the higher layers of the sand substrate. If the particle's size is less than the filter's voids, van der Waals forces will drive it to adhere to the medium. Huisman's method [26] could potentially be used to numerically calculate the diameter of the gap between the filter medium pores. It is therefore possible to establish the MP size that can travel through the filter material [27–32, 40–42]. The efficiency of RSF in eliminating MPs is dependent on the pore size distribution of the sand medium (Table 6.1).

6.5 APPLICABILITY OF RAPID SAND FILTRATION TECHNIQUE

Rapid sand filtration systems are widely employed water treatment procedures with multiple applications, particularly in municipal and industrial water supply systems. These are frequently utilized in municipal water treatment plants to remove suspended particles, turbidity, and microbes from raw water, making it safe to drink. These are also used in various industries to treat water used in industrial processes, ensuring that suspended particles and other impurities are eliminated to protect machinery and maintain quality requirements [43–49]. Rapid sand filtration technique improves water quality in cooling towers and other cooling water systems by eliminating contaminants that might cause scaling, fouling, or corrosion. Rapid sand filtration systems are also effective in treating water from lakes, rivers, and reservoirs, which frequently contain suspended particles, algae, and organic matter. RSF could potentially be used to remove metals like Fe and Mn from groundwater, which are frequent contaminants in some areas and cause staining, taste, and odor problems in water [50]. Applicability of RSFs also requires large energy inputs, extremely sophisticated technical installations, and highly trained personnel for both building and operation. Thus, rapid sand filtration systems are effective in eliminating particles and microbes

TABLE 6.1
Key Components of the Rapid Sand Filtration Technique for Remediation of Microplastics, Including their Role and Significance

S.No.	Element	Description	Role and Significance
1.	Filter Media	More often silica is used with specific particle size	Physically remove particles from water through straining, and depth filtration, act as an adsorbent surface, support biological growth, and influence the hydraulic properties of the filtration system.
2.	Filter Bed	The depth of the filter bed typically ranges from 0.6 to 1.8 meters	The depth and composition of the filter media are important for effective particle removal. Filter beds must have enough porosity to let water pass through while trapping particulate matter.
3.	Underdrain System	Underdrain pipes or troughs are located at the bottom of the filter bed	Collects filtered water and prevents the sand from leaving the filter bed during operation or backwashing.
4.	Control Valves and Flow Meter	Flow Control Valves: regulate the flow rate of water entering and leaving the filter. Flow meters are used to monitor the flow rates of influent and effluent water	Proper control must be maintained to ensure appropriate filtration rates and avoid damage to the filter bed.
5.	Backwash System	Designed to clean and maintain the filter media by removing accumulated contaminants	primary role – to restore filter's functionality – reverses the flow of water to dislodge and remove trapped particles from the filter media, provides operational flexibility, and extends life of the filter media.
6.	Coagulants and Flocculants	Work together to combine smaller particles into bigger clusters, to be extracted by filtration. Standard coagulants – aluminum sulfate, and ferric chloride.	Coagulants – neutralize electrostatic charges that keep particles suspended in water. Flocculants – bind these particles together. Coagulation and flocculation processes significantly reduce water turbidity.
7.	Monitoring and Control Systems	Systems include sensors, automated controls, data logging, and real-time analytics. All work together to ensure efficient and effective operation of the filtration system.	Real-time data on the filtering process's state is provided via monitoring devices, continuously measuring key parameters like water turbidity and flow rate across the filter bed. Control systems optimize the performance of the filtration process.

Rapid Sand Filtration Technique

from water, making them useful in a variety of industries, including municipal, industrial, and agricultural water treatment.

6.6 ADVANTAGES OF RAPID SAND FILTRATION TECHNIQUES

The rapid sand filtration systems used for the remediation of MPs present several advantages that make them useful for water treatment including:

1. They are effective at lowering water turbidity and eliminating suspended particles; they frequently achieve values below 0.1 NTU, which is good for safety and aesthetics.
2. High filter rate (4'000 – 12'000 liters per hour per square meter of surface) makes them suitable for municipal water treatment plants and industrial applications. Small land requirements make them more suitable for urban or space-constrained environments.
3. Operational flexibility can be achieved by adjusting the filtration rate based on demand and water quality. Growing water needs can also be readily accommodated by rapidly scaling up RSFs by adding more units or expanding the area of the filter bed.
4. Automated monitoring and control systems simplify operations and guarantee consistent water quality.
5. Backwashing the filter media makes it possible to regularly remove trapped particles, preserving filter effectiveness and prolonging the media's lifespan. Backwashing takes only a few minutes, and the filters may be placed back into operation immediately.
6. RSFs provide more economical initial capital expenditures and operating expenses when compared to more sophisticated treatment methods like membrane filtration.
7. RSF helps to safeguard the environment and public health by efficiently eliminating pollutants. They can reduce the strain on subsequent disinfection procedures and contribute to the production of water for consumption, thereby reducing the total amount of chemicals used in water treatment.

6.7 DISADVANTAGES OF RAPID SAND FILTRATION TECHNIQUE

Although rapid sand filtration is a popular and successful technique for treating water, some disadvantages and restrictions should be taken into account. The following are the main drawbacks of rapid sand filtration:

1. RSFs are primarily designed to remove suspended particles and are not effective in removing heavy metals, nitrates, pesticides, bacteria, viruses, and organic matter (requires pre- and post-treatment).
2. To maintain efficiency, regular cleaning (backwashing) is required every 24 to 72 hours.
3. The efficacy of the sand and gravel layers may decrease with time due to degradation or fouling, necessitating replacement or renovation.

4. Skilled personnel are required for RSF operation and maintenance, particularly for operations like media replacement, backwashing, and system calibration. Inadequate upkeep might result in decreased efficiency and water quality.
5. To guarantee optimal performance, characteristics including turbidity, flow rates, and pressure differentials must be continuously monitored. This calls for complex instrumentation and control systems.

6.8 COMPARISON OF RAPID AND SLOW SAND FILTRATION TECHNIQUES

Rapid sand filtration technique and slow sand filtration (SSF) technique of water purification are two different techniques in which sand is used as a filter medium, but there are notable differences in their construction, functionality, and uses. While the SSF technique primarily depends on biological activity, the rapid sand filtration technique mostly depends on physical straining. Rapid sand filtration systems have a far greater filtering rate than SSF systems. It is around 4 to 21 $m^3/h.m^2$ for rapid sand filtration systems and 0.1 to 0.4 $m^3/h.m^2$ for SSF systems [32, 51, 52]. Whereas fine sand is used as filter media in SSF systems, coarse sand and a layer of gravel support the filter media in rapid sand filtration systems. Pre-treatment procedures like flocculation and coagulation are necessary in rapid sand filtration systems but not in SSF systems. Thus, due to these differences in their design and functionality their field of application is also different. Rapid sand filtration systems are used in the areas where large volumes of water need to be treated quickly like municipal water treatment plants and industrial applications. SSF systems on the other hand are employed in situations requiring long-term, low-maintenance operations as in rural or small-scale water treatment systems.

6.9 CONCLUSION

Rapid sand filtration is a tried and tested technique for removing MPs from water and wastewater. It is an efficient means for addressing one of the most significant environmental problems of contemporary times due to its potential to collect a broad variety of particle sizes, including MPs. The removal of certain MP sizes can be made possible through the use of pre-sedimentation, coagulation, flocculation, sedimentation, and filtering. Physical straining, adsorption and aggregation, customization, optimization, cost-effectiveness, and scalability are the main factors that make rapid sand filtration successful in removing MPs.

REFERENCES

[1] M. Umar, C. Singdahl-Larsen, & S.B. Ranneklev, "Microplastics removal from a plastic recycling industrial wastewater using sand filtration." *Water*, 15(5), 896, Feb. 2023, doi: 10.3390/w15050896.

Rapid Sand Filtration Technique

[2] S. Wolff, F. Weber, J. Kerpen, M. Winklhofer, M. Engelhart, & L. Barkmann, "Elimination of microplastics by downstream sand filters in wastewater treatment." *Water*, *13*(1), 33, Dec. 2020, doi: 10.3390/w13010033.

[3] B.D. Marsono, A. Yuniarto, A. Purnomo, & E.S. Soedjono, "Comparison performances of microfiltration and rapid sand filter operated in water treatment plant". In *IOP Conference Series: Earth and Environmental Science* (Vol. 1111, No. 1, p. 012048), IOP Publishing, 2022, doi: 10.1088/1755-1315/1111/1/012048.

[4] M. Eriksen, S. Mason, S. Wilson, C. Box, A. Zellers, W. Edwards, H. Farley, & S. Amato, "Microplastics pollution in the surface waters of the Laurentian Great Lakes." *Marine Pollution Bulletin*, 77, 177–182, Dec. 2013, doi: 10.1016/j.marpolbul.2013.10.007.

[5] S. Kawamura, *"Integrated Design of Water Treatment Facilities."* John Wiley & Sons, Inc., New York, NY, 2000.

[6] M.J. Adelman, M.L. Weber-Shirk, A.N. Cordero, S.L. Coffey, W.J. Maher, D. Guelig, D., ... & L.W. Lion, "Stacked filters: Novel approach to rapid sand filtration." *Journal of Environmental Engineering*, 138(10), 999–1008, Feb. 2012, doi:10.1061/(ASCE) EE.1943-7870.0000562.

[7] C.R. O'Melia, & D.K. Crapps, "Some chemical aspects of rapid sand filtration." *Journal-American Water Works Association*, 56(10), 1326–1344, Jun. 1964, doi: 10.1002/j.1551-8833.1964.tb01340.x.

[8] S.M. Mintenig, M.G. Löder, S. Primpke, & G. Gerdts, "Low numbers of microplastics detected in drinking water from groundwater sources." *Science of Total Environment*, 648, 631–635, Aug. 2018, doi: 10.1016/j.scitotenv.2018.08.178.

[9] D. Vries, C. Bertelkamp, F.S. Kegel, B. Hofs, J. Dusseldorp, J.H. Bruins, ... & B. Van den Akker, "Iron and manganese removal: Recent advances in modelling treatment efficiency by rapid sand filtration." *Water Research*, 109, 35–45, Nov. 2016, doi: 10.1016/ j.watres.2016.11.032.

[10] E. Sembiring, et al. *"Performance of RSF to Remove Microplastics, Water Supply."* Environmental Engineering Department, Institut Teknologi Bandung, Bandung, Indonesia, 2021, doi: 10.2166/ws.2021.060.

[11] L. Metcalf & H.P. Eddy, G. , *"Wastewater Engineering Treatment, Disposal, and Reuse."* 3rd edn. McGraw-Hill, Inc., New York, NY, 1991, doi: 10.1016/ 0309-1708(80)90067-6.

[12] S. Morganaa, L. Ghigliottia, N. Estevez-Calvara, R. Stifanesea, A. Wieckzorekb, T. Doyle, J.S. Christiansenc, M. Faimalia, & F. Garaventaa, "Microplastics in the Arctic: A case study with sub-surface water and fish samples off Northeast Greenland." *Environmental Pollution*, 242 (part B), 1078–1086, Aug. 2018, doi: 10.1016/ j.envpol.2018.08.001.

[13] B.E. Oßmann, G. Sarau, H. Holtmannspötter, M. Pischetsrieder, S.H. Christiansen, & W. Dicke, "Small-sized microplastics and pigmented particles in bottled mineral water." *Water Research*, 141, 307–316, May. 2018, doi: 10.1016/j.watres.2018.05.027.

[14] D. Schmitt, & C. Shinault, *"Rapid Sand Filtration."* Virginia Tech, Blacksburg, 1996.

[15] B. Deboch, & K. Faris, "Evaluation on the Efficiency of Rapid Sand Filtration." *Addis Abada (Ethiopia): 25th WECD Conference*, 1999.

[16] R. Sabale, & S. Mujawar, "Improved rapid sand filter for performance enhancement." *International Journal of Science and Research* (IJSR), 3(10), 1031-1033, 2014.

[17] M. Fajar, E. Sembiring, & M. Handajani, "The effect of Filter Media size and loading Rate to filter performance of removing microplastics using rapid sand filter." *Journal of Engineering and Technological Sciences*, 54(5), Jul. 2022, doi: 10.5614/j.eng. technol.sci.2022.54.5.12.

[18] F. Brikke, & M. Bredero, "Linking Technology Choice with Operation and Maintenance in the context of community water supply and sanitation. A reference Document for Planners and Project Staff." World Health Organization and IRC Water and Sanitation Centre, Geneva , 2003.

[19] L. Su, Y. Xue, L. Li, D. Yang, P. Kolandhasamy, D. Li, & S. Huahong, "Microplastics in Taihu Lake, China." *Environmental Pollution*, 216, 711–719, Jun. 2016, doi: 10.1016/j.envpol.2016.06.036.

[20] J. Bayo, J. López-Castellanos, & S. Olmos, "Membrane bioreactor and rapid sand filtration for the removal of microplastics in an urban wastewater treatment plant." *Marine Pollution Bulletin*, 156, 111211, May. 2020, doi: 10.1016/j.marpolbul.2020.111211.

[21] C. Di Marcantonio, C. Bertelkamp, N. van Bel, T.E. Pronk, P.H. Timmers, P. van der Wielen, & A.M. Brunner, A. M, "Organic micropollutant removal in full-scale rapid sand filters used for drinking water treatment in The Netherlands and Belgium." *Chemosphere*, 260, 127630, doi: 10.1016/j.chemosphere.2020.127630.

[22] V. Hidalgo-Ruz, L. Gutow, R.C. Thompson, & M. Thiel, "Microplastics in the marine environment: A review of the methods used for identification and quantification." *Environmental Science Technology*, 46, 3060–3075, Feb. 2012, doi: 10.1021/es2031505.

[23] M. Siegfried, A.A. Koelmans, E. Besseling, & C. Kroeze, "Export of microplastics from land to sea. A modelling approach." *Water Research*, 127, 249–257, 2017, doi: 10.1016/j.watres.2017.10.011.

[24] S. Gupta, P.K. Mishra, & D. Khare, "Performance and suitability of an economically efficient Rapid Sand Filter for filtration of roof-top harvested rainwater." *Materials Today: Proceedings*, Jun. 2023, doi: 10.1016/j.matpr.2023.06.203.

[25] G. Tchobanoglous, F.L. Burton, & H.D. Stensel, "*Wastewater Engineering Treatment & Reuse*." 4th edn. McGraw-Hill Book Company, New Delhi, India (2003), Feb. 2012, doi: 10.1021/es2031505.

[26] P.F. Iqbal, K.M. Athar, M.S. Muzammil, & P.A. Waqar, "Treatment of greywater by rapid sand filter." *International Journal of Interdisciplinary Innovative Research and Development*, 1(4), 17–21.

[27] I. Chubarenko, A. Bagaev, M. Zobkov, & E. Esiukova, "On some physical and dynamical properties of microplastic particles in marine environment." *Marine Pollution Bulletin*, 108 (1–2), 105–112, April. 2016, doi: 10.1016/j.marpolbul.2016.04.048.

[28] D. Eerkes-Medrano, H.A. Leslie, & B. Quinn, "Microplastics in drinking water: a review and assessment." *Current Opinion in Environmental Science & Health*, 7, 69–75, Mar. 2019, doi: 10.1016/j.coesh.2018.12.001.

[29] T.J. Hoellein, A.J. Shogren, J.L. Tank, P. Risteca, & J.J. Kelly, "Microplastic deposition velocity in streams follows patterns for naturally occurring allochthonous particles." *Scientific Report*, 9, 3740, Feb. 2019, doi: 10.1038/s41598-019-40126-3.

[30] F.C. Alam, E. Sembiring, B.S. Muntalif, & V. Suendo, "Microplastic distribution in surface water and sediment river around slum and industrial area case study: Ciwalengke. River, Majalaya district, Indonesia." *Chemosphere*, 224, 637–645, Feb. 2019, doi: 10.1016/j.chemosphere.2019.02.188.

[31] B. Ma, W. Xue, C. Hu, H. Liu, J. Qu, & L. Li, "Characteristics of microplastic removal via coagulation and ultrafiltration during drinking water treatment." *Chemical Engineering Journal*, 359, 159–167, Nov. 2019, doi: 10.1016/j.cej.2018.11.155.

[32] J. Talvitie, A. Mikola, A. Koistinen, & O. dan Setala, "Solutions to microplastics pollution removal of microplastics from wastewater effluent with advanced wastewater treatment technologies." *Water Research*, 123, 401–407, July. 2017, doi: 10.1016/j.watres.2017.07.005.

[33] A.C. Twort, D.D. Ratnayaka, & M.J. Brandt, *"Water Supply."* Butterworth-Heinemann, Oxford, UK, 2006.

[34] AWWA. *"Water Quality Treatment."* AWWA. Mc Graw Hill, New York, NY, 1990.

[35] M.A. Browne, P. Crump, S.J. Niven, E. Teuten, A. Tonkin, T. Galloway, & R. Thompson, "Accumulation of microplastics on shorelines worldwide: Sources and sinks." *Environmental Science Technology,* 45 (21), 9175–9179, Sep. 2011, doi: 10.1021/es201811s.

[36] G.J. Churchman, W.P. Gates, B.K.G. Theng, & G. Yuang, "Clay and Clay Minerals for Pollution Control." In *Handbook of Clay.* Elsevier, Amsterdam, the Netherlands, Edited by: Faiza Bergaya, Gerhard Lagaly, Nov. 2013, doi: 10.1016/B978-0-08-098259-5.00021-4.

[37] C.B. Crawford, & B. Quinn, *"Microplastics Pollutants."* 1st edn. Elsevier Science, Amsterdam, the Netherlands, 2017.

[38] M. Cole, P. Lindeque, C. Halsband, & T.S. Galloway, "Microplastics as contaminants in the marine environment: A review." *Marine Pollution Bulletin*, 62 (12), 2588–2597, 2011.

[39] J.C. Crittenden, R.R. Trussell, D.W. Hand, K.J. Howe, & G. Tchobanoglous, *"Water Treatment Principles and Design."* 3rd edn. John Wiley & Sons, Inc., New York, NY, 2012.

[40] J.C. Van Dijk, J.H.C.M. Oomen, "Slow Sand Filtration for Community Water Supply in Developing Countries." *A Design and Construction Manual*, 1978.

[41] R. L. Droste, *"Theory and Practice of Water and Wastewater Treatment."* John Wiley & Sons, New York, NY 1997.

[42] D. Eerkes-Medrano, R.C. Thompson, & D.C. Aldridge, "Microplastics in freshwater systems: A review of the emerging threats, identification of knowledge gaps and prioritization of research needs." *Water Research*, 75, 63–82, Feb. 2015, doi: 10.1016/j.watres.2015.02.012.

[43] T.S. Galloway, "Micro-and Nano-Plastics and Human Health." In: *Marine Anthropogenic Litter* (Bergmann, M., Gutow, L. & Klages, M, eds), pp. 343–366. Springer, Dordrecht, the Netherlands, 2015, doi: 10.1007/978-3-319-16510-3.

[44] A.A. Horton, A. Walton, D.J. Spurgeon, E. Lahive, & C. Svendsen, "Microplastics in freshwater and terrestrial environments: evaluating the current understanding to identify the knowledge gaps and future research priorities." *Science of the Total Environments*, 586, 127–141, Feb. 2017, doi: 10.1016/j.scitotenv.2017.01.190.

[45] S. Klein, E. Worch, & T.P. Knepper, "Occurrence and spatial distribution of microplastics in river shore sediment of the Rhine-Main Area, Germany." *Environmental Science and Technology*, 49, 6070–6076, Apr. 2015, doi: 10.1021/acs.est.5b00492.

[46] P.J. Kole, A.J. Lohr, F.G.A.J. Belleghem, & A.M.J. Ragas, "Wear and tear of tyres: A stealthy source of microplastics in the environment." *International Journal of Environmental Research and Public Health*, 14, 1265, Oct. 2017, doi: 10.3390/ijerph14101265.

[47] M. Pivokonsky, L. Cermakova, K. Novotna, P. Peer, T. Cajthaml, & V. Janda, "Occurrence of microplastics in raw and treated drinking water." *Science of Total Environment,* 643, 1644–1651, Aug. 2018, doi: 10.1016/j.scitotenv.2018.08.102.

[48] D. Schymanski, C. Goldbeck, H.U. Humpf, & P. Fürst, "Analysis of microplastics in water by micro-Raman spectroscopy: Release of plastic particles from different packaging into mineral water." *Water Research*, 129, 154–162, Nov. 2018, doi: 10.1016/j.watres.2017.11.011.

[49] E.W. Steel, & T.J. McGhee, *"Water Supply and Sewerage."* 5th edn. McGraw-Hill, New York, NY, 1985.

[50] P. Stundt, P. Schulze, & F. Syversen, "Sources of Microplastics-Pollution to the Marine Environment." Mepex for the Norwegian Environment Agency Report no M-321/2015, 2015.

[51] C.C Dorea, & B.A. Clarke, "Chemically enhanced gravel pre-filtration for slow sand filters: Advantages and pitfalls." *Water Science and Technology: Water Supply*, 6(1), 121–128, Jan. 2006, doi: 10.2166/ws.2006.029.

[52] K. Zhang, W. Gong, J. Lv, X. Xiong, & C. Wu, "Accumulation of floating microplastics behind the three Gorges Dam." *Environmental Pollution*, 204, 117–123, 2015, doi: 10.1016/j.envpol.2015.04.023.

7 Bioremediation of Microplastics

Anjali Yadav, Shubhvardhan Singh,
Bhawana Jangir, Jaya Dwivedi, and
Manish Srivastava

7.1 INTRODUCTION

Plastics are widely used in a variety of sectors around the world due to their accessibility, affordability, and versatility, providing substantial convenience to humanity. Despite its practical application, it has suddenly emerged as a major global environmental problem [1–3]. The environment has unavoidably become overrun with a vast number of discarded plastics due to the increasing demand and use of plastic products [4]. Plastic waste becomes an environmental contamination that cannot disintegrate in the natural environment even after 100 years [5,6]. Brahney et al. [7] anticipated that by 2025 the amount of plastic in the environment would have increased to 11 billion tons. However, the present rate of plastic recovery is a little higher than 5%.

Plastic particles smaller than 5 mm are known as microplastics, or MPs [8]. Because of their detrimental effects on the biosphere, particularly in marine environments, there is growing concern about MPs. Current research indicates that MPs can be found in sediments, rivers, wastewater, airways, and soil in addition to the ocean [9–11]. Fish, zooplankton, phytoplankton, and seabirds are just a few of the many marine animals that can eat MPs, which build up in tissues and circulatory systems and can be harmful. Additionally, they have been discovered in deep marine sediments and polar glaciers [4]. MPs are small, abundant, and easily absorbed by the human body via the food chain. Thus, compared to bigger plastic objects, MPs offer a far higher risk to humans [12]. In conclusion, the global frequency of MP pollution has increased, raising concerns about the eventual fate of these ubiquitous particles. MPs cause damage to a variety of creatures as well as soil and water systems when they build up in the environment. Research indicates that a diverse range of creatures consume MPs, leading to various issues such as reduced growth rates, physiological stress, false satiety, and reproductive abnormalities. Furthermore, MPs contain a range of substances that have the potential to contaminate soil and water [13–15]. MPs also cause several difficulties in sea areas, influencing aquatic plant growth and fish reproductive systems, for example [16]. Because MPs have such a significant

DOI: 10.1201/9781003486947-7

environmental impact, understanding how to reduce their numbers has become a hot topic. Recent research endeavors have delved into the dispersion, ingestion, destiny, behavior, prevalence, and consequences of MPs. However, the effectiveness of proposed MP removal and remediation strategies remains questionable. Furthermore, there has been no comprehensive summary of the breakdown of MPs.

7.2 ORIGIN OF MPS

MPs are difficult to trace due to their small size. It is imperative to comprehend their origins to eliminate plastic materials and MPs in marine habitats effectively. Microplastics are routinely released into the environment through:

- Wastewater treatment plants [17]
- Drainage systems [18]
- Debris from ships and recreational activities
- Breakdown of agricultural polyethylene foils through dissolution
- Laundry and cloth cleaning [19]
- Car tire abrasions
- Fertilizer runoff [20]

The origins, particle sizes, and the transmission of MPs via ocean waves and currents all contribute to the variability within aquatic ecosystems.

7.3 SOURCES OF MPS

In the environment, MPs can be classified as primary or secondary MPs, depending on the production process and breakdown of bigger plastic products [21–23]. Primary and secondary MPs are abundant in the ecosystem. MPs can enter the ecosystem via a variety of channels, including hospitals, factories, municipal solid waste, and everyday activities. The distribution of MPs throughout the environment is influenced by various factors including size, density, shape, and chemical composition. Several sources and examples of MPs in the environment are shown in Figure 7.1.

7.3.1 PRIMARY MPS

Primary MPs originate as large plastic items manufactured for specific purposes. Examples of mechanical exfoliants encompass a diverse array of products such as adhesive scrub sheets, microfiber clothing, and various cosmetic items including hair care products, handwash, eyeliner, fragrances, facewash, beauty talc, nail polish, pesticides, toothpaste, and sunscreen [24,25]. Photo, thermal, or biodegradation processes transform primary plastics into MPs, which can manifest as pellets, microbeads, fragments, and fibers. MP particles composed of polypropylene (PP), polyethylene (PE), and polystyrene (PS) have been found to be present in primary plastic objects, such as food containers, textiles, and plastic bottles. MP particles derived from textiles include polyethylene terephthalate (PET), polyethersulfone (PES), and polyamide (PA). MP beads, measuring less than 0.5 mm in diameter, are

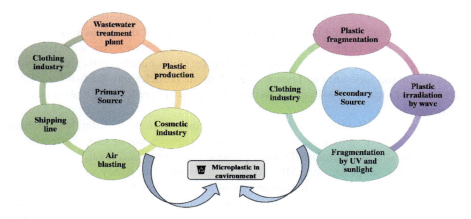

FIGURE 7.1 Origins of MPs.

present in numerous personal care products available in the market today [22,25]. MPs are conveyed through sewage systems to municipal wastewater treatment facilities. The origins of primary MPs, including the quantities discharged from each category, largely remain unidentified, representing a significant knowledge gap.

7.3.2 Secondary MPs

Secondary MPs are created when larger plastic artifacts break into smaller particles due to physical, chemical, or biological processes [23,25]. As macroscopic plastic objects break down into smaller bits, the harm to animals from MPs is predicted to worsen, and the spreading of these particles into marine and terrestrial habitats. As particle size decreases, the diversity of animals capable of consuming rubbish increases. Smaller plastic particles are therefore more readily absorbed, enhancing the susceptibility and adsorption, leaching, and desorption capabilities of MPs [21–23]. To determine the origins of secondary MPs, it is critical to identify the existence of MPs and outline the processes by which they break down in various contexts. As both macro and MPs exhibit continuous movement within the environment and undergo dynamic degradation, the precise determination of their origins poses a challenge.

7.4 EFFECT OF MICROPLASTICS

7.4.1 On Human Health

Fish are regarded as a potential protein source for human consumption. Consequently, as the top consumers in the food chain, humans inadvertently ingest MPs present in various fish and crustaceans they consume. Hence, this mode of ingestion poses a potential health hazard and has garnered growing attention in recent times [26,27]. MPs can be detected within the organs (such as the intestine and tissue) of numerous aquatic organisms, including fish, crabs, and bivalves [28]. The morphology, color, size, and mass of MPs all have an impact on how bio-accessible they are to aquatic species. The study found that MPs accumulated on mussel tissues, as well as latex-based spheres on rainbow trout skin, particularly in the phagocytes underlying the epidermal cells, in a variety of sites. Thus, it is clear that epithelial cells play a significant role in the adherence and penetration of plastic particles into the fish body [29]. According to research, MPs are present in sea salt, which reflects their abundance in seafood.

An analysis of around 17 salt samples acquired from 8 different countries showed that there were, on average, 0–10 plastic particles per kg [30]. Furthermore, fragmented and filamentous MPs were found in both honey and sugar.

One study suggested that MPs could be transported to honey via atmospheric winds, while another found low quantities of MPs in honey [31]. Canned foods like sprat and sardines contain MPs. Four of the twenty canned food brands that were evaluated had at least one to three MP particles in them. Concerns have been raised in recent years after major MP levels were found in canned food. Experts recommend imposing laws and limits to eliminate these health dangers [32].

MPs, owing to their ability to adsorb a variety of toxic compounds from the environment, represent a potential hazard to human health. Polyethylene terephthalate, or PET, is the most widely used plastic polymer in the manufacturing of pipes, food packaging, insulating materials for buildings, and drinking water bottles. Prolonged exposure to PET poses health risks due to its carcinogenic properties. Another common plastic polymer is PS, which is found in many different types of plastic cookware. Certain plastics, such as PVC (polyvinyl chloride) and PS, can emit toxic monomers that have the potential to cause cancer in humans. Furthermore, certain PVC compounds can build up in the blood of humans.

On the other hand, nanoplastics made of PS can cause alterations in the morphology of intestinal cancer cells, decrease their viability, cause inflammation, and modify human gene expression. MPs can transmit several diseases, chemical parasites, and fatal illnesses to people, making inhaling them exceedingly harmful. Absorption of MPs can potentially affect the human lungs and gastrointestinal tract, with endocytosis playing a role. Furthermore, the circulatory and gastric systems can carry MPs to other organs. Several critical parameters, including the bound protein corona, surface functionalization, particle size, surface loading, and hydrophobicity, have a significant impact in this situation. Minute plastic particles, including nanoplastics derived from polystyrene, can travel through the gastrointestinal tract's circulatory system, dispersing broadly throughout the bloodstream and organs. A potential connection between MPs and lung cancer has been suggested by the discovery of several synthetic fibers in the lung tissue of cancer patients.

Nonylphenol, triclosan, brominated flame retardants (BFRs), bisphenol A (BPA), phthalates, and organotin compounds are common additives used in plastic manufacture. There is speculation that after the ingestion of organisms containing MPs, BPA and nonylphenol might permeate biological cells [33]. Most plastics contain the hazardous compound BPA, utilized as an antioxidant or plasticizer. Being in contact with these plastic items, such as food containers, can lead to food contamination through the leaching of BPA. It induces a range of health issues, including liver dysfunction, decreased insulin sensitivity, alterations in the reproductive system,

Bioremediation of Microplastics

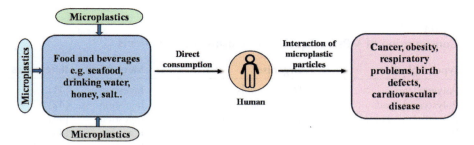

FIGURE 7.2 Effect on human health.

cognitive impairment, and pregnancy complications such as uterine abnormalities. Additionally, BPA interferes with thyroid hormone production, which impacts pancreatic beta cell function and may cause cardiovascular problems as well as obesity [34,35]. Phthalates are plasticizers found in plastics that have been connected to health problems like cancer, birth defects, and irregularities in reproduction [36]. MP additives based on polymers have been shown to cause cancer, which is one reason why MPs are linked to a wide range of diseases.

Furthermore, the ability of MPs to transport other contaminants, like heavy metals and harmful microbes, complicates the assessment of their long-term impacts on human health after exposure.

Numerous medical ailments including birth defects, lung cancer, obesity, cardiovascular diseases, respiratory disorders, gastrointestinal disorders, asthma, and viral infections have all been linked to MP pollution (Figure 7.2).

7.4.2 On Marine Environment

MPs are frequently mistaken for prey by aquatic animals, and some species intentionally consume them [37]. MPs cause chemical and physical paralysis when ingested by aquatic organisms. Plastic attachment to cell surfaces can impair flexibility and cause clogs in the digestive tract; it can also cause hepatic stress and impaired development. The MPs may contain chemical substances, such as diethylhexylphtalate, that are hazardous to aquatic creatures. Numerous MPs that have been found in aquatic species' digestive tracts are listed in Table 7.1. Moreover, numerous chemicals and metals may remain associated with MPs, potentially exacerbating their detrimental effects on aquatic life [38]. Previous studies have documented the neurotoxic effects of MPs, assessing acetylcholinesterase activity under controlled laboratory conditions. MPs can also induce oxidative stress, leading to lipid peroxidation of cellular membranes. The identification of MPs in various commercially important edible fish species prompts concerns regarding their possible transfer to higher organisms. Campanale et al. [39] examined the impact of MPs on human health, encompassing respiratory issues, gastrointestinal accumulation, and circulation within the bloodstream. The effect of MPs on aquatic organisms is shown in Figure 7.3, and Figure 7.4 represents the transfer of MPs across the food chain.

TABLE 7.1
Consumption of MPs by Aquatic Organisms and the Effects

Type of plastic	Organism	Mechanism	References
Polyethylene	*Mytilus edulis*	Ingestion	[41]
Polystyrene beads	*Artemia nauplii*	Ingestion	[42]
Polyethylene, polypropylene	*Balaenoptera Physalus*	Ingestion	[43]
Polyethylene	*Mytilus edulis*	Ingestion	[44–46]
Polylactic acid, Polyethylene	*Ostrea edulis*, *Arenicola marina*	Ingestion	[47,48]
Polystyrene	*Calanus helgolandicus*, *Centriscuscristatus*, *Euphasiapacifa* (Copepod)	Ingestion	[49,50]
Polystyrene microbeads	*Paramecium* sp. strain RB1	Ingestion	[51]

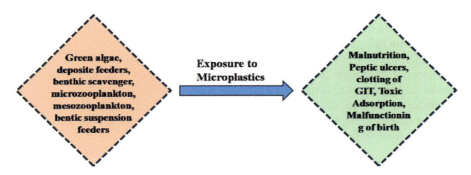

FIGURE 7.3 Impact of microplastic on aquatic organisms.

Several studies have demonstrated the onset of genotoxicity, oxidative stress, physical and chemical toxicity, and abnormal behavior in response to exposure to MPs in a variety of test organisms, such as mollusks, fish, and crustaceans. Sussarellu et al. [40] discovered that polystyrene MPs have a deleterious influence on oyster reproduction and nutrition, resulting in lower egg and sperm counts. Additionally, it has been demonstrated that penguins are susceptible to MPs in water. Additionally, MPs exert a notable influence on the zooplankton community. The acid digestion method was used to study the occurrence of MPs in two economically significant zooplankton species, *Euphasia Pacific* and *Neocalanuscristatus*, in the North Atlantic.

7.5 BIOREMEDIATION OF MPS

Bioremediation entails mineralizing polymers with biotic elements such as algae, fungi, and bacteria. This process comprises the growth of biofilms on plastic surfaces, which triggers the breakdown of the plastic structure via depolymerization catalyzed by certain enzymes. Monomers, oligomers, and dimers may be produced, as well as

Bioremediation of Microplastics

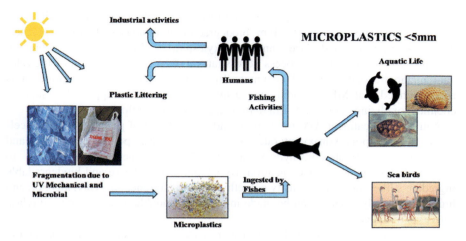

FIGURE 7.4 Transfer of MPs across the food chain.

carbon dioxide, water, and byproducts of depolymerization. Biotransformation can occur, resulting in the production of persistent organic pollutants that are hazardous to the environment. In aerobic environments, biodegraded polymers yield water and carbon dioxide, but in anaerobic conditions, they generate carbon dioxide and methane [52,53].

7.5.1 Bacterial Degradation of Microplastics

Several studies have employed microorganisms to break down MPs. Bacteria that can break down MPs have been found in a variety of environments, including contaminated sediments, wastewater, sludge, compost, marine sediments, municipal landfills, and extreme climates including mangroves and Antarctic soils. Furthermore, MP-degrading bacteria have been identified from earthworm stomach microflora. It is widely acknowledged that bacteria present in contaminated environments frequently have the enzyme machinery needed for MP degradation. This degrading process can be carried out utilizing either pure bacterial cultures or bacterial consortiums.

Nonetheless, pure cultures have certain advantages in the degrading process, including the opportunity to study the metabolic pathways involved. It pure cultures may also help evaluate the effects of environmental parameters such as pH, substrate characteristics, temperature, and surfactants on the degradation process. However, the greatest disadvantage is the noticeably slow rate of disintegration. As a result, novel and imaginative solutions are required to optimize conditions and boost the efficacy of decomposing bacterial isolates, hence speeding up the process.

When a single bacteria performs biodegradation, toxic byproducts are often produced, but these can be effectively managed in a stable microbial community. As a result, it is often recommended to use a bacterial consortium. The main physico-chemical breakdown process shortens the polymer chain and modifies the functional groups of MPs to increase their susceptibility to microbial enzyme activity. The following enzymes are involved in biodegradation: amidases, lipases, laccases, cutinases, esterases, hydrolases, and carboxylesterases [54]. Therefore, achieving an effective biodegradation process necessitates a deep comprehension of the relevant metabolic pathways and enzymes. To improve plastic biodegradation, it is recommended that MPs be pretreated using physicochemical approaches such as chemical oxidizing agents, UV irradiation, and thermooxidation.

Nitric acid treatment, UV irradiation, and the addition of polymers such as cellulose esters, polyhydroxybutyrate, starch derivatives, polycaprolactone, and poly(3-hydroxybutyrate-co-3-hydroxyvalerate) have all been demonstrated to improve polypropylene biodegradation. By adding pro-oxidative and biodegradable ingredients like starch, low-density polyethylene, polystyrene, polyvinyl alcohol, and high-density polyethylene can be made more biodegradable and have their amylase activity stimulated [55].

Cacciari et al. [56] conducted pioneering research on microorganisms proficient in degrading MPs. They utilized a consortium comprising *Pseudomonas chlororaphis*, *Vibrio sp.,* and *Pseudomonas stutzeri* to degrade polypropylene. It was also discovered that the addition of starch enhanced the biodegradation capability.

Arkatkar et al. [57] and Fontanella et al. [58] subsequently showed the biodegradation of polypropylene. A consortium consisting of *Bacillus subtilis, B. fexus, Pseudomonas stutzeri*, and *Rhodococcusrhodochrous* was employed by Arkatkar et al. [57], while a similar consortium was used by Fontanella et al. [58]. Kowalczyk et al. [59] observed that the isolation of *Achromobacterxylosoxidans* resulted in the formation of a biofilm.

After 40 days of incubation, Auta et al. [60] found that two bacterial strains taken from mangrove sediments, Bacillus and Rhodococcus, had polypropylene breakdown efficiencies of 4.0% and 6.4%, respectively. They also noted that MPs can be broken down by *Bacillus cereus* and *Bacillus gottheilii*. For polyethylene terephthalate, polypropylene, polystyrene, and polyethylene, B. gottheilii produced weight loss rates of 3.0%, 3.6%, 5.8%, and 6.2%, respectively.

Bacterial species such as Bacillus, Pseudomonas, Chelatococcus, and *Lysinibacillus fusiformis* were found in a variety of settings, including mangrove ecosystems, compost, cow dung, and plastic-contaminated soil. The intestinal microbiomes of other arthropods, including *Tenebrio molitor*(mealworms), *Plodia interpunctella* (Indian meal moth), and *Galleria mellonella*, have also been found to include microbes that can degrade MPs.

Enterobacter asburiae YT1 and *Bacillus sp.* YP1 are the two bacterial strains that Yang et al. [61] discovered from the intestinal tracts of waxworms. These strains demonstrated the ability to degrade polyethylene by reducing its hydrophobic characteristics and changing the surface of polypropylene. *Exiguobacterium sp.*, a bacterial species that was isolated by Yang et al. [62] from mealworm digestive tracts, was shown to be capable of breaking down polystyrene and forming biofilms. *Microbacteriumparaoxydans* and *Pseudomonas aeruginosa* demonstrated significant biodegradation efficiency of low-density polyethylene, with around 61.0% and 50.5% degradation, respectively [63]. Research by Huerta Lwanga et al. [64] focused on earthworms' degradation of low-density polyethylene.

Bioremediation of Microplastics

It was discovered through testing that specific gut isolates belonging to the Actinobacteria and Firmicutes genera were able to break down low-density polyethylene MP. They also emitted volatile chemicals such as tricosane, docosane, and eicosane. A consortium of *Enterobacter* and *Pseudomonas* isolated from cow manure showed an increased weight decrease of up to 15% for 120 days [65]. Many marine hydrocarbonoclastic bacteria, including *Alcanivoraxborkumensis*, show competence in degrading alkanes, branching aliphatic compounds, alkyl cycloalkanes, and isoprenoid hydrocarbons [66]. It has previously been shown that the isolate under research could form biofilms on low-density polyethylene with the aid of pyruvate, hexadecane, and yeast extract [67]. Moreover, the hydrophilicity of the cell membrane is changed in the presence of alkanes, leading to the creation of biosurfactants that interact with the plastic surface to form C=O and COOH/OH functional groups. Several *actinomycetes*, including *Streptomyces* and *Rhodococcusruber*, were involved in the degradation of polyethylene [68].

Matjašič et al. [69] discovered that 21% of the bacteria engaged in MP degradation were Pseudomonas, 15% were Bacillus, and 17% were a combination of both. Additional bacteria involved in MP biodegradation comprised *Rhodococcusrhodochrous, Rhodococcusruber, Enterobacter asburiae, Ideonellasakaiensis, Nocardia asteroids, Streptomyces badius, Comamonasacidovorans, Exiguobacterium sp., Thermomonosporafusca, Pseudomonas chlororaphis, Bacillus sp., Clostridium thermocellum*, and *Pseudomonas putida AJ*.

Tables 7.2 and 7.3 offer a comprehensive compilation of diverse bacteria and *actinomycetes* associated with MP degradation.

In conclusion, it has been foundthat bacteria that may degrade MPs exist in microbiota, contaminated sediments, municipal landfills, sludge, compost, and harsh conditions.

TABLE 7.2
Bacterial Species Responsible for MP Degradation

Source of microbe	Bacterial strains that have been separated	Percentage of degradation	References
Polluted soil samples	*Lysinibacillus sp.*	4 and 9%	[70]
Compost	*Bacillus cereus*	-	[71]
Municipal landfll sediment	*Bacillus* sp. and *Paenibacillus*sp	14.7 %	[72]
Mangrove sediment	*Bacillusgottheilii*	6.2%, 3.0%, 3.6%, 5.8%	[73]
Compost	*Bacillus thuringiensis*	12%	[74]
Mangrove sediments	*Sporosarcinaglobispora*	11%	[75]
Sewage treatment plants	Microbial consortia	47%, 58% and 56%	[76]
Municipal solid waste	*Stenotrophomonaspanacihumi* PA3-2	20.3 ± 1.39%	[77]
Sandy beaches	*Agios Onoufrios* and Kalathas	0.19%	[78]

TABLE 7.3
Degradation of Microplastics by Actinomycetes

Source	Actinomycetes strains	% of degradation	References
Antarctic soil	*Pseudomonas* sp. ADL15	17.3%	[79]
Soils from waste coal	Bacterial consortia	17.03%	[80]
Not reported	*Microbacteriumparaoxydans*	61.0%	[63]
Mangrove sediment	*Rhodococcus*	6.4%	[60]

Bacteria [81] have been evaluated for their MP breakdown capability using both pure cultures and microbial consortia. Bacterial consortia are notable for their increased stability and efficiency in the community.

7.5.2 FUNGI

Fungi may be able to use MPs as their only source of carbon and energy by producing enzymes.

Numerous plastic substrates, such as polyester polyurethane, LDPE (low-density polyethylene), PVC, polystyrene sulfonate, HDPE, PP, PET, and PA, have demonstrated the effectiveness of fungi. Fungal degradation ability stems from a combination of external enzymatic activities that release hydrolases for the breakdown of MPs, as well as intracellular detoxification processes and fungal adaptability. Sanchez [82] proposed several techniques to improve biodegradation activities, including employing antioxidant stabilizers during plastic manufacture, decreasing biocides, and adding pro-oxidant species. It was demonstrated that fungi break down PE more efficiently than bacteria. They serve as suitable candidates for polymer breakdown because they can generate hydrophobin and adhere hyphae to hydrophobic substrates. Fungi are extremely adaptable to a variety of conditions and have been found in both terrestrial and marine settings, including deep-sea and anoxic ecosystems. In the meantime, only a limited number of research have looked at the enzymes responsible for MP breakdown.

Consequently, further research should explore conducting experiments in oceanic environments, particularly below the sea surface, where fungi are more abundant [83]. Although the biodegradability of bacteria and fungi has been evaluated by weight loss measurement, the American Society for Testing and Materials Standards recommends that tracking CO_2 evolution is a more effective method.

Nevertheless, photosynthetic organisms or micropollutants drawn on polymers can both produce CO_2. To address this issue, Amaral-Zettler et al. [83] propose integrating CO_2 assessment with tracing methodologies such as isotope tagging to measure the extent of assimilation. Until yet, only PET has been efficiently depolymerized by enzymes for polymers with hydrolyzable linkages such as esters or amides, whereas PA and PU have yet to be reported [84]. On the contrary, investigating the biodegradation of polyolefin chains is hindered by their lack of functional groups and high molecular weight, requiring abiotic treatment methods for polyolefins. Moreover,

Bioremediation of Microplastics

the biodegradation is hindered by its hydrophobic nature and cross-linked chemical structure. Natural bacteria are unlikely to break down MPs, but they can establish "plastisphere" microbial communities by interacting with them. Nevertheless, photosynthetic organisms or micropollutants drawn on polymers can both produce CO_2. As a result, advanced oxidation processes (AOPs) such as photocatalysis or activation with H_2O_2/PMS/PDS act as abiotic treatments, with the produced reactive oxygen species (ROS) introducing oxygen-containing functional groups, reducing molecular weight, and decreasing plastic hydrophobicity.

7.5.3 ALGAE

In wastewater streams, microalgae have been seen to adhere to the surface of plastic, which starts the breakdown of plastic by ligninolytic and exopolysaccharide enzyme production. Algae have demonstrated the ability to transform plastic waste into biomass and metabolites during the mineralization process, which has positive environmental effects. The polystyrene (PS) MPs purchased from Chengdu Huaxia Chemical Reagent Co. were incubated using *Chlamydomonas reinhardtii* from the freshwater algal culture collection of the Chinese Academy of Sciences' Institute of Hydrobiology. The MPs were then activated in a liquid SE medium. SEM pictures showed both homo- and hetero-aggregation within microalgae cells as well as between MPs and microalgae. Because the hetero-aggregation mechanism helps particles settle when the cells are floating in the aqueous phase, it may help microalgae reduce PS MP contamination. Li et al. [85] and another study [86] found that various algal species thrive on degrading PE trash collected from wastewater in Kota, Rajasthan.

Microcystis aeruginosa, Amphora ovalis, Monoraphidiumvontortum, Phormidium tenue, Oscillatoria tenuis, Selenastrumminutum, Chlorella vulgaris, Closterium costatum, and *Scenedesmus acuminatus* are some of these species [87]. *Naviculapupula, Anabaena spiroides,* and *Scenedesmus dimorphus* were discovered among a group of green photosynthetic microalgae growing on abandoned PE bags collected from freshwater reservoirs such as pools, ponds, and ditches in Vanagaram, Poonamallee, and Maduravoyal, Tamil Nadu. The reported rates of degradation for LDPE sheets were 3.74±0.26%, 8.18±0.66%, and 4.44±0.82% [88]. After 42 days of incubation, PE strips treated with *Phormidium lucidum* and *Oscillatoria subbrevis* showed a 30% weight decrease. The cyanobacteria were separated from PE substrates colonized by algae and collected from sewage water in Assam, India [89].

A study was conducted to examine the biodegradation capacity of*Chlorella vulgaris*in the presence of BPA, a chemical often used in the plastic industry. *Chlorella vulgaris* was incubated with BPAat a concentration of 20 mg/l for 7 days, resulting in a degradation rate of more than 50%. Furthermore, from microalgal samples extracted from PE bags and water samples in Malaysia, Bhuyar et al. [90] discovered a group of microalgae, including *Chlorella sp.* and *Cyanobacteria sp.* The study found that incubating this consortium with LDPE samples resulted in a 37.91% reduction in carbon weight.

7.5.4 PLANTS

Using phytoremediation techniques, MP pollutants in water can be eliminated with the assistance of plants. Phytostabilization, phytoextraction, and phytofiltration are examples of phytoremediation methods used to treat MP-contaminated water or soil. MPs are absorbed by the soil's roots and transferred to above-ground tissues during the phytoextraction process. Phytoextraction involves the absorption or adsorption of MPs from groundwater and liquid waste streams by plant roots or seedlings. Because of their size, MPs cannot enter plant tissues because they cannot break through the cell wall.

On the other hand, research by [91] using fluorescent nanobeads to track the endocytosis process in tobacco BY-2 cells showed that nanoplastics might permeate plant tissues. Lemna minor showedan incidence of 0.23 ±0.05-0.47 ±0.14 firmly attached MP particles per milligram of plant fresh weight and 0.65 ±0.33-1.93 ±0.53 loosely attached MP particles per milligram of plant fresh weight when exposed to polyethylene MPs from a body scrub [92]. Silver birch, or *Betula pendula Roth*, was grown as a young plant in a soil combination that contained fluorescently dye-labeled MPs.

Analysis revealed that 5-17% of the tree root portions contained MP particles, suggesting the capability of root tissue to absorb such particles. Polyamide was used as the MP source for this experiment. After exposure to MPs of different sizes, MP accumulation was observed on the testa of *Lepidium sativum* seeds within 24 hours and on the seedlings within 8 hours of growth [93].

A build-up of 100 nm fluorescent polystyrene MP beads was seen in the root tips of *Vicia faba* by Jiang et al. [94]. Additionally, [95] identified the bioaccumulation of plastic in the *rhizosphere* of maize plants. When assessing the carbon content of the *rhizosphere* of maize plants exposed to MPs, 30% of it was attributed to polyethylene. Gao et al. conducted a study [96] that showed polyethylene's adherence to the root surface of purple and green lettuce plants. Concerns have been expressed in the phytoremediation of MPs about the creation of released MPs in unexpected plants or their ingestion by animals, both of which could be harmful. This emphasizes the necessity for extensive research on MP degradation before considering it a safe technique for regulating MP pollution.

7.5.5 MICROORGANISM

The amount of MP in many species is rapidly increasing. Plastic particles saturate the ecosystem, reaching even the most remote parts of the world. As a result, biodegradable MPs could be a viable solution for reducing plastic waste. Enzymes from these microorganisms play a crucial part in driving this process, which breaks down plastic waste into methane, CO_2, biomass, water, and numerous inorganic chemicals. Because plastic polymers have different physical and chemical properties, the specific types of plastic polymers present affect how quickly MPs biodegrade. Temperature, sunshine, atmospheric humidity, and UV radiation are some of the environmental factors that affect the process of biotic degradation [96].

Bioremediation of Microplastics

7.5.6 Microbial Degradation

Compared to other abiotic treatment methods, biodegradation has several benefits, such as being less expensive, requiring less energy, and leaving a less carbon imprint. Plastics can be used by microorganisms as sources of nitrogen and carbon to help them survive and develop.

It has been shown that bacteria and fungi, among other microbes, are capable of degrading specific polymers. This feature has substantial potential for applications in MP management and remediation, even though it is often not designed for MP degradation.

7.6 ADVANTAGES OF BIOREMEDIATION

Microorganisms such as algae, bacteria, and fungi have been studied for their ability to remove xenobiotics from the environment, using pure enzymes, immobilized forms, and biosorption approaches. Because of their environmental adaptability and capacity to thrive in a wide range of conditions, fungi have demonstrated extraordinary competence in detoxifying xenobiotics. Often used for biodegradation, microorganisms can withstand harsh conditions and high amounts of pollution caused by humans. Moreover, a variety of microbes are employed to attain the best possible biodegradation. Microorganisms use aerobic, anoxic, and anaerobic processes to detoxify xenobiotics, which include oxidation and reduction reactions. Secretory enzymes, such as hydrolases and carboxylesterases, break down a variety of substrates with efficiency and selectivity while remaining stable in organic solvents. The employment of a certain type of bacteria in bioremediation improves biodegradation efficiency in a variety of ways. The synergistic collaboration of numerous metabolic pathways and oxidoreductases yields exceptional degradation efficiency. Finally, bioremediation produces either biotransformed or mineralized biomass, demonstrating an environmentally friendly strategy for xenobiotic elimination. Researchers employ metabolomic and genetic techniques to predict the metabolism of certain pollutants during degradation. These techniques will allow the investigation of prospective techniques to bioengineer particular microorganisms for effective degradation [97,98].

7.7 CHALLENGES

Not all bacteria have the ability to break down a variety of MPs. MPs at low concentrations present hurdles for remediation of the environment due to limited gene expression, while high quantities may inhibit bacterial development. Although MP biodegradation has been studied extensively, little is still known about metabolic pathways, biocatalysts, and gene expression. Nonspecific gene expression may have an impact on the effectiveness of fungus-based bioremediation of MPs. Temperature, inoculum size, pH, and NaCl concentration all have an impact on MP breakdown efficiency, creating issues for long-term sustainability. Bioremediation on a wide scale faces challenges such as high purifying costs, limited catalytic activity, and insufficient stability of enzymes involved in metabolization. Accurately detecting intermediate metabolites is difficult due to the complexity and transience of microbial metabolic processes. This may lead to the misidentification of certain metabolites, obscuring relevant metabolic pathways. Some oxidation reactions yield dihydrodiol epoxides, which can have both advantageous and toxic effects. There is a pressing demand for innovative experimental approaches to swiftly and accurately detect reaction end products using modern methodologies [99].

7.8 CONCLUSION AND FUTURE OUTLOOK

MPs are an increasing environmental threat, damaging marine and terrestrial ecosystems. They release hazardous substances that can spread throughout contaminated environments, posing serious toxicity hazards to living beings. Given the widespread use of plastic in our daily lives, it is critical to reduce its consumption and focus on efforts to recycle it.

Recycling and environmental remediation of plastic waste depend on its efficient management. Bioremediation approaches have demonstrated efficacy in the removal of plastic waste, detritus, and MPs in recent decades. To reduce the accumulation of MPs, it is critical to address the fundamental sources of plastic consumption in a variety of applications. Furthermore, the development and acceptance of bioplastics provide a possible alternative to traditional plastics. Reducing the number of MPs and larger plastic polymers in the environment requires the use of integrated eco-technologies for remediation, better waste management techniques, and the adoption of circular economy ideas. Further study into molecular bioremediation shows promise for improving the remediation of MP contamination in impacted areas. Future research on the effects of MPs of varying sizes as pollutants, as well as the development of various eco-friendly remediation technologies, will help guide policy decisions about their large-scale management in the environment. Cutting-edge biotechnological technologies can help analyze the synergistic impacts of MP breakdown and remediation methods. Considering the importance of MPs and their effects on the health of humans and animals, it is expected that many researchers will continue to focus on them in the coming years. A significant component of this researchwill focus on developing novel technologies for the identification and characterization of MPs. Investigating new removal strategies is another critical aspect of understanding MPs, a topic that is expected to receive more attention in the future. The research to produce materials for their environmental elimination, such as magnetic sorbents, covalent organic frameworks, metal-organic frameworks, molecularly imprinted polymers, and similar substances, is another important aspect of MPs.

REFERENCES

[1] Ali, Imran, Tengda Ding, Changsheng Peng, Iffat Naz, Huibin Sun, Juying Li, and Jingfu Liu. "Micro-and nanoplastics in wastewater treatment plants: Occurrence, removal, fate, impacts and remediation technologies–a critical review." *Chemical Engineering Journal* 423 (2021): 130205. https://doi.org/10.1016/j.cej.2021.130205

[2] Ali, Sameh S., Tamer Elsamahy, Rania Al-Tohamy, Daochen Zhu, Yehia A-G. Mahmoud, Eleni Koutra, Metwally A. Metwally, Michael Kornaros, and Jianzhong

Sun. "Plastic wastes biodegradation: Mechanisms, challenges and future prospects." *Science of the Total Environment* 780 (2021): 146590. https://doi.org/10.1016/j.scitotenv.2021.146590

[3] Cheng, Yan Laam, Jong-Gook Kim, Hye-Bin Kim, Jeong Hwan Choi, Yiu Fai Tsang, and Kitae Baek. "Occurrence and removal of microplastics in wastewater treatment plants and drinking water purification facilities: A review." *Chemical Engineering Journal* 410 (2021): 128381. https://doi.org/10.1016/j.cej.2020.128381

[4] Kumar, Rakesh, Prabhakar Sharma, Camelia Manna, and Monika Jain. "Abundance, interaction, ingestion, ecological concerns, and mitigation policies of microplastic pollution in riverine ecosystem: A review." *Science of the Total Environment* 782 (2021): 146695. https://doi.org/10.1016/j.scitotenv.2021.146695

[5] Ricardo, Ivo A., Edna A. Alberto, Afonso H. Silva Júnior, Domingos Lusitâneo P. Macuvele, Natan Padoin, Cintia Soares, Humberto Gracher Riella, Maria Clara V. M. Starling, and Alam G. Trovo. "A critical review on microplastics, interaction with organic and inorganic pollutants, impacts and effectiveness of advanced oxidation processes applied for their removal from aqueous matrices." *Chemical Engineering Journal* 424 (2021): 130282. https://doi.org/10.1016/j.cej.2021.130282

[6] Tiwari, Neha, Deenan Santhiya, and Jai Gopal Sharma. "Microbial remediation of micro-nano plastics: Current knowledge and future trends." *Environmental Pollution* 265 (2020): 115044. https://doi.org/10.1016/j.envpol.2020.115044

[7] Brahney, Janice, Margaret Hallerud, EricHeim, Maura Hahnenberger, and Suja Sukumaran. "Plastic rain in protected areas of the United States." *Science* 368, no. 6496 (2020): 1257–1260. https://doi.org/10.1126/science.aaz5819

[8] Ryberg, Morten W., Michael Z. Hauschild, Feng Wang, Sandra Averous-Monnery, and Alexis Laurent. "Global environmental losses of plastics across their value chains." *Resources, Conservation and Recycling* 151 (2019): 104459. https://doi.org/10.1016/j.resconrec.2019.104459

[9] Tiwari, M., T. D. Rathod, P. Y. Ajmal, R. C. Bhangare, and S. K. Sahu. "Distribution and characterization of microplastics in beach sand from three different Indian coastal environments." *Marine Pollution Bulletin* 140 (2019): 262–273. https://doi.org/10.1016/j.marpolbul.2019.01.055

[10] Patchaiyappan, Arunkumar, Syed Zaki Ahmed, Kaushik Dowarah, Shanmuganathan Jayakumar, and Suja P. Devipriya. "Occurrence, distribution and composition of microplastics in the sediments of South Andaman beaches." *Marine Pollution Bulletin* 156 (2020): 111227. https://doi.org/10.1016/j.marpolbul.2020.111227

[11] Geng, Xianhui, Jun Wang, Yan Zhang, and Yong Jiang. "How do microplastics affect the marine microbial loop? Predation of microplastics by microzooplankton." *Science of the Total Environment* 758 (2021): 144030. https://doi.org/10.1016/j.scitotenv.2020.144030

[12] Gaylarde, Christine C., José Antonio Baptista Neto, and Estefan Monteiro da Fonseca. "Nanoplastics in aquatic systems-are they more hazardous than microplastics?" *Environmental Pollution* 272 (2021): 115950. https://doi.org/10.1016/j.scitotenv.2021.146695

[13] Liu, Guangzhou, Zhilin Zhu, Yuxin Yang, Yiran Sun, Fei Yu, and Jie Ma. "Sorption behavior and mechanism of hydrophilic organic chemicals to virgin and aged microplastics in freshwater and seawater." *Environmental Pollution* 246 (2019): 26–33. https://doi.org/10.1016/j.envpol.2018.11.100

[14] Ma, Jie, Jinghua Zhao, Zhilin Zhu, Liqing Li, and Fei Yu. "Effect of microplastic size on the adsorption behavior and mechanism of triclosan on polyvinyl chloride." *Environmental Pollution* 254 (2019): 113104. https://doi.org/10.1016/j.envpol.2019.113104

[15] Yu, Fei, Changfu Yang, Zhilin Zhu, Xueting Bai, and Jie Ma. "Adsorption behavior of organic pollutants and metals on micro/nanoplastics in the aquatic environment." *Science of the Total Environment* 694 (2019): 133643. https://doi.org/10.1016/j.scitotenv.2019.133643

[16] Guo, Xuan, and JianlongWang. "The chemical behaviors of microplastics in marine environment: A review." *Marine Pollution Bulletin* 142 (2019): 1–14. https://doi.org/10.1016/j.marpolbul.2019.03.019

[17] Murphy, Fionn, Ciaran Ewins, Frederic Carbonnier, and Brian Quinn. "Wastewater treatment works (WwTW) as a source of microplastics in the aquatic environment." *Environmental Science &Technology* 50, no. 11 (2016): 5800–5808. https://doi.org/10.1021/acs.est.5b05416

[18] Wagner, Stephan, Thorsten Hüffer, Philipp Klöckner, Maren Wehrhahn, Thilo Hofmann, and Thorsten Reemtsma. "Tire wear particles in the aquatic environment- a review on generation, analysis, occurrence, fate and effects." *Water Research* 139 (2018): 83–100. https://doi.org/10.1016/j.watres.2018.03.051

[19] Mintenig, Svenja M., Ivo Int-Veen, Martin G. J. Löder, Sebastian Primpke, and Gunnar Gerdts. "Identification of microplastic in effluents of waste water treatment plants using focal plane array-based micro-Fourier-transform infrared imaging." *Water Research* 108 (2017): 365–372. https://doi.org/10.1016/j.watres.2016.11.015

[20] Dubaish, Fatehi, and Gerd Liebezeit. "Suspended microplastics and black carbon particles in the Jade system, southern North Sea." *Water, Air, &Soil Pollution* 224 (2013): 1–8. https://doi.org/10.1007/s11270-012-1352-9

[21] Ahmed, Riaz, Ansley K. Hamid, Samuel A. Krebsbach, Jianzhou He, and Dengjun Wang. "Critical review of microplastics removal from the environment." *Chemosphere* 293 (2022): 133557. https://doi.org/10.1016/j.chemosphere.2022.133557

[22] Gong, Jian, and Pei Xie. "Research progress in sources, analytical methods, eco-environmental effects, and control measures of microplastics." *Chemosphere* 254 (2020): 126790. https://doi.org/10.1016/j.chemosphere.2020.126790

[23] Wang, Chunhui, Jian Zhao, and Baoshan Xing. "Environmental source, fate, and toxicity of microplastics." *Journal of Hazardous Materials* 407 (2021): 124357. https://doi.org/10.1016/j.jhazmat.2020.124357

[24] Coyle, Róisín, Gary Hardiman, and Kieran O'Driscoll. "Microplastics in the marine environment: A review of their sources, distribution processes, uptake and exchange in ecosystems." *Case Studies in Chemical and Environmental Engineering* 2 (2020): 100010. https://doi.org/10.1016/j.cscee.2020.100010

[25] Auta, Helen Shnada, Chijioke U. Emenike, and Shahul Hamid Fauziah. "Distribution and importance of microplastics in the marine environment: A review of the sources, fate, effects, and potential solutions." *Environment International* 102 (2017): 165–176. https://doi.org/10.1016/j.envint.2017.02.013

[26] Wright, Stephanie L., and Frank J. Kelly. "Plastic and human health: A micro issue?" *Environmental Science &Technology* 51, no. 12 (2017): 6634–6647. https://doi.org/10.1021/acs.est.7b00423

[27] De-la-Torre, Gabriel Enrique. "Microplastics: An emerging threat to food security and human health." *Journal of Food Science and Technology* 57, no. 5 (2020): 1601–1608. https://doi.org/10.1007/s13197-019-04138-1

[28] Naji, Abolfazl, Marzieh Nuri, and A. Dick Vethaak. "Microplastics contamination in molluscs from the northern part of the Persian Gulf." *Environmental Pollution* 235 (2018): 113–120. https://doi.org/10.1016/j.envpol.2017.12.046

[29] Su, Lei, Huiwen Cai, Prabhu Kolandhasamy, Chenxi Wu, Chelsea M. Rochman, and Huahong Shi. "Using the Asian clam as an indicator of microplastic pollution

in freshwater ecosystems." *Environmental Pollution* 234 (2018): 347–355. https://doi.org/10.1016/j.envpol.2017.11.075

[30] Iñiguez, Maria E., Juan A. Conesa, and Andres Fullana. "Microplastics in Spanish table salt." *Scientific Reports* 7, no. 1 (2017): 8620. https://doi.org/10.1038/s41598-017-09128-x

[31] Schymanski, Darena, Christophe Goldbeck, Hans-Ulrich Humpf, and Peter Fürst. "Analysis of microplastics in water by micro-Raman spectroscopy: Release of plastic particles from different packaging into mineral water." *Water Research* 129 (2018): 154–162. https://doi.org/10.1016/j.watres.2017.11.011

[32] Karami, Ali, Abolfazl Golieskardi, Cheng Keong Choo, Vincent Larat, Samaneh Karbalaei, and Babak Salamatinia. "Microplastic and mesoplastic contamination in canned sardines and sprats." *Science of the Total Environment* 612 (2018): 1380–1386. https://doi.org/10.1016/j.scitotenv.2017.09.005

[33] Koelmans, Albert A., Ellen Besseling, and Edwin M. Foekema. "Leaching of plastic additives to marine organisms." *Environmental Pollution* 187 (2014): 49–54. https://doi.org/10.1016/j.envpol.2013.12.013

[34] Rani, Manviri, Won Joon Shim, Gi Myung Han, Mi Jang, Najat Ahmed Al-Odaini, Young Kyong Song, and Sang Hee Hong. "Qualitative analysis of additives in plastic marine debris and its new products." *Archives of Environmental Contamination and Toxicology* 69 (2015): 352–366. https://doi.org/10.1007/s00244-015-0224-x

[35] Ohore, Okugbe E., and Songhe Zhang. "Endocrine disrupting effects of bisphenol A exposure and recent advances on its removal by water treatment systems. A review." *Scientific African* 5 (2019): e00135. https://doi.org/10.1016/j.sciaf.2019.e00135

[36] Gómez, Cristina, and Hectorgg Gallart-Ayala. "Metabolomics: A tool to characterize the effect of phthalates and bisphenol A." *Environmental Reviews* 26, no. 4 (2018): 351–357. https://doi.org/10.1139/er-2018-0010

[37] Lönnstedt, Oona M., and Peter Eklöv. "RETRACTED: Environmentally relevant concentrations of microplastic particles influence larval fish ecology." *Science* 352, no. 6290 (2016): 1213–1216. https://doi.org/10.1126/science.aad8828

[38] Mammo, F. K.,I. D. Amoah, K. M.Gani, L. Pillay, S. K. Ratha, F. Bux, and S. Kumari. "Microplastics in the environment: Interactions with microbes and chemical contaminants." *Science of The Total Environment* 743 (2020): 140518. https://doi.org/10.1016/j.scitotenv.2020.140518

[39] Campanale, Claudia, Carmine Massarelli, Ilaria Savino, Vito Locaputo, and Vito Felice Uricchio. "A detailed review study on potential effects of microplastics and additives of concern on human health." *International Journal of Environmental Research and Public Health* 17, no. 4 (2020): 1212. https://doi.org/10.3390/ijerph17041212

[40] Sussarellu, Rossana, Marc Suquet, Yoann Thomas, Christophe Lambert, Caroline Fabioux, Marie Eve Julie Pernet, Nelly Le Goïc et al. "Oyster reproduction is affected by exposure to polystyrene microplastics." *Proceedings of the National Academy of Sciences* 113, no. 9 (2016): 2430–2435. https://doi.org/10.1073/pnas.1519019113

[41] Van Cauwenberghe, Lisbeth, Lisa Devriese, François Galgani, Johan Robbens, and Colin R. Janssen. "Microplastics in sediments: A review of techniques, occurrence and effects." *Marine Environmental Research* 111 (2015): 5–17. https://doi.org/10.1016/j.marenvres.2015.06.007

[42] De Lange, Hendrika J., Joost Lahr, Joost J. C. Van der Pol, and Jack H. Faber. "Ecological vulnerability in wildlife: Application of a species-ranking method to food chains and habitats." *Environmental Toxicology and Chemistry* 29, no. 12 (2010): 2875–2880. https://doi.org/10.1002/etc.336

[43] Fossi, Maria Cristina, Letizia Marsili, Matteo Baini, Matteo Giannetti, Daniele Coppola, Cristiana Guerranti, Ilaria Caliani et al. "Fin whales and microplastics: The Mediterranean Sea and the Sea of Cortez scenarios." *Environmental Pollution* 209 (2016): 68–78. https://doi.org/10.1016/j.envpol.2015.11.022

[44] Avio, Carlo Giacomo, Stefania Gorbi, Massimo Milan, Maura Benedetti, Daniele Fattorini, Giuseppe d'Errico, Marianna Pauletto, Luca Bargelloni, and Francesco Regoli. "Pollutants bioavailability and toxicological risk from microplastics to marine mussels." *Environmental Pollution* 198 (2015): 211–222. https://doi.org/10.1016/j.envpol.2014.12.021

[45] Chua, Evan M., Jeff Shimeta, Dayanthi Nugegoda, Paul D. Morrison, and Bradley O. Clarke. "Assimilation of polybrominated diphenyl ethers from microplastics by the marine amphipod, Allorchestescompressa." *Environmental Science &Technology* 48, no. 14 (2014): 8127–8134. https://doi.org/10.1021/es405717z

[46] Von Moos, Nadia, Patricia Burkhardt-Holm, and Angela Köhler. "Uptake and effects of microplastics on cells and tissue of the blue mussel Mytilus edulis L. after an experimental exposure." *Environmental Science &Technology* 46, no. 20 (2012): 11327–11335. https://doi.org/10.1021/es302332w

[47] Besseling, Ellen, Anna Wegner, Edwin M. Foekema, Martine J. Van Den Heuvel-Greve, and Albert A. Koelmans. "Effects of microplastic on fitness and PCB bioaccumulation by the lugworm Arenicola marina (L.)." *Environmental Science &Technology* 47, no. 1 (2013): 593–600. https://doi.org/10.1021/es302763x

[48] Green, Dannielle Senga. "Effects of microplastics on European flat oysters, Ostrea edulis and their associated benthic communities." *Environmental Pollution* 216 (2016): 95–103. https://doi.org/10.1016/j.envpol.2016.05.043

[49] Cole, Matthew, Penelope K. Lindeque, Elaine Fileman, James Clark, Ceri Lewis, Claudia Halsband, and Tamara S. Galloway. "Microplastics alter the properties and sinking rates of zooplankton faecal pellets." *Environmental Science &Technology* 50, no. 6 (2016): 3239–3246. https://doi.org/10.1021/acs.est.5b05905

[50] Desforges, Jean-Pierre W.,Moira Galbraith, and Peter S. Ross. "Ingestion of microplastics by zooplankton in the Northeast Pacific Ocean." *Archives of Environmental Contamination and Toxicology* 69 (2015): 320–330. https://doi.org/10.1007/s00244-015-0172-5

[51] Bulannga, Rendani B., and Stefan Schmidt. "Uptake and accumulation of microplastic particles by two freshwater ciliates isolated from a local river in South Africa." *Environmental Research* 204 (2022): 112123. https://doi.org/10.1016/j.envres.2021.112123

[52] Zhang, Fan, Yuting Zhao, Dandan Wang, Mengqin Yan, Jing Zhang, Pengyan Zhang, Tonggui Ding, Lei Chen, and Chao Chen. "Current technologies for plastic waste treatment: A review." *Journal of Cleaner Production* 282 (2021): 124523. https://doi.org/10.1016/j.jclepro.2020.124523

[53] Giacomucci, Lucia, Noura Raddadi, Michelina Soccio, Nadia Lotti, and Fabio Fava. "Biodegradation of polyvinyl chloride plastic films by enriched anaerobic marine consortia." *Marine Environmental Research* 158 (2020): 104949. https://doi.org/10.1016/j.marenvres.2020.104949

[54] Amobonye, Ayodeji, Prashant Bhagwat, Suren Singh, and Santhosh Pillai. "Plastic biodegradation: Frontline microbes and their enzymes." *Science of the Total Environment* 759 (2021): 143536. https://doi.org/10.1016/j.scitotenv.2020.143536

[55] Zadjelovic, Vinko, Audam Chhun, Mussa Quareshy, Eleonora Silvano, Juan R. Hernandez-Fernaud, María M. Aguilo-Ferretjans, Rafael Bosch, Cristina Dorador, Matthew I. Gibson, and Joseph A. Christie-Oleza. "Beyond oil

degradation: Enzymatic potential of Alcanivorax to degrade natural and synthetic polyesters." *Environmental Microbiology* 22, no. 4 (2020): 1356–1369. https://doi.org/10.1111/1462-2920.14947

[56] Cacciari, I., P. Quatrini, G. Zirletta, E. Mincione, V. Vinciguerra, Paolo Lupattelli, and G. Giovannozzi Sermanni. "Isotactic polypropylene biodegradation by a microbial community: Physicochemical characterization of metabolites produced." *Applied and Environmental Microbiology* 59, no. 11 (1993): 3695–3700. https://doi.org/10.1128/aem.59.11.3695-3700.1993

[57] Arkatkar, Ambika, Asha A. Juwarkar, Sumit Bhaduri, Parasu Veera Uppara, and Mukesh Doble. "Growth of Pseudomonas and Bacillus biofilms on pretreated polypropylene surface." *International Biodeterioration & Biodegradation* 64, no. 6 (2010): 530–536. https://doi.org/10.1016/j.ibiod.2010.06.002

[58] Fontanella, Stéphane, Sylvie Bonhomme, Jean-Michel Brusson, Silvio Pitteri, Guy Samuel, Gérard Pichon, Jacques Lacoste, Dominique Fromageot, Jacques Lemaire, and Anne-Marie Delort. "Comparison of biodegradability of various polypropylene films containing pro-oxidant additives based on Mn, Mn/Fe or Co." *Polymer Degradation and Stability* 98, no. 4 (2013): 875–884. https://doi.org/10.1016/j.polymdegradstab.2013.01.002

[59] Kowalczyk, Anna, Marek Chyc, Przemysław Ryszka, and Dariusz Latowski. "Achromobacterxylosoxidans as a new microorganism strain colonizing high-density polyethylene as a key step to its biodegradation." *Environmental Science and Pollution Research* 23 (2016): 11349–11356. https://doi.org/10.1007/s11356-016-6563-y

[60] McClanahan, T. R., E. Sala, P. J. Mumby, and S. Jones. "Phosphorus and nitrogen enrichment do not enhance brown frondose"macroalgae"."*Marine Pollution Bulletin* 48, no. 1 (2004): 196–199. 10.1016/j.marpolbul.2003.10.004

[61] Yang, Jun, Yu Yang, Wei-Min Wu, Jiao Zhao, and Lei Jiang. "Evidence of polyethylene biodegradation by bacterial strains from the guts of plastic-eating waxworms." *Environmental Science &Technology* 48, no. 23 (2014): 13776–13784. https://doi.org/10.1021/es504038a

[62] Yang, Yu, Jun Yang, Wei-Min Wu, Jiao Zhao, Yiling Song, Longcheng Gao, Ruifu Yang, and Lei Jiang. "Biodegradation and mineralization of polystyrene by plastic-eating mealworms: Part 2. Role of gut microorganisms." *Environmental Science &Technology* 49, no. 20 (2015): 12087–12093. https://doi.org/10.1021/acs.est.5b02663

[63] Rajandas, Heera, Sivachandran Parimannan, Kathiresan Sathasivam, Manickam Ravichandran, and Lee Su Yin. "A novel FTIR-ATR spectroscopy based technique for the estimation of low-density polyethylene biodegradation." *Polymer Testing* 31, no. 8 (2012): 1094–1099. https://doi.org/10.1016/j.polymertesting.2012.07.015

[64] Lwanga, Esperanza Huerta, Binita Thapa, Xiaomei Yang, Henny Gertsen, Tamás Salánki, Violette Geissen, and Paolina Garbeva. "Decay of low-density polyethylene by bacteria extracted from earthworm's guts: A potential for soil restoration." *Science of the Total Environment* 624 (2018): 753–757. https://doi.org/10.1016/j.scitotenv.2017.12.144

[65] Skariyachan, Sinosh, Neha Taskeen, Alice Preethi Kishore, Bhavya Venkata Krishna, and Gautami Naidu. "Novel consortia of Enterobacter and Pseudomonas formulated from cow dung exhibited enhanced biodegradation of polyethylene and polypropylene." *Journal of Environmental Management* 284 (2021): 112030. https://doi.org/10.1016/j.jenvman.2021.112030

[66] Davoodi, Seyyed Mohammadreza, Saba Miri, Mehrdad Taheran, Satinder Kaur Brar, Rosa Galvez-Cloutier, and Richard Martel. "Bioremediation of unconventional oil contaminated ecosystems under natural and assisted conditions: A review."

Environmental Science &Technology 54, no. 4 (2020): 2054–2067. https://doi.org/10.1021/acs.est.9b00906

[67] Delacuvellerie, Alice, Valentine Cyriaque, Sylvie Gobert, Samira Benali, and Ruddy Wattiez. "The plastisphere in marine ecosystem hosts potential specific microbial degraders including Alcanivoraxborkumensis as a key player for the low-density polyethylene degradation." *Journal of Hazardous Materials* 380 (2019): 120899. https://doi.org/10.1016/j.jhazmat.2019.120899

[68] Sivan, Alex. "New perspectives in plastic biodegradation." *Current Opinion in Biotechnology* 22, no. 3 (2011): 422–426. https://doi.org/10.1016/j.copbio.2011.01.013

[69] Matjašič, Tjaša, Tatjana Simčič, Neja Medvešček, Oliver Bajt, Tanja Dreo, and Nataša Mori. "Critical evaluation of biodegradation studies on synthetic plastics through a systematic literature review." *Science of the Total Environment* 752 (2021): 141959. https://doi.org/10.1016/j.scitotenv.2020.141959

[70] Jeon, Jong-Min, So-Jin Park, Tae-Rim Choi, Jeong-Hoon Park, Yung-Hun Yang, and Jeong-Jun Yoon. "Biodegradation of polyethylene and polypropylene by Lysinibacillus species JJY0216 isolated from soil grove." *Polymer Degradation and Stability* 191 (2021): 109662. https://doi.org/10.1016/j.polymdegradstab.2021.109662

[71] Jain, Kimi, H. Bhunia, and M. Sudhakara Reddy. "Degradation of polypropylene-poly-L-lactide blends by Bacillus isolates: A microcosm and field evaluation." *Bioremediation Journal* 26, no. 1 (2022): 64–75. https://doi.org/10.1080/10889868.2021.1886037

[72] Park, Seon Yeong, and Chang Gyun Kim. "Biodegradation of micro-polyethylene particles by bacterial colonization of a mixed microbial consortium isolated from a landfill site." *Chemosphere* 222 (2019): 527–533. https://doi.org/10.1016/j.chemosphere.2019.01.159

[73] Auta, H. S., C. U. Emenike, and S. H. Fauziah. "Screening of Bacillus strains isolated from mangrove ecosystems in Peninsular Malaysia for microplastic degradation." *Environmental Pollution* 231 (2017): 1552–1559. https://doi.org/10.1016/j.envpol.2017.09.043

[74] Jain, Kimi, H. Bhunia, and M. Sudhakara Reddy. "Degradation of polypropylene-poly-L-lactide blend by bacteria isolated from compost." *Bioremediation Journal* 22, no. 3–4 (2018): 73–90. https://doi.org/10.1080/10889868.2018.1516620

[75] Auta, S. H., C. U. Emenike, and S. H. Fauziah. "Screening for polypropylene degradation potential of bacteria isolated from mangrove ecosystems in Peninsular Malaysia." (2017). https://doi.org/10.17706/ijbbb.2017.7.4.245-251

[76] Skariyachan, Sinosh, Amulya A. Patil, Apoorva Shankar, Meghna Manjunath, Nikhil Bachappanavar, and S. Kiran. "Enhanced polymer degradation of polyethylene and polypropylene by novel thermophilic consortia of Brevibacillussps. and Aneurinibacillus sp. screened from waste management landfills and sewage treatment plants." *Polymer Degradation and Stability* 149 (2018): 52–68. https://doi.org/10.1016/j.polymdegradstab.2018.01.018

[77] Jeon, Hyun Jeong, and Mal Nam Kim. "Isolation of mesophilic bacterium for biodegradation of polypropylene." *International Biodeterioration & Biodegradation* 115 (2016): 244–249. https://doi.org/10.1016/j.ibiod.2016.08.025

[78] Syranidou, Evdokia, Katerina Karkanorachaki, Filippo Amorotti, Martina Franchini, Eftychia Repouskou, Maria Kaliva, Maria Vamvakaki et al. "Biodegradation of weathered polystyrene films in seawater microcosms." *Scientific Reports* 7, no. 1 (2017): 17991. https://doi.org/10.1038/s41598-017-18366-y

[79] Habib, Syahir, Anastasia Iruthayam, Mohd Yunus Abd Shukor, Siti Aisyah Alias, Jerzy Smykla, and Nur Adeela Yasid. "Biodeterioration of untreated polypropylene

microplastic particles by Antarctic bacteria." *Polymers* 12, no. 11 (2020): 2616. https://doi.org/10.3390/polym12112616

[80] Nowak, Bożena, Jolanta Pająk, and Jagna Karcz. *Biodegradation of pre-aged modified polyethylene films*. 2012. http://hdl.handle.net/20.500.12128/8652

[81] Moog, Daniel, Johanna Schmitt, Jana Senger, Jan Zarzycki, Karl-Heinz Rexer, Uwe Linne, Tobias Erb, and Uwe G. Maier. "Using a marine microalga as a chassis for polyethylene terephthalate (PET) degradation." *Microbial Cell Factories* 18 (2019): 1–15. https://doi.org/10.1186/s12934-019-1220-z

[82] Sánchez, Carmen. "Fungal potential for the degradation of petroleum-based polymers: An overview of macro-and microplastics biodegradation." *Biotechnology Advances* 40 (2020): 107501. https://doi.org/10.1016/j.biotechadv.2019.107501

[83] Amaral-Zettler, Linda A., Erik R. Zettler, and Tracy J. Mincer. "Ecology of the plastisphere." *Nature Reviews Microbiology* 18, no. 3 (2020): 139–151. https://doi.org/10.1038/s41579-019-0308-0

[84] Wei, Ren, Till Tiso, Jürgen Bertling, Kevin O'Connor, Lars M. Blank, and Uwe T. Bornscheuer. "Possibilities and limitations of biotechnological plastic degradation and recycling." *Nature Catalysis* 3, no. 11 (2020): 867–871. https://doi.org/10.1038/s41929-020-00521-w

[85] Li, Shuangxi, Panpan Wang, Chao Zhang, Xiangjun Zhou, Zhihong Yin, Tianyi Hu, Dan Hu, Chenchen Liu, and Liandong Zhu. "Influence of polystyrene microplastics on the growth, photosynthetic efficiency and aggregation of freshwater microalgae Chlamydomonas reinhardtii." *Science of the Total Environment* 714 (2020): 136767. https://doi.org/10.1016/j.scitotenv.2020.136767

[86] Sharma, M., A. Dubey, and A. Pareek. "Algal flora on degrading polythene waste." *CIBTech Journalof Microbiology* 3 (2014): 43–47.

[87] Kumar, R. Vimal, G. R. Kanna, and Sanniyasi Elumalai. "Biodegradation of polyethylene by green photosynthetic microalgae." *Journal of Bioremediation & Biodegradation* 8, no. 381 (2017): 2. https://doi.org/10.4172/2155-6199.1000381

[88] Sarmah, Pampi, and Jayashree Rout. "Efficient biodegradation of low-density polyethylene by cyanobacteria isolated from submerged polyethylene surface in domestic sewage water." *Environmental Science and Pollution Research* 25 (2018): 33508–33520. https://doi.org/10.1007/s11356-018-3079-7

[89] Gulnaz, Osman, and Sadik Dincer. "Biodegradation of bisphenol A by Chlorella vulgaris and Aeromonas hydrophilia." *Journal of Applied Biological Sciences* 3, no. 2 (2009): 79–84.

[90] Bhuyar, Prakash, Sathyavathi Sundararaju, Ho Xuan Feng, Mohd Hasbi Ab Rahim, Sudhakar Muniyasamy, Gaanty Pragas Maniam, and Natanamurugaraj Govindan. "Evaluation of Microalgae's Plastic Biodeterioration Property by a Consortium of Chlorella sp. and Cyanobacteria sp." *Environmental Research, Engineering and Management* 77, no. 3 (2021): 86–98. https://doi.org/10.5755/j01.erem.77.3.25317

[91] Bandmann, Vera, Jasmin Daniela Müller, Tim Köhler, and Ulrike Homann. "Uptake of fluorescent nano beads into BY2-cells involves clathrin-dependent and clathrin-independent endocytosis." *FEBS Letters* 586, no. 20 (2012): 3626–3632. https://doi.org/10.1016/j.febslet.2012.08.008

[92] Rozman, Ula, Anita Jemec Kokalj, Andraž Dolar, Damjana Drobne, and Gabriela Kalčíková. "Long-term interactions between microplastics and floating macrophyte Lemna minor: The potential for phytoremediation of microplastics in the aquatic environment." *Science of the Total Environment* 831 (2022): 154866. https://doi.org/10.1016/j.scitotenv.2022.154866

[93] Austen, Kat, Joana MacLean, Daniel Balanzategui, and Franz Hölker. "Microplastic inclusion in birch tree roots." *Science of the Total Environment* 808 (2022): 152085. https://doi.org/10.1016/j.scitotenv.2021.152085

[94] Jiang, Xiaofeng, Hao Chen, Yuanchen Liao, Ziqi Ye, Mei Li, and Göran Klobučar. "Ecotoxicity and genotoxicity of polystyrene microplastics on higher plant Vicia faba." *Environmental Pollution* 250 (2019): 831–838. https://doi.org/10.1016/j.envpol.2019.04.055

[95] Urbina, Mauricio A., Francisco Correa, Felipe Aburto, and Juan Pedro Ferrio. "Adsorption of polyethylene microbeads and physiological effects on hydroponic maize." *Science of the Total Environment* 741 (2020): 140216. https://doi.org/10.1016/j.scitotenv.2020.140216

[96] Dey, Thuhin K., Md Elias Uddin, and Mamun Jamal. "Detection and removal of microplastics in wastewater: Evolution and impact." *Environmental Science and Pollution Research* 28 (2021): 16925–16947. https://doi.org/10.1007/s11356-021-12943-5

[97] Ben Ali, Rihab, Sabrine Ben Ouada, Christophe Leboulanger, Jihene Ammar, Sami Sayadi, and Hatem Ben Ouada. "Bisphenol A removal by the Chlorophyta Picocystis sp.: Optimization and kinetic study." *International Journal of Phytoremediation* 23, no. 8 (2021): 818–828. https://doi.org/10.1080/15226514.2020.1859985

[98] Nakamura, Aline M., Marco Antonio Seiki Kadowaki, André Godoy, Alessandro S. Nascimento, and Igor Polikarpov. "Low-resolution envelope, biophysical analysis and biochemical characterization of a short-chain specific and halotolerant carboxylesterase from Bacillus licheniformis." *International Journal of Biological Macromolecules* 120 (2018): 1893–1905. https://doi.org/10.1016/j.ijbiomac.2018.10.003

[99] Uthra, Karupanagounder Thangaraj, Vellapandian Chitra, Narayanasamy Damodharan, Anitha Devadoss, Moritz F. Kuehnel, Antonio Jose Exposito, Sanjay Nagarajan, Sudhagar Pitchaimuthu, and Gururaja Perumal Pazhani. "Microplastic emerging pollutants-impact on microbiological diversity, diarrhea, antibiotic resistance, and bioremediation." *Environmental Science: Advances* 2 (2023): 1469–1487. https://doi.org/10.1039/D3VA00084B

8 Electrocoagulation for Remediation of Microplastics

Anita Choudhary, Anshul Yadav,
Priyanka Ghanghas, and Kavita Poonia

8.1 INTRODUCTION

Long-chain organic polymers that makeup plastics are in high demand across several industries, leading to a steady increase in global production of plastics year after year [1]. They are used in a varied variation of single-use products intextiles, furniture, electronics, cars, food, and beverage packaging insulating materials and even medical equipment [2]. However, there are serious environmental concerns with how these materials are disposed off. They fragment into smaller particles known as microplastics (MPs) over time as a result of a sequence of physico-chemical reactions. Nearly every ecosystem has been impacted by these tiny pieces including freshwater bodies estuaries vast oceans and even the farthest reaches of Arctic ice [3–5].

Wastewater discharged from cities is a significant factor in the ubiquity of MPs in natural environments. Regretfully the wastewater treatment facilities that are in place today lack the necessary apparatus to capture and remove these MPs. Thus, these particles may avoid treatment plants and enter nearby water bodies where they eventually assemble [6]. This emphasizes how urgently improved methods and technologies are needed to decrease the quantity of MPs released into the environment and control their effects on ecosystems. Numerous studies have examined a range of methods intended to decrease the amount of MPs in aquatic environments. The methods of EC sedimentation and sand filtering have been studied in great detail. When positively charged hydrolysates from coagulants are introduced during coagulation the negative (-) charges on the electric double layer of colloidal MPs are neutralized making it easier to remove the MPs [7]. The productivity of coagulants in eliminating micropollutants from water has been assessed by research using materials such as iron aluminum and polyamine [8]. Notably, research has shown that adding coagulants based on aluminum effectively neutralizes the charge on polyethylene terephthalate (PET) MPs [9, 10]. On the other hand, adding too much of a polyamine-based coagulant can cause the colloidal system to re-stabilize [11].

DOI: 10.1201/9781003486947-8

147

Moreover, fast sand filtration has proven to be extremely effective in eliminating more than 70% of MPs from wastewater [12]. Similar to this, granular activated carbon filtration has shown promise in removing material at a rate of about 70% [13]. Relative to traditional secondary-activated sludge treatment methods research on membrane bioreactors has demonstrated a more consistent ability to retain MPs [14]. However, it is important to remember that membranes have a major drawback in that they are prone to fouling and clogging. Therefore, there is an urgent need to create new affordable techniques specifically designed for the targeted elimination of MPs from water management facilities. This emphasizes how crucial it is to keep up the investigation and development in the field of synthetic polymers. Numerous current techniques for reuse and treating water have renewable issues and a certain amount of them rely on the use of hazardous chemicals. Furthermore, because organic waste from industries is obstinate dealing with it is a difficult task. An innovative water treatment technique called electrocoagulation (EC),which has great sustainability and an economical operating strategy, is a promising future solution. This covers among other things pollutants like bacteria oils and heavy metals. Its ability to effectively handle various impurities renders it a valuable instrument in the endeavor to clean water properties. In addition, communities and industries looking for cost-effective and ecologically responsible water treatment solutions find it appealing due to its sustainable features and economical operation. Despite a thorough examination of the process, there is a discernible lack of specialized research that is specifically focused on the removal of MPs. Surprisingly this method when used properly achieves removal rates of over 90% proving its effectiveness in two real-world wastewater scenarios [15,16] and carefully monitored lab environments. All of these results point to the significant potential of EC as a powerful tool in the ongoing fight against MP pollution providing an effective and workable solution for industrial and environmental applications. Metal ions like Al^{3+} and Fe^{2+} are liberated by votive conductors into a marine rivulet that contains hydroxide ions [17]. The ensuing describes the reactions that occur at the anode/cathode surface in an EC route in the resolution: where the metal anode iron or aluminum is indicated by the letter Q. Iron and aluminum hydroxides produced in the process act as micro-coagulants causing colloidal contaminants to become unstable and form flocs. To remove the formed flocs of MPs sedimentation or flotation can be applied. A 90% efficacy has been achieved by electro-coagulating polyethylene microbead in a stirred tank batch reactor with an aluminum electrode. When it comes to removing MPs an aluminum electrode seems to be more effective than an iron electrode. With removal efficiencies of more than 90% aluminum anode can get rid of MPs such as polyethylene, polypropylene, poly-methyl methacrylate, and cellulose acetate. According to one study there is a positive correlation between the amount of organic matter and ionic strength and removal effectiveness [18]. Previous studies examined the effects of critical variables on the effectiveness of MP removal with initial pH electrolyte concentration and current density. During the EC step in these investigations, solid plate electrodes were used. Regardless of the extensive research on electrocoagulation-based MP removal, the size and impact of size on removal rates has been largely unexplored. Furthermore, the majority of studies focused only

Electrocoagulation for Remediation of Microplastics 149

on one electrode configuration ignoring the possibility that electrode geometry and arrangement could have an impact on the process. Furthermore, there was a conspicuous lack of particular investigation concerning the efficacy of EC in eliminating MPs made of polystyrene, a polymer that is widely present in the environment. The impact of polystyrene MP size, electrode configuration, and geometry was thus thoroughly investigated in this study, which was a ground breaking step. The Procedure for Order of Preference by Similarity to Ideal Solution (TOPSIS), a multi-criteria executive tool, was used to improve operational factors. This methodological decision was taken in order to rankorder and thoroughly assess the best configurations for the removal of MPs.

8.2 RELEVANCE AND CHEMISTRY OF MICROPLASTICS

Approximately 8million tons of plastic debris are thought to arrive in the ocean annually harming nearly 700 marine species. Unwanted MP releases into the atmosphere can exacerbate environmental stressors and have harmful effects on plant ecosystems and lower trophic levels if they are not managed properly [19]. According to current trends, it is estimated that by 2060 the amount of secondary MPs in the environment will increase by 155–265 million tons with MPs accounting for 13.2% of this total [20, 21].

MPs are produced from polypropylene, polyethylene, poly(oxyethyleneoxyter ephthaloyl), polystyrene, and polychloroethene [22–24]. Plastics can be classified according to their size: nano-plastics (less than 100 nm), MPs (0.1 μm-10 mm), mesoplastics (5–25 mm), and macro-plastics (>25 mm). MPs are difficult to measure precisely because of their variability in polymer type, size, shape, persistence, and matrix [25]. MPs are typically created indirectly from larger plastics or directly from industrial sources [26].

The primary concerns about the widespread presence of MPs in the environment are related to their effects on marine and human health [27]. Due to their small size, MPs can easily move into marine life as contaminated materials and effluents without being detected. They can be found in many of our everyday foods. Additionally, MPs act as a grid or transporter for other hazardous conservation contaminants and have a high adsorption capacity. There are negative health effects when MP is exposed through ingestion, inhalation, or skin contact. Chronic pneumonia or bronchitis, hepatic stress, granuloma formation, bronchial reactions, punctured lung, inhaling lesions, altered instinctive microbial configuration and absorption, and perceptive decline have all been related to human exposure to MPs [28].

The primary biodegradation methods for MP remediation used today rely on bacteria, fungi, and enzymes [29]. Nevertheless, the removal of MP has also been accomplished by photo-degradation thermal treatments and chemical treatments [30–32]. According to Wu *et al.,* alternative low-toxicity generation strategies for MP remediation have been developed more recently including photo-catalysis and plasma-based oxidation. But MPs typically have a high resistance to photo- and biodegradation, which leads to low removal efficiency [33]. Because of these qualities tracking and getting rid of MP waste is a difficult but necessary task. The identification of MPs is therefore crucial for enhancing regulatory procedures and MP transport monitoring.

Recent publications [34–36] have analyzed reviews on the subject of MP monitoring and measurement. Additionally, an electrochemical review on the use of boron-doped diamond electrodes for the discovery of minor carbon-based molecules known as plastic leachates has been published. In this chapter, we discuss the technological chances for MP pollutants as well as recent findings on the use of electrochemical methodologies for remediation and direct MP monitoring.

8.3 ELECTROCOAGULATION

8.3.1 THEORETICAL BACKGROUND

The "electrolysis" principle is the cornerstone of EC. "Electrolysis" is the term used to describe the process of breaking down substances using voltage. In 1820, Michael Faraday presented the idea of electrolysis. The procedure ensues in an electrolyte solution and allows ions to flow between electrodes. In the EC cell, (+) ions flow towards and decrease in quantity towards the cathode. At the same time, (-) ions travel towards the anode and experience oxidation [37, 38]. In this procedure, wastewater was electrolyzed by combining it with saltwater. The primary aim of the EC method in its initial phases was to yield chlorine for the disinfection and odor reduction of manure waste. Water purification using EC has been practiced for a very long time. It was introduced in London in 1889 and was employed in a sewage action plant for ten years [39]. The EC method of purifying ship bilge water was created and patented by A. E. Dietrich in 1906. Similar to this, J.T. Harries was granted a patent in 1909 in the United States for electrolysis technologythatused Fe and Al as electrodes to clean wastewater [40]. However, because electricity and investment were so expensive at the time, EC was not commonly used for water purification. When EC was shown to be more active than outdated organic coagulation at eliminating both organic and inorganic pollutants from surface and groundwater, it began to gain popularity. In the United States, EC was first applied to large amounts of drinking water in 1984 [41]. Emulsified oils, bacteria, heavy metals, and total suspended solids (TSS) can all be eliminated from water using the sophisticated EC process. It is an amalgam of well-known methods including flotation, coagulation, and electrochemistry. On the other hand, the literature on combining the three techniques is comparatively thin [42]. Despite being a merciless and efficacious sewage management technique, the electrochemical process had partial acceptance and success in the 20th era. The early areas of improvement were wastewater throughput rates and electrical power consumption reduction [43]. Recent technological developments have enabled EC to operate at low currents, which qualifies it for use in wind turbines, solar cells, and fuel cells [44]. They are considered eco-friendly because no chemical reagents are added, the use of electrode material is minimal, and metal coagulants are readily available, non-toxic, and cheaper to produce than other metals [45].

8.3.2 Mechanism of Removal

The sequence of events and processes in the electrochemical and chemical, as well as the mechanisms of coagulation-flocculation, are comparable. Hexaaquoaluminium $(Al(H_2O)_6)^{3+}$ is the hydrated form of the aluminum anode that forms when it dissolves into Al^{3+} ions in an Al-based electrochemical reaction. Following that, this ion undergoes a series of quick deprotonation events that result in the production of resolvable monomeric type $(Al(OH)(H_2O)_5)^{2+}$, polymeric type $(Al_{13}O_4(OH)_{24})^{7+})$, dense $Al(OH)_3$, and aluminate ions $(Al(OH)_4)$ [46,47]. There are various processes for removing organic materials, including adsorption, complexation, trapping, neutralization, or destabilization of charges [48]. Which mechanism predominates dependson a number of aspects, includingthe pH, coagulant type and dosage, charge, size, and hydrophobicity of the pollutant, as well as other components of the water matrix. Recalling that the general appliance for coagulation-based pollution removal is often complex, with various appliances occupied together to enhance exclusion effectiveness, is imperative [48]. The term "natural organic matter" (NOM) describes organic material that includes hydrophilic acids, proteins, lipids, carbohydrates, and a variety of humic components (such as phenolic compounds, fulvic and humic acids). A range of NOM-containing fluids, such as peat water [49, 50], river water [51, 52], lake water [52–54], and synthetic wastewater [55, 56], have all been reported to be treated by EC. Because there is less salinity and chloride ions present, low NOM levels in aquatic can stop bifurcation, caustic embrittlement, and the production of Cl during electrolysis. Fe^{2+} can be transformed into $Fe(OH)_3(s)$ through "sweep coagulation" when iron electrodes are used. This process involves raising the pH and DOC in oxygenated water. $Al(OH)_3$ is the principal oxide produced through Al-based EC in normal water, and the occurrence of NOM can obstruct its crystal-like precipitation, much like it does with aluminum chloride in chemical precipitation. The anodic breakdown of aluminum/iron didnot emit oxygen, indicating that all electrons were involved in the process [57].

8.3.3 Removal of MPs

Malleable polymer finds its application in several trades due to its affordability, strength, durability, and low weight. MPs with atom dimensions < 5 mm in width are being produced by the massive volume of plastic debris being released into the environment [58–60]. MPs are present throughout the world in a variety of sizes and are mostly produced by fragmentation of plastic trash, plastic manufacture, and domestic sewage discharge. Due to their inability to effectively remove MPs, wastewater treatment plants play a major role in their release into the environment, whereupon they accumulate in aquatic ecosystems. Because of their polymeric structure and ease of transportation between multiple ecosystems, MPs cannot disintegrate. Planktonic and invertebrate species are particularly prone to accumulating MPs in aquatic environments. These particles can subsequently arrive in the human body through the food chain and cause physical injury, reduced nutritional value, and disease exposure [61,

62]. Furthermore, chemical additions included in raw plastics include phthalates, bisphenol A, and polybrominated diphenyl ethers, which can be harmful to living things if ingested [61].

The separation of MPs from water and wastewater has considerable promise using coagulation-based procedures [60]. The straightforward operation, great efficiency, and low cost of the EC process make it a potentially advantageous method for removing MPs [60, 63]. EC is a thoroughly researched field, butthe elimination of MPs has received very little attention in studies. From real wastewater [18, 64, 65] and artificial wastewater [16], these investigations show that EC may efficiently eliminate MPs (i.e., > 90% clearance under ideal reaction circumstances).

In situ coagulants produced by EC using metal electrodes have the ability to breakdown MPs and absorb charged contaminants. The primary methods by which EC removes MPs are as follows:

(1) Three processes occur, sorption, electrical neutralization, and flotation, in which the adsorption of negatively charged ionic MPs is made possible by the flocculants' positive surface charge. During the EC process, H_2 and O_2 bubbles may also be produced, which may aid in bringing the flocs to the top so that skimming can remove them [58].
(2) MPs can be oxidized into small molecules of non-toxic substrates through an oxidation reaction involving Fe^{4+} formed on the anode in conjunction with metal-ion precipitation and ROS, such as O_2, $O_2^{\bullet-}$, H_2O_2, and OH^- [58].

8.3.4 PROS AND CONS OF ELECTROCOAGULATION

Compared to conventional physio-chemical treatment techniques, the EC method has a number of benefits and drawbacks [17, 66]. For treating wastewater, EC is superior to chemical coagulation (CC) because it allows for easier automation, faster and more effective separation of organic matter, chemical-free pH control, cheaper operating costs, and less secondary contamination. In addition, it creates larger, more stable flocs; gas bubbles aid in the removal of pollutants; and it has the ability to extract very tiny colloidal particles. Additional chemical and electrochemical processes, like electro-Fenton (EF) and photo-catalysis (PC), use more chemicals, hence EC is less expensive than these methods since it requires less equipment and upkeep [67, 68]. Furthermore, EC is capable of operating in a variety of environments, including those with high salt and pH levels. On the other hand, EF can only function in a limited acidic pH range, whereas PC is responsive to ecological variables such as pH, temperature, and the occurrence of carbon-based substances [18, 69, 70]. EC is not without its drawbacks, though, such as the need for cathode passivation, the need to replace sacrificial anodes regularly, the requirement for subsequent treatment due to elevated alloy particle agglomeration, high energy expense in areas with finite power, and possible mud increase power through a continuous process. Moreover, rather than relying on biotic developments, which are inadequate todecompose ecological contaminations and dispersible biotic compounds, EC only uses electrochemical ways to remove pollutants [67]. But for treating water including organic contaminants, alternative techniques such as PC [67, 71], EF [68], bacterial electroplating cells,

Electrocoagulation for Remediation of Microplastics 153

and microbial energy cells [70] might work better. These techniques may be more effective since they use chemicals and light to aid in treatment or bio-electrochemical processes.

8.4 ELECTROCHEMICAL REMEDIATION OF MICROPLASTICS

MPs can be removed ruined or recycled from the atmosphere as part of remediation. EC isa cheap and green technique for removing impurities and pollutants from water [64]. The method consists of an electric field producing cations (anodic dissolution) on a metal electrode that binds to MPs to congeal them into large aggregates that can be filtered out [72–73]. In order to maximize remediation efficiency and minimize electrical power utilization it is imperative to select the optimal applied electric field during this process to facilitate EC [16]. The EC ranges and power of the electric field strength relies on the distance between the anode and cathode electrodes. The strength of the electric field increases with electrode proximity. When total organic carbon from municipal wastewater was electro-coagulated the electric field strength was greatest at 0.5 cm between electrodes and lowermost at 2 cm. It is therefore also feasible to say that the di-electrophoresis force is maximum at 0. 5 cm. Generally speaking, MP aggregation is caused by the interaction of (+) charged metallic ions and their complexes with (-) charged MPs (Figure 8.1a) [74]. Aluminum anode and cathode were employed in parallel using wastewater and a laboratory-scale stir-tank batch apparatus. The efficacy of the method in treating MP wastewater was assessed by testing the properties of pH and viscosity of the wastewater as well as the applied current density. Using EC, the MPs were effectively eliminated in simulated wastewater up to 90–100% of the time. Additionally, the method worked well with wastewater with a pH range from 3 to 10 indicating that the low-cost method can potentially eliminate other types of MPs from a range of residential and commercial wastewater. The anode/cathode material in this process is essential to the remediation efficiency. For instance, it was discovered that Al electrodes were more effective than Fe electrodes [75]. MP removal efficiency with the Al electrode was 90%, compared to 59–85% for PE, polymethyl methacrylate (PMMA), cellulose acetate (CA), and polypropylene (PP) MPs with the Fe electrode. Furthermore, iron corrosion may have contributed to the thick precipitation and quick alluviation of the MP flocs generated by the Fe anode. Therefore, iron corrodes more quickly than aluminum, which results in less effective application. Because the aluminum anode produced flocs with a small floc size MP remediation could occur faster and with a stronger adsorption capacity due to a larger surface area. The EC method's phosphate removal efficiency was 100% with Al electrodes and 84.7%with Fe electrodes, respectively. Similar to the Fe electrode the Al electrode showed a reduced electrode mass depletion under the same experimental conditions [76]. Because aluminum had a higher rate of COD turbidity and phosphate removal than iron it was chosen as the electrode. The original pH (7.8) 100 A/m^2 and 10 min EC time were found to be the ideal operating conditions yielding removal competencies of phosphorus turbidity and COD of 72, 98, and 98%, respectively. Table 8.1 presents an assessment of the cathode conductor electrolyte category reaction period and effectiveness for different methods of EC [77]. Using iron and copper as the anode and cathode materials, respectively, obviously results

TABLE 8.1

Comparison of Different Electro-coagulation Setups in Terms of Electrode Material, Electrolyte, Reaction Period, and Efficacy [77]

Anode material	Cathode material	Electrolyte	Time	Efficiency (%)	Ref.
Al	Cu	Na_2SO_4 (0.05M)	4h	92-98	Shen et al. (2022) [16]
Fe	Cu	Na_2SO_4 (0.05M)	4h	59-85	Shen et al. (2022) [16]
Al	Al	NaCl (2g/L)	1h	99	Perren et al.(2018) [15]
Al	Al	N/A	30 min	90	Kim and Park (2021) [18]
Al	Al	N/A	4h	94	Elkhatib et al. (2021) [64]

in reduced efficiency. The most effective EC was achieved with Al as the electrode material.

Although the MP flocs created by the aluminum terminal had a strong adsorption volume and fast speed, the MP flocs formed by the Fe anode had solid precipitation and rapid alluviation. To remove PE, PP, PMMA, and cellulose MPs EC was further developed (Figure 8.1a). Additionally, the removal rates varied from 50 to 99% based on the types of MPs with PP showing more effective rates compared to the slowest rates with PMMA. Energy consumption necessary for the real-world implementation of this remediation strategy must be taken into account using the EC removal procedure for MP. Metal ions like Al(III) (Figure 8.1b) neutralize the charge of the plastic MPs during EC forming froth flotation or sedimentation as a result [15]. In the future, the precipitates or froths produced during EC may be employed as sensor signaling probes and as markers of the presence of MP.

Electrochemical oxidation of polymeric materials can also lead to MP degradation. Reacting through the polymeric chemical bonds in MPs are a variety of reactive oxygen species produced at high concentrations during an electrochemical process such as hydrogen peroxide and hydroxyl radical. Harsh oxidation mediators like Ag, Co, or Ce ions, which all have redox potentials above 1.7 V, are necessary for the electro-oxidation of MP [78]. Because it can break down and detect different types of pollutants the boron-doped diamond electrode (BDD) has become more popular than metallic electrodes [79]. Additionally, BDD and electro-oxidation processes have beenused to degrade MPs in water, specifically PS MPs. In contrast to smaller particles MPs were directly broken down (~90%) into gaseous products like CO_2 within a few hours of electrolysis. A C-O bond is formed when radicals initiate the breaking of the polymer C-H bond, which is the mechanism by which PS degrades. Hydroxyl radicals are created at the anode surface of water. Effectively the PS polymeric C-C bonds arealso broken. Eventually, the polymeric chain decomposition into CO_2 and water result from further oxidation. A sensor that detects MP presence may be developed using the ensuing CO_2 production. Numerous effective factors were

Electrocoagulation for Remediation of Microplastics 155

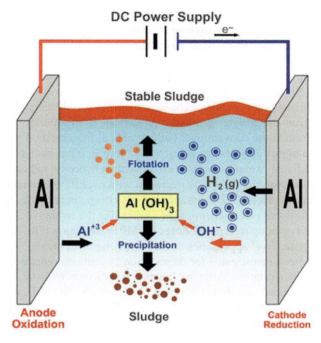

FIGURE 8.1a Electrochemical treatment and monitoring approaches for micro-plastics (A) Schematic of the electrocoagulation process used to remove micro-plastics from water [74].

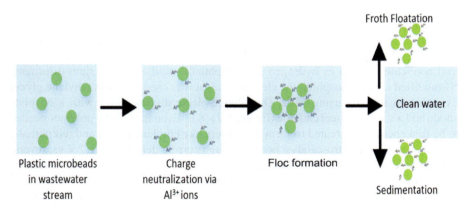

FIGURE 8.1b Schematic of electrocoagulation of microbeads from wastewater [15].

examined to optimize the electro-oxidation process and enhance MP degradation. These factors include current intensity, anode surface, electrolyte category, electrolyte application, and reaction period. Other anode electrode materials such as titanium electrode used as the cathode and metal oxide (iridium oxide) were also tested for

MP breakdown. The reaction rate for the degradation of MPs was also observed to be higher using the BDD compared to the abovementioned anode electrodes. The MP removal efficiency (%) of BDD electrodes was found to be higher than that of mixed metallic oxide and Ir oxide anodes. Because the BDD can produce hydroxyl radicals at a higher rate this has been linked to the higher removal efficiency [80]. Nanoplastics (NPs), which are persistent plastics in the environment, can also be produced by MPs as they weather. The active agents for NP degradation were sulphate radicals from electrolytes and hydroxyl radicals from water discharge. The procedure outperformed the conventional (sulfate radical free) method by 2.6 times and demonstrated a higher degradation efficiency of 86.6%. It has been reported that sulfate radical-mediated oxidation is more effective than hydroxyl radical-mediated oxidation [81]. Moreover, the NPs were degraded with over 85% efficiency using BDD or carbon felt electrodes. This technique has also been successfully used to anodic oxidize PS. Additional electrochemical processes like the oxidative dehydrogenation of alkanes could open up new possibilities for MP remediation [78]. Still not much is known about these novel approaches to the electrochemical breakdown of polymers. MPs have also been investigated for degradation by cathode reduction treatments in addition to anodic oxidation. After six hours at a persistent realistic latent of -0.7 V vs. Ag/AgCl, the electrocatalytic treatment of MPs with (TiO_2/C) cathode produced a high dechlorination productivity of 75% for PVC. The degradation process started with the simultaneous anodic oxidation of the hydroxyl radicals and cathodic reduction of PVC [82]. Halogenated polymers can be eliminated using this process as opposed to anodic oxidation. To monitor MP existence in an instrument application indirectly the development of halogenated side products like HCl or perchlorates can be measured. Conversion of the product is a feasible alternative even though deletion and poverty are typically the primary objectives of MP remediation. In addition to producing new value-added chemicals, recycling and repurposing MPs can also eliminate themfrom the environment. Recently Hua *et al.* achieved the electrocatalytic upcycling of poly(oxyethyleneoxyterephthaloyl) to chemicals with valuable industrial applications such as hydrogen gas terephthalic acid and potassium diformate. Ni-modified cobalt phosphide the electro-catalyst was used to assess ethylene glycol oxidation to formate using MEA (membrane-electrode assembly) with a Ni plate anode and a stainless-steel cathode (Figure 8.2) [83]. Even though the polymer needed to be first chemically treated to form monomers like ethylene glycol and terephthalic acid the use of an electro-catalyst produced high current densities at 1.7 V vs. reversible hydrogen electrode, possessing superior faradaic efficiency (80%) and selectivity (80%) for formate product. Although the focus of this work was on the value-added chemical formation from MPs one could envision designing and developing sensors to identify MPs present in a sample of interest using indicators such as terephthalic acid hydrogen gas or formic acid.

In general, compared to traditional methodologies electrochemical processes for MP remediation take hours, which makes these procedures valuable for real-world applications. Even though the MP crisis may be resolved by using electrochemical remediation techniques there are still challenges to address. The high applied potentials or chemical oxidation needed for current electrochemical processes could result in unfavorable side reactions. Because of the inadequate definition of the

Electrocoagulation for Remediation of Microplastics

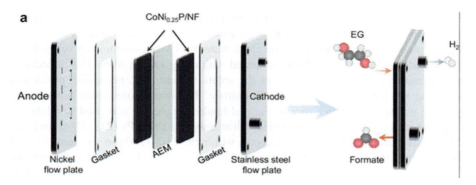

FIGURE 8.2 Membrane-electrode assembly for electro-oxidation of polymer breakdown products, specifically ethane 1,2 diol oxidation into methanoate [83].

current protocols for isolation, characterization, and measurement MP remediation is challenging. Standardized protocols and a versatile platform to remove, degrade, and recycle MPs make the remediation process quicker and more scalable. The creation of new electrochemical techniques and the proposal and growth of nanomaterials electrode resources and catalysts are some potential future directions for electrochemical remediation of MPs [84]. It is evident nevertheless that electrochemical rectification produces a range of chemical entities that serve as oblique markers of MPs and can be used as sensing probes to detect the presence of MPs. The next section goes over the importance of electrochemical sensor-based MP detection for environmental monitoring.

8.5 ELECTROCHEMICAL SENSORS FOR MICROPLASTICS

According to Oliveira *et al.*, leachates from MPs are among the environmental contaminants for which electrochemical sensors have been increasingly used, but they are only now beginning to receive notice for their limited applications in MP detection. Due to the short response, time, low cost, easy process, and portability of electrolytic cells, the growth of electrochemical detectors for MPs is of particular interest. The electrochemical methods in contrast to other conventional methods do not require MP purification or isolation beforehand and can be willingly extended to in-field testing of a variety of sample types. Currently, electrochemical technique and label-free Electrochemical Impedence Spectroscopy (EIS) are used to monitor and sense MP. Flow cytometry and an EIS-based sensor were used to detect PE MPs [85]. The sensor contained a movement cell with flow cytometry for particle exposure and Au-plated route panels with all Au-plated conductors for EIS recognition. This method was developed with the assumption that, at low frequencies, the real part of the impedance should change proportionately to the particle volume. As a result, at low frequencies, the MP elements flowing over electrodes cause a modification in interference that is proportionate to the MP particle volume. This system allowed for the separation of biological and plastic particles at any measured frequency. This podium can be used for MP discovery in complex media because interestingly biological particles produced a negative change in impedance whereas MPs instigated a positive change. To assess the sensor's sensitivity and volume to identify and differentiate among different MP sizes the clean water flow was continuously verified and compared to water that had been tampered with using known MPs. The impedance wasaltered in the presence of MP in the solution under test, producing peaks and enabling the measurement of atom magnitude. The linear connection between the particle width and the dice root of the real hindrance change was used to relate the impedance change to the particle size. Ultimately, the PE MPs (212–1000 μm) and MP beads (including biological structures within the 210–1200 μm size range) could be measured and quantified by this sensor. The sensor yielded a 90% recovery rate for MPs in the 100–300 um size range and a 1% false-positive rate when it came to classifying biological materials as MPs, according to Colson and Michel (2021) [85]. The next wave of sensors should concentrate on differentiating between MP particle types other than PE in addition to MP detection in the field even though EIS was effectively used as a label-free technique for particle finding with MPs. Simizu *et al.* also employed particle impact electrochemistry to detect spherical PE MPs (1–22 μm) [86]. One popular method for studying particles suspended in solution is the particle-electrode impact method. Chronoamperometry measurements demonstrated that a fast current response was produced upon particle impact with the carbon fiber microwave electrode. In the electrochemical analysis setup, which contained of an entire three-electrode system maintained at a precise potential to detect a desired reaction, particle-electrode collision produced the signal change that was measured. To ascertain MP, an analysis was conducted on the transient current response, or spike, that resulted from the particles colliding with the electrode. Because of the decrease in oxygen in PE MP particles, there was a current spike in the chronoamperogram that resulted from an MP collision with the working electrode. By applying this technique, a strong correlation with an R_2 value of 0.96 was found between the MP particle concentrations and spikes frequency. The MP measurement was generally more accurate and consistent when using this discovery technique in comparison to more conventional methods. It is possible to detect conductive particles using the same method that was used to detect electrically insulating MPs.

Serial faradaic ion concentration polarization, or fICP, is an alternate method for MP detection [87]. The fICP uses their electrophoretic mobilities to sort the MPs. Their electrophoretic mobilities affect how the MPs interact with the EFG. Particles with higher electrophoretic flexibility concentrate at a low electric field location. Conversely, at higher electric fields, smaller electrophoretic mobilities are concentrated. The MPs were sorted into distinct chambers using a trifurcated trunk, and the bipolar electrodes (BPE) were positioned close to the trunk. The BPE indicates its location by positioning the ion depletion zones (IDZ) and EFG across the bifurcated channel opening. With the larger electrophoretic mobility particles going into the lower channel and the smaller ones into the upper channel, the MPs are sorted as a result. The Faradaic Ion Concentration Polarization (FICP) controlled PS MP flow in a trifurcated microchannel. Continuous focusing, sorting, and separation of MPs was made possible by this technique. MPs interacting with electric field gradients made it possible to sort them according to their electrophoretic mobilities, which were then identified using optical and epifluorescence microscopy. As the MPs approached the cathodic end of the bipolar electrodes, their trajectory was guided towards the trifurcated micro-channels through the electric field inclines. These micro-channels organized the MPs based on their sizes and electrophoretic mobilities. This technique further simplifies the technology by enabling the continuous and real-time monitoring of MPs in water systems as well as the on-demand isolation or pre-concentration of different MPs deprived of the essential for membrane-based separation. Crucially integrating FICP with other methods would provide more latitude in the development and design of sensors; however, MP detection has not yet taken benefit of this potential. More recently, the development of a cost-effective, high-efficiency technique was made possible by a versatile resistive beat instrument [88]. This silver-wire-based microfluidic sensor detected changes when an analyte (MP) passed through a thin contraction. To illustrate the usefulness of this process for counting and tracking MPs with biological particles, the expedient was employed to track algae and MPs in tea bags. Apart from the amperometric and impedimetric sensors of MPs, other instances of dual-mode sensors have been described [89] and a sensor created that employs exoelectrogenic bacteria, voltammetry, and impedance to detect PE MP. Electroactive bacterial films have demonstrated great potential as a low-energy approach for microbial electrochemical systems. To ascertain the response and study the morphology, EPS, electrochemical characteristics, and microbial structure of the biofilms in response to MPs, the biofilms were exposed to MPs. In this three-electrode setup, the reference electrode was Ag/AgCl, the cathode counter electrode was Ti interlaced wire mesh, and the working electrode was a carbon fiber brush anode. The elevated internal resistance was caused by the PE-MP attaching to the film. The impedance of the Microbial Fuel Cells (MFCs) and Microbial Electrolysis Cells (MECs) was measured in EIS using Nyquist plots before and after binding. The cells' resistance was increased by the presence of PE-MP. The resistance to charge transfer was responsible for the majority of the resistance as determined by the equivalent circuit model. Furthermore, an increase in the population of dead cells may be brought on by the poisonous properties of PE-MPs, which would significantly increase the resistance to charge transfer (Rct). The recent density in microbial petroleum cells stayed constant when PE MP was extant. However, the current signal declined as MP conc. increased in the microbial electrolysis cell, and this pattern lasted for more than 42 days. The binding of PE MP to the film increased internal resistance and caused a decline in the signal. Therefore, the concentration of MP would need to be determined using microbial electrolysis cells in order to identify PE MP using exoelectrogenic biofilms. A microbial electrolysis cell might eventually be used to separate MP types and sizes in order to broaden its application to a real-world setting. Moreover, MP remediation in microbial electrochemical wastewater management systems could be accomplished using the same technique.

Even though MP remediation has made use of electrochemical techniques, little is known about MP sensors. Although effective electrochemical sensors for MP have been demonstrated, more sensor schemes and growth are necessary for MP monitoring. Some MP remediation approaches are ideal for creating novel challenging instruments. To reduce MP interactions with biologicals or other biologicals, for example, or to detect MP coagulation, MP remediation may be used indirectly for sensing applications such as MP byproduct sensing through anodic oxidation.

The challenges that still need to be overcome to advance electrochemical sensors for MPs are MP documentation and variation, standard operating measures for MPs separation and description, conservational monitoring and following, and MP discovery in the variability of illustration kinds (such as soil runoffs, different watersheds, pool wastewater, manufacturing wastes, drainage area, clouds, etc.). The electrochemical strategies might also deliver a way to continuously and in real-time monitor the path and fate of MPs in their surroundings. The identification and survey of elastic elements ensuing from MP is critical since the degradation of MP into smaller NPs is an important aspect of MP chemistry. Due to the easy adsorption and interaction of MPs with other biomolecules, it will be possible to monitor and remediate MP environments in new ways by using electrochemical methods to understand these interactions.

8.6 CHALLENGES

Despite the long history of EC use the literature currently does not provide a systematic method for both the design and operation of EC. However, because current research mostly focuses on the removal of specific impurities from sewage using successive trials scaling up to meet industry demands is challenging [17]. Constant models will improve the understanding of the design parameters and facilitate the prediction of EC reactor efficiency during construction allowing technology development to progress past its current impracticability.

Thus, the ability of such technology to achieve business goals—like reducing operating and maintenance costs and turning a profit on investment as quickly as possible—will dictate whether or not it is commercialized. Additionally, the technology faces competition from tried-and-true water treatment techniques like membrane systems and adsorption. The probability of success would increaseif the EC approach was combined with existing technologies. Since it has been demonstrated that excessive use of the electrodes can cause their homogeneous planar shape to change into a heterogeneous one the passivation process is a significant barrier to the acceptance of EC [90]. The inactive layer that develops on the outer surface of electrodes increases electricity intake but reduces the effectiveness of pollutant removal. Efficient current transfer between both electrodes is limited by the growth of an inert oxide coating on the cathode surface upon the application of direct current (DC). Conversely alternating current (AC) effectively addresses the drawbacks of the conventional direct current-EC (DC-EC) method by enabling regular switching of the cathode and anode. Because AC requires less electrode passivation than DC it is generally less expensive. AC has a sufficient operating lifespan due to the interruption in cathode passivation and anode degradation.

8.7 CONCLUSION

As previously noted, the EC technique works incredibly well with a wide range of drug kinds and classifications. Enhancing and improving removal efficiency can be achieved by further investigating this technique and optimizing its parameters. Additionally, for the best outcomes, this technique can be used in conjunction with

Electrocoagulation for Remediation of Microplastics

other operating techniques. However, there are some shortcomings that must be addressed. While the lack of research on EC effects on a wide range of medications is understandable EC can have varying effects on different members of the same family. Among the penicillin family, for instance, it may be very effective against Amoxicillin antibiotics but ineffective against AMP. They belong to the same family but have different results due to their spatial arrangement, surface charges, and atom-to-atom bonding energies. The impact of EC on all second-generation cephalosporins, nearly all anticonvulsants, most glucocorticoids, and benzylpenicillin (Penicillin G), nafcillin, oxacillin, etc., is not well studied. Because their family is mentioned in the current articles these points are brought up. As a result, further research is required to identify the ineffectiveness of additional medications and to treat them. By including a comprehensive list of pollutants and differentiating between their categories the current study builds on earlier research. Other recently discovered organic pollutants like those resulting from farm runoff were also discussed.

REFERENCES

[1] Anthony L. Andrady. "The plastic in micro-plastics: A review." *Marine Pollution Bulletin* 111 (2017): 12–22. doi:10.1016/j.marpolbul.2017.01.082

[2] Robert C. Hale, Meredith E. Seeley, Mark J. La Guardia, Lei Mai, Eddy Y. Zeng. "A Global Perspective on Micro-plastics." *Journal of Geophysical Research: Oceans* 125 (2020): 1–40. doi:10.1029/2018JC014719

[3] Fares John Biginagwa, BahatiSosthenesMayoma, Yvonne Shashoua, KristianSyberg, Farhan R. Khan. "First evidence of micro-plastics in the African Great Lakes: Recovery from Lake Victoria Nile perch and Nile tilapia."*Journal of Great Lakes Research* 42 (2016): 146–149.doi:10.1016/j.jglr.2015.10.012

[4] Mark A. Browne, Tamara S. Galloway, Richard C.Thompson. "Spatial patterns of plastic debris along Estuarine shorelines."*Environmental Science & Technology* 44 (2010): 3404–3409.doi:10.1021/es903784e

[5] Mark A.Browne, Tamara S.Galloway, Richard C.Thompson. "Accumulation of micro-plastic on shorelines worldwide: Sources and sinks."*Environmental Science and Technology* 45 (2011): 9175–9179. doi:10.1021/es201811s

[6] Steve A.Carr, JinLiu, Arnold G.Tesoro. "Transport and fate of micro-plastic particles in wastewater treatment plants."*Water Research* 15 (2016): 174–182. doi:10.1016/j.watres.2016.01.002

[7] Guanyu Zhou, Qingguo Wang, JiaLi, QiansongLi, HaoXu, Qian Ye, YunqiWang, ShihuShu, JingZhang. "Removal of polystyrene and polyethylene micro-plastics using PAC and $FeCl_3$ coagulation: Performance and mechanism". *Science of the Total Environment* 15 (2021): 752. doi:10.1016/j.scitotenv.2020.141837

[8] Katriina Rajala, OutiGrönfors, MehrdadHesampour, Anna Mikola. "Removal of micro-plastics from secondary wastewater treatment plant effluent by coagulation/flocculation with iron, aluminum and polyamine-based chemicals."*Water Research* 15 (2020): 183. doi:10.1016/j.watres.2020.116045

[9] Lu Sen, Liu Libing, YangQinxue, DemissieHailu, JiaoRuyuan, AnGuangyu, WangDongsheng. "Removal characteristics and mechanism of micro-plastics and tetracycline composite pollutants by coagulation process."*Science of the Total Environment* 54 (2021): 786.doi:10.1016/j.scitotenv.2021.147508

[10] Yujian Zhang, Guanyu Zhou, JiapengYu, XinyiXing, ZhiweiYang, XinyuWang, QingguoWang, JingZhang. "Enhanced removal of polyethylene terephthalate

micro-plastics through poly-aluminum chloride coagulation with three typical coagulant aids." *Science of The Total Environment* 800 (2021): 149589. doi:10.1016/j.scitotenv.2021.149589

[11] HaerulHidayaturrahman, Tae Gwan Lee. "A study on characteristics of micro-plastic in wastewater of South Korea: Identification, quantification, and fate of microplastics during treatment process." *Marine Pollution Bulletin* 146 (2019): 696–702. doi:10.1016/j.marpolbul.2019.06.071

[12] MartinPivokonský, LenkaPivokonská, KateřinaNovotná, LenkaČermáková, MartinaKlimtová. "Occurrence and fate of micro-plastics at two different drinking water treatment plants within a river catchment." *Science of The Total Environment* 741 (2020): 140236. doi:10.1016/j.scitotenv.2020.140236

[13] MirkaLares, Mohamed Chaker Ncibi, MarkusSillanpää, MikaSillanpää. "Occurrence, identification and removal of micro-plastic particles and fibers in conventional activated sludge process and advanced MBR technology." *Water Research* 133 (2018): 236–246. doi:10.1016/j.watres.2018.01.049

[14] Iqra Nabi, Aziz-Ur-Rahim Bacha, LiwuZhang. "A review on micro-plastics separation techniques from environmental media." *Journal of Cleaner Production* 337 (2022): 130458. doi:10.1016/j.jclepro.2022.130458

[15] William Perren, Arkadiusz Wojtasik, QiongCai. "Removal of microbeads from wastewater using electrocoagulation." *ACS Omega* 3 (2018): 3357–3364. doi:10.1021/acsomega.7b02037.

[16] Maocai Shen, Yaxing Zhang, EydhahAlmatrafi, TongHu, BiaoSong, ZhuotongZeng, GuangmingZeng. "Efficient removal of micro-plastics from wastewater by an electrocoagulation process." *Chemical Engineering Journal* 428 (2022): 131161. doi:10.1016/j.cej.2021.131161

[17] Dina T.Moussa, Muftah H. El-Naas, MustafaNasser, Mohammed J. Al-Marri. "A comprehensive review of electrocoagulation for water treatment: Potentials and challenges." *Journal of Environmental Management* 186 (2017): 24–41. doi:10.1016/j.jenvman.2016.10.032

[18] KeugTaeKim, SanghwaPark. "Enhancing micro-plastics removal from wastewater using electro-coagulation and granule-activated carbon with thermal regeneration." *Processes* 9 (2021): 617. doi:10.3390/pr9040617

[19] Veronica Piazza, AbdusalamUheida, ChiaraGambardella, FrancescaGaraventa, MarcoFaimali. "Ecosafety screening of photo-fenton process for the degradation of microplastics in water". *Frontiers in Marine Science* 8 (2022): 791431. doi:10.3389/fmars.2021.791431

[20] Laurent Lebreton, Anthony Andrady. "Future scenarios of global plastic waste generation and disposal." *Palgrave Communication* 5 (2019): 1. doi:10.1057/s41599-018-0212-7

[21] Zahra Sobhani, Yongjia Lei, YouhongTang, Liwei Wu, Xiang Zhang, Ravi Naidu. "Micro-plastics generated when opening plastic packaging." *Scientific Report* 10(2020): 4841. doi:10.1038/s41598-020-61146-4

[22] Anthony L.Andrady. "Micro-plastics in the marine environment." *Marine Pollution Bulletin* 62 (2011): 596–1605. doi:10.1016/j.marpolbul.2011.05.030

[23] MatthewCole, PennieLindeque, ClaudiaHalsband, Tamara S. Galloway."Microplastics as contaminants in the marine environment: A review." *Marine Pollution Bulletin* 62 (2011): 2588–2597. doi:10.1016/j.marpolbul.2011.09.025

[24] Thiago M.B.F. Oliveira, Francisco W.P. Ribeiro, SimoneMorais, Pedrode Lima-Neto, Adriana Correia. "Removal and sensing of emerging pollutants released from (micro)

plastic degradation: Strategies based on boron-doped diamond electrodes."*Current Opinion in Electrochemistry* 31 (2022): 100866. doi:10.1016/j.coelec.2021.100866

[25] MartheKiendrebeogo, M. R. KarimiEstahbanati, Ali KhosravanipourMostafazadeh, PatrickDrougi, R. D.Tyagi."Treatment of micro-plastics in water by anodic oxidation: A case study for polystyrene." *Environmental Pollution* 269 (2021):116168. doi:10.1016/j.envpol.2020.116168

[26] KurunthachalamKannan, KrishnamoorthiVimalkumar. "A review of human exposure to micro-plastics and insights into micro-plastics as obesogens." *Frontiers in Endocrinology* 12 (2021):724989. doi:10.3389/fendo.2021.724989

[27] Joana CorreiaPrata, João P. da Costa, Isabel Lopes, Armando CDuarte, Teresa Rocha-Santos. "Environmental exposure to micro-plastics: An overview on possible human health effects." *Science Total Environment* 702 (2020): 134455. doi:10.1016/j.scitotenv.2019.134455

[28] JoanaCorreiaPrata. "Airborne micro-plastics: Consequences to human health." *Environmental Pollution* 234 (2018):115–126. doi:10.1016/j.envpol.2017. 11.043

[29] ElenaCorona,CeciliaMartin,RomanaMarasco, Carlos M.Duarte."Passive and active removal of marine micro-plastics by a mushroom coral (danafungiascruposa)." *Frontiers Marine Science* 7(2020):128. doi:10.3389/fmars.2020.00128

[30] SonaHermanová, Martin Pumera."Micro-machines for micro-plastics treatment." *ACS Nanoscience* 2 (2022):225–232. doi:10.1021/acsnanoscienceau.1c00058

[31] Weimin Wu, Jun Yang, Criag S.Criddle."Micro-plastics pollution and reduction strategies." *Frontiers of Environmental Science Engineering* 11 (2017): 1–14. doi:10.1007/s11783-017-0897-7

[32] Kunsheng Hu, Weinjie Tian, YangyangYang, Gang Nie, Peng Zhou, Yuxian Wang. "Micro-plastics remediation in aqueous systems: Strategies and technologies." *Water Research* 198 (2021): 117144. doi:10.1016/j.watres.2021.117144

[33] KunshengHu, PengZhou, YangyangYang, TonyHall, GangNie, XiaoguangDuan."Degradation of micro-plastics by a thermal fenton reaction." *ACS ES&T Engineering* 2 (2022): 110–120. doi:10.1021/acsestengg.1c00323

[34] XinGuo, HelenLin, ShupingXu, Lili He. "Recent advances in spectroscopic techniques for the analysis of micro-plastics in food." *Journal of Agricultural and Food Chemistry* 70 (2022): 1410–1422. doi:10.1021/acs.jafc.1c06085

[35] Natalia P.Ivleva. "Chemical analysis of micro-plastics and nano-plastics: challenges, advanced methods, and perspectives." *Chemical Reviews* 121 (2021): 11886–11936. doi:10.1021/acs.chemrev.1c00178

[36] ChenxuYu, PaulTakhistov, EvangelynAlocilja, Jose ReyesdeCorcuera, MargaretW. Frey, CarmenL. Gomes,MengshiLin. "Bioanalytical approaches for the detection, characterization, and risk assessment of micro/nano-plastics in agriculture and food systems." *Analytical and Bioanalytical Chemistry* 414(2022):4591–4612. doi:10.1007/s00216-022-04069-5

[37] Omprakash Sahu, Bidyut Mazumdar, P. K.Chaudhari. "Treatment of wastewater by electrocoagulation: A review." *Environmental Science and Pollution Research* 21 (2014): 2397–2413. doi:10.1007/s11356-013-2208-6

[38] B. K. Zaied, Mamunur Rashid, Mohd Nasrullah, A. W. Zularisam, Deepak Pant, Lakhveer Singh. "A comprehensive review on contaminants removal from pharmaceutical wastewater by electrocoagulation process." *Science of the Total Environment* 726 (2020): 138095. doi:10.1016/j.scitotenv.2020.138095

[39] Eilen A.Vik, Dale A. Carlson, Arild S. Eikum, Egil T. Gjessing. "Electrocoagulation of potable water." *Water Research* 18, no. 11 (1984): 1355–1360. doi.org/10.1016/0043-1354(84)90003-4

[40] AhmedSamirNaje, Saad A. Abbas. "Electrocoagulation technology in wastewater treatment: A review of methods and applications."*Civil and Environmental Research* 3, no. 11 (2013): 29–42.

[41] M. Bharath, B. M. Krishna, B.Manoj Kumar. "A review of electrocoagulation process for wastewater treatment."*International Journal of ChemTech Research* 11, no. 3 (2018): 289–302.

[42] A. Shahedi, A. K.Darban, F. Taghipour, A. J. C. O. I. E.Jamshidi-Zanjani. "A review on industrial wastewater treatment via electrocoagulation processes."*Current Opinion in Electrochemistry* 22 (2020): 154–169. doi:10.1016/j.coelec.2020.05.009

[43] K. Padmaja, Jyotsna Cherukuri, M. AnjiReddy. "A comparative study of the efficiency of chemical coagulation and electrocoagulation methods in the treatment of pharmaceutical effluent."*Journal of Water Process Engineering* 34 (2020): 101153. doi:10.1016/j.jwpe.2020.101153

[44] Pranjal P.Das, MukeshSharma, Mihir K. Purkait. "Recent progress on electrocoagulation process for wastewater treatment: A review."*Separation and Purification Technology* 292 (2022): 121058. doi:10.1016/j.seppur.2022.121058

[45] P. V. Nidheesh, T. S.Anantha Singh. "Arsenic removal by electrocoagulation process: Recent trends and removal mechanism."*Chemosphere* 181 (2017): 418–432. doi:10.1016/j.chemosphere.2017.04.082

[46] AdelaideDura. *Electrocoagulation for water treatment: The removal of pollutants using aluminium alloys, stainless steels and iron anodes.* National University of Ireland, Maynooth (Ireland), 2013.

[47] B. M.Belongia,P. D.Haworth, J. C.Baygents, S. Raghavan. "Treatment of alumina and silica chemical mechanical polishing waste by electro-decantation and electrocoagulation."*Journal of the Electrochemical Society* 146, no. 11 (1999): 4124–4130. doi:10.1149/1.1392602

[48] SergiGarcia-Segura, MariaMaesia S. G.,Eiband, JailsonVieirade Melo, and CarlosAlbertoMartínez-Huitle. "Electrocoagulation and advanced electrocoagulation processes: A general review about the fundamentals, emerging applications and its association with other technologies."*Journal of Electroanalytical Chemistry* 801 (2017): 267–299.

[49] NazeriAbdulRahman,CalvinJoseJol,AlleneAlbaniaLinus,VerawatyIsmail. "Emerging application of electrocoagulation for tropical peat water treatment: A review."*Chemical Engineering and Processing-Process Intensification* 165 (2021): 108449. doi:10.1016/j.cep.2021.108449

[50] NazeriAbdulRahman, NurhidayahKumar Muhammad Firdaus Kumar, Umang JataGilan, Elisa ElizebethJihed, AdarshPhillip, Allene AlbaniaLinus, DasimaNen, and VerawatyIsmail. "Kinetic study & statistical modelling of Sarawak Peat Water Electrocoagulation System using copper and aluminium electrodes."*Journal of Applied Science & Process Engineering* 7, no. 1 (2020): 439–456. doi:10.33736/jaspe.2195.2020

[51] ShwetaKumari, R. NareshKumar. "River water treatment using electrocoagulation for removal of acetaminophen and natural organic matter."*Chemosphere* 273 (2021): 128571. doi:10.1016/j.chemosphere.2020.128571

[52] ShivamSnehi, HarirajSingh, TanwiPriya, Brijesh KumarMishra. "Understanding the natural organic matter removal mechanism from mine and surface water through the electrocoagulation method."*Journal of Water Supply: Research and Technology—AQUA* 68, no. 7 (2019): 523–534. doi:10.2166/aqua.2019.167

[53] Sean T.McBeath, AminNouri-Khorasani, MadjidMohseni, David P. Wilkinson. "In-situ determination of current density distribution and fluid modeling of an

electrocoagulation process and its effects on natural organic matter removal for drinking water treatment."*Water Research* 171 (2020): 115404. doi:10.1016/j.watres.2019.115404

[54] FerideUlu, Erhan Gengec, MehmetKobya. "Removal of natural organic matter from Lake Terkos by ECprocess: Studying on removal mechanism by floc size and zeta potential measurement and characterization by HPSECmethod."*Journal of Water Process Engineering* 31 (2019): 100831. doi:10.1016/j.jwpe.2019.100831

[55] AbdellatifEl-Ghenymy, Mohammad Alsheyab, AhmedKhodary, IgnasiSirés, AhmedAbdel-Wahab. "Corrosion behavior of pure titanium anodes in saline medium and their performance for humic acid removal by electrocoagulation."*Chemosphere* 246 (2020): 125674. doi:10.1016/j.chemosphere.2019.125674

[56] GonaHasani, Afshin Maleki, HiuaDaraei, RezaGhanbari, MahdiSafari, GordonMcKay, KaanYetilmezsoy, FatihIlhan, NaderMarzban. "A comparative optimization and performance analysis of four different electrocoagulation-flotation processes for humic acid removal from aqueous solutions."*Process Safety and Environmental Protection* 121 (2019): 103–117. doi:10.1016/j.psep.2018.10.025

[57] ShankararamanChellam, Mutiara AyuSari. "Aluminum electrocoagulation as pretreatment during microfiltration of surface water containing NOM: A review of fouling, NOM, DBP, and virus control."*Journal of Hazardous Materials* 304 (2016): 490–501. doi:10.1016/j.jhazmat.2015.10.054

[58] FangyuanLiu, ChunpengZhang, HuilinLi, Nnanake-Abasi O. Offiong, YuhangBi, RuiZhou, HejunRen. "A systematic review of electrocoagulation technology applied for microplastics removal in aquatic environment."*Chemical Engineering Journal* 456 (2023): 141078. doi:10.1016/j.cej.2022.141078

[59] MengqiaoLuo, ZhaoyangWang, ShuaiFang, BoSong, PengweiCao, HaoLiu, YixuanYang. "Removal and toxic forecast of microplastics treated by electrocoagulation: Influence of dissolved organic matter."*Chemosphere* 308 (2022): 136309. doi:10.1016/j.chemosphere.2022.136309

[60] IqraNabi, LiwuZhang. "A review on microplastics separation techniques from environmental media."*Journal of Cleaner Production* 337 (2022): 130458. doi:10.1016/j.jclepro.2022.130458

[61] PadervandMohsen, LichtfouseEric, RobertDidier, ChuanyiWang. "Removal of microplastics from the environment. A review."*Environmental Chemistry Letters* 18, no. 3 (2020): 807–828.

[62] ShentanLiu, HailiangSong, SizeWei, FeiYang, XianningLi. "Bio-cathode materials evaluation and configuration optimization for power output of vertical subsurface flow constructed wetland—Microbial fuel cell systems."*Bioresource Technology* 166 (2014): 575–583. doi:10.1016/j.biortech.2014.05.104

[63] MaocaiShen, Biao Song, YuanZhu, GuangmingZeng, YaxinZhang, YuanyuanYang, XiaofengWen, MingChen, HuanYi. "Removal of microplastics via drinking water treatment: Current knowledge and future directions."*Chemosphere* 251 (2020): 126612. doi:10.1016/j.chemosphere.2020.126612

[64] DouniaElkhatib, Vinka Oyanedel-Craver, ElvisCarissimi. "Electrocoagulation applied for the removal of micro-plastics from wastewater treatment facilities."*Separation and Purification Technology* 276 (2021): 118877.

[65] CeyhunAkarsu, Halil Kumbur, Ahmet ErkanKideys. "Removal of micro-plastics from wastewater through electrocoagulation-electro-flotation and membrane filtration processes."*Water Science and Technology* 84, no. 7 (2021): 1648–1662. doi:10.2166/wst.2021.356

[66] AbdulkarimAlmukdad, MhdAmmar Hafiz, Ahmed T. Yasir, RadwanAlfahel, Alaa H. Hawari. "Unlocking the application potential of electrocoagulation process

through hybrid processes."*Journal of Water Process Engineering* 40 (2021): 101956. doi:10.1016/j.jwpe.2021.101956

[67] AtousaGhaffarianKhorram, NargesFallah. "Comparison of electrocoagulation and photocatalytic process for treatment of industrial dyeing wastewater: Energy consumption analysis." *Environmental Progress & Sustainable Energy* 39, no. 1 (2020): 13288. doi:10.1002/ep.13288

[68] AbdurrahmanAkyol, Orhan Taner Can, ErhanDemirbas, and MehmetKobya. "A comparative study of electrocoagulation and electro-Fenton for treatment of wastewater from liquid organic fertilizer plant." *Separation and Purification Technology* 112 (2013): 11–19. doi:10.1016/j.seppur.2013.03.036

[69] TingGao, JingqiLin, KeZhang, MohsenPadervand, YifanZhang, WeiZhang, MenglinShi, ChuanyiWang. "Porous defective Bi/Bi3NbO7 nanosheets for efficient photocatalytic NO removal under visible light." *Processes* 11, no. 1 (2022): 115. doi:10.3390/pr11010115

[70] MonaliPriyadarshini, AzhanAhmad, SovikDas, Makarand M. Ghangrekar. "Application of innovative electrochemical and microbial electrochemical technologies for the efficacious removal of emerging contaminants from wastewater: A review." *Journal of Environmental Chemical Engineering* 10, no. 5 (2022): 108230. doi:10.1016/j.jece.2022.108230

[71] MohsenPadervand, Baker Rhimi, ChuanyiWang. "One-pot synthesis of novel ternary Fe3N/Fe2O3/C3N4 photo-catalyst for efficient removal of rhodamine B and CO2 reduction." *Journal of Alloys and Compounds* 852 (2021): 156955. doi:10.1016/j.jallcom.2020.156955

[72] B.Zeboudji, NadjibDrouiche, HakimLouinici, NabilMameri, NoreddineGhaffour. "The influence of parameters affecting boron removal by electrocoagulation process." *Separation Science and Technology*48(2013): 1280–1288. doi:10.1080/01496395.2012.731125

[73] Lalita Sharma, Shashank Prabhakar, Vijay Tiwari, Atul Dhar, Aditi Halder. "Optimization of ECparameters using Fe and Al electrodes for hydrogen production and wastewater treatment." *Environmental Advances* 3 (2021): 100029. doi:10.1016/j.envadv.2020.100029

[74] EdwarAguilar-Ascón, Armando Solari-Godiño, MiguelCueva-Martínez, WalterNeyra-Ascón, MiguelAlbrecht-Ruíz. "Characterization of sludge resulting from chemical coagulation and electrocoagulation of pumping water from fishmeal factories." *Processes* 11 (2023): 567. doi:10.3390/pr11020567

[75] FuatOzyonar, BunyaminKaragozoglu. "Operating cost analysis and treatment of domestic wastewater by electrocoagulation using aluminum electrodes." *Polish Journal of Environmental Studies* 20 (2011):173–179.

[76] M. Behbahani, M. R. AlaviMoghaddam, M. Arami. "A comparison between aluminum and iron electrodes on removal of phosphate from aqueous solutions by electrocoagulation process." *International Journal of Environmental Research* 5 (2011):403–412. doi:10.22059/ijer.2011.325

[77] Sanela Martic, Meaghan Tabobondung, Stephanie Gao, Tyra Lewis."Emerging electrochemical tools for microplastics remediation and sensing." *Frontier in Sensors* 3 (2022): 958633. doi:10.3389/fsens.2022.958633

[78] R. S.Weber, K.K. Ramasamy. "Electrochemical oxidation of lignin and waste plastic." *ACS Omega* 5 (2020): 27735–27740. doi:10.1021/acsomega.0c03989

[79] Annabel Fernandez, Christopher Pereira, Violeta Koziol, Maria José Pacheco, Lurdes Ciriaco, Ana Lopes. (2020). "Emerging contaminants removal from effluents

Electrocoagulation for Remediation of Microplastics 167

with complex matrices by electro-oxidation." *Science of Total Environment* 740 (2020): 140153. doi:10.1016/j.scitotenv.2020.140153

[80] GovindarajDivyapriya, P. V. Nidheesh. "Electrochemically generated sulfate radicals by boron doped diamond and its environmental applications". *Current Opinion Solid State and Material Science* 25 (2021): 100921. doi:10.1016/j.cossms.2021.100921

[81] P. V.Nidheesh, RamanRajan."Removal of rhodamine B from a water medium using hydroxyl and sulphate radicals generated by iron loaded activated carbon." *RSCAdvances* 6 (2016):5330–5340. doi:10.1039/C5RA19987E

[82] Fie Miao, YanfengLiu, MingmingGao, Xin Yu, PengweiXiao, Mie Wang. "Degradation of polyvinyl chloride micro-plastics via an electro-Fenton-like system with a TiO_2/graphitecathode."*Journal of Hazardous Materials* 399 (2020):123023.doi:10.1016/j.jhazmat.2020.123023

[83] Hua Zhou, Yue Ren, ZhenhuaLi, MingXu, Ye Wang, Ge, Ruixiang, Qinghui Zeng. "Electrocatalytic upcycling of polyethylene terephthalate to commodity chemicals and H_2 fuel." *Nature Communications* 12 (2021):4679. doi:10.1038/s41467-021-25048-x

[84] Gayathri Chellasamy, Rose Mary Kiriyanthan, TheivanayagamMaharajan, A Radha. "Remediation of micro-plastics using bio-nanomaterials: A review." *Environmental Research* 208 (2022): 112724. doi:10.1016/j.envres.2022.112724

[85] BacettC. Colson, AnnaMichel. "Flow-through quantification of micro-plastics using impedance spectroscopy." *ACS Sensors* 6 (2021): 238–244. doi:10.1021/acssensors.0c02223

[86] Kenichi Shimizu, Stanislav Sokolv, Enno Katelhon, Jennifer Holter, Neil P.Young, Richard G.Compton. "In situ detection of micro-plastics: Single micro-particle-electrode impacts. " *Electro-analysis* 29 (2017): 2200–2207. doi:10.1002/elan.201700213

[87] Collin D. Davies, Richard M. Crooks. "Focusing, sorting, and separating micro-plastics by serial faradaic ion concentration polarization." *Chemical Science* 11 (2020): 5547–5558. doi:10.1039/D0SC01931C

[88] Marcus Pollard, Eugenie Hunsicker, Mark Platt."A tunable three-dimensional printed microfluidic resistive pulse sensor for the characterization of algae and micro-plastics." *ACS Sensors* 5 (2020):2578–2586. doi:10.1021/acssensors.0c00987

[89] Song Wang, Mingyi Xu, Biao Jin, Urban J. Wünsch, Yanyan Su. "Electrochemical and microbiological response of exoelectrogenic biofilm to polyethylene micro-plastics in water." *Water Research* 211 (2022): 118046. doi:10.1016/j. watres.2022.118046

[90] Sriram Boinpally, Abhinav Kolla, Jyoti Kainthola, Ruthviz Kodali. "Astate-of-the-artreviewoftheelectrocoagulationtechnologyforwastewatertreatment." *Water Cycle* 4 (2023): 26–36. doi:10.1016/j.watcyc.2023.01.001

9 Photocatalytic Degradation and Remediation of Microplastics

Manjinder Kour and Sapana Jadoun

9.1 INTRODUCTION

In recent years, microplastics, tiny plastic particles typically measuring less than 5 mm, have emerged as a pervasive environmental concern [1]. They originate from various sources, including industrial processes, the degradation of larger plastic debris, and the direct release of microplastic products. Depending on their origin, these particles are classified into primary and secondary microplastics [2]. The classification of microplastics is depicted in Figure 9.1.

Primary microplastics are intentionally manufactured at a microscopic scale for specific purposes, such as microbeads in personal care products or pellets used in industrial processes. As shown in Figure 9.2, they stem from the manufacturing processes involved in producing skincare items and personal hygiene products [3, 4]. Secondary microplastics are formed by breaking larger plastic particles through biological, physical, and chemical processes [4]. These are primarily fragmented in coastal areas due to intense UV radiation, oxygen exposure, wave abrasion, and turbulent air currents.

Microplastics are further classified from these primary categories based on their polymer compositions (Figure 9.3). For instance, some belong to biodegradable categories, such as polyethylene, polylactic acid, and polystyrene. In contrast, others fall into nonbiodegradable classifications like polyurethane, polyvinyl chloride, polyamide, polypropylene, polymethyl methacrylate, and polyesters [5, 6].

The importance of plastic in our everyday existence can be credited to its remarkable resilience and flexibility, cost-effectiveness, and lightweight characteristics [7, 8]. The prevalence of microplastic contamination has surfaced as a novel environmental issue owing to our dependency on plastic materials. The global production of plastics has witnessed exponential growth, escalating from 1.5 million tons to approximately 359 million tons over the past seven decades [9]. This surge in plastic production, coupled with insufficient waste management practices, has led to the widespread distribution of microplastics across terrestrial, freshwater, and marine ecosystems [10–12].

DOI: 10.1201/9781003486947-9

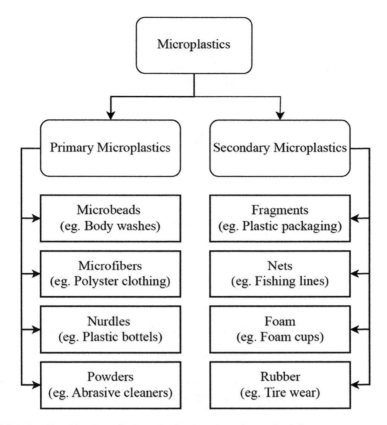

FIGURE 9.1 Classification of microplastics based on size and origin.

Land-based sources account for most microplastic pollution, encompassing plastic waste from various sectors such as packaging, construction, and textiles. In contrast, ocean-based sources, including maritime activities and coastal tourism, contribute to a smaller yet significant proportion of microplastic contamination. Microplastics come in various shapes, including fibers, fragments, spheres, and films, each with distinct characteristics influencing their environmental behavior and interactions with organisms [13]. Additionally, they exhibit a wide range of polymer compositions, with polyethylene (PE), polypropylene (PP), and polystyrene (PS) being among the most common. Surface modifications of microplastics, such as oxidation, biofouling, and sorption of organic and inorganic compounds, further complicate their behavior. The widespread presence of microplastics in our environment, from land to water and marine ecosystems, has raised concerns about their impact on nature [14–16]. Many species eat these small particles, which move through food chains and pose health threats to all life types – marine, freshwater, and terrestrial [17]. Microplastics can cause reduced growth and reproduction and decreased survival in living things. They can block digestive tracts, and transfer toxins to these species [18]. Microplastics can

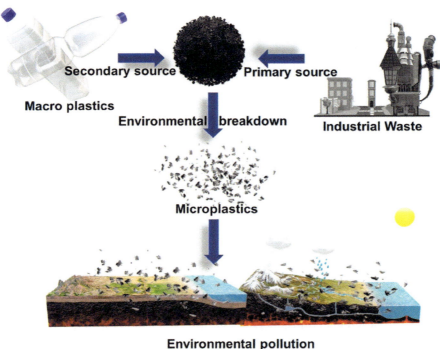

FIGURE 9.2 Overview of general sources of microplastics in the environment, encompassing primary and secondary sources. Reproduced from Ref. (5) with permission.

also ferry damaging microbes and pollutants into the environment and could pose a human health risk. These reports highlight the direct harm that microplastics cause to human health and indirect implications.

Microplastics can cause complicated disruptions to the delicate balance between plants, animals, and the environment as they cascade through ecosystems. When they accumulate in sediment at the seabed, they disturb ecosystems and obstruct the cycles of carbon storage and regeneration and nutrient cycles [19]. They can create perfect surfaces for invasive species to thrive. They can modify the composition of microbial communities. And they can affect the availability of nutrients and organic matter. This linkage between microplastic impacts attests to how urgently we must develop sustainable, practical solutions to degrade microplastic and preserve ecosystem health.

Moreover, recent studies show that microplastics are effective vectors of toxic pollutants, adding further environmental and health degradation to the mix [20]. Numerous typical interaction mechanisms can occur between microplastics and pollutants, facilitating the transport of these pollutants within the ecosystem. Microplastics carry organic and inorganic contaminants through many interaction mechanisms [2, 21–23]. These have been reported to absorb contaminants such as

Photocatalytic Degradation and Remediation 171

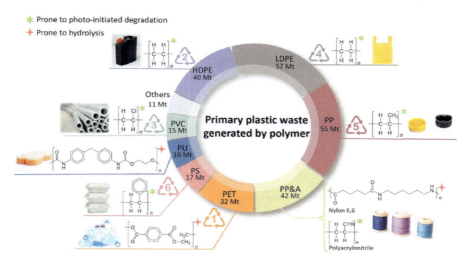

FIGURE 9.3 Commonly documented polymers in microplastic studies and their share of primary plastic waste in 2015. Abbreviations: LDPE (low-density polyethylene), HDPE (high-density polyethylene), PP&A (polyester, polyamide, and acrylic) [Reproduced from Ref. (5) with permission].

persistent organic pollutants, heavy metals, and antibiotics, leading to their bioaccumulation in marine and terrestrial ecosystems [24, 25]. Hydrophobic interactions are the predominant sorption mechanism, especially for organic pollutants. However, other forces such as electrostatic forces, van der Waals forces, hydrogen bonding, and π-π interactions also contribute [26–28]. The physicochemical properties of microplastics, such as size, structure, and functional groups, along with environmental factors like pH, temperature, and salinity, influence the sorption of pollutants [2]. Furthermore, microplastics have been found to affect the growth and metabolism of organisms, highlighting the complex interplay between microplastics and ecosystem health. As shown in Figure 9.4, ingesting microplastics presents significant health risks to human beings, encompassing a range of adverse effects on various organ systems [29–32].

Potential consequences include cancer, immunotoxicity, intestinal diseases, pulmonary diseases, and cardiovascular diseases. Microplastics, commonly found in food and water, can translocate across the digestive epithelium, entering the bloodstream and affecting organs throughout the body [33, 34]. Additionally, they can exacerbate inflammatory responses, disrupt gut microbiota composition, and increase gut permeability, leading to intestinal diseases [35]. Furthermore, microplastics can induce oxidative stress, inflammation, and cellular toxicity, impacting immune function and worsening inflammatory diseases. Pregnant individuals and their offspring are particularly vulnerable, as exposure during gestation and lactation can result in metabolic abnormalities, neural development issues, and immune system dysregulation. Moreover, plasticizer additives such as bisphenol A and phthalates,

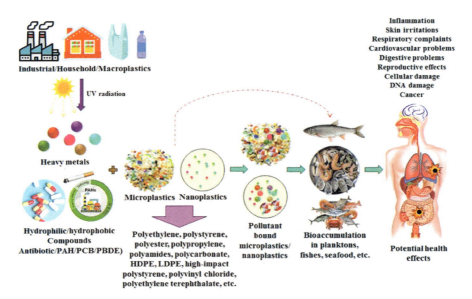

FIGURE 9.4 Bioaccumulation patterns of microplastics and nanoplastics within trophic levels and their harmful effect on human health. Source [Reproduced from Ref. (5) with permission].

leached from plastic materials, have been associated with cardiovascular diseases and developmental abnormalities. Despite ongoing research, more comprehensive studies are needed to fully understand the mechanisms underlying the detrimental effects of microplastics on human health and to develop effective mitigation strategies [36].

Generic approaches to microplastic remediation, such as mechanical removal, coagulation, rapid sand filtration, membrane bioreactor technology, adsorption, and waste management, have limitations in addressing microplastics' complex distribution and persistence across diverse environments [37]. Furthermore, these methods often entail high costs, operational challenges, and potential ecological disturbances, necessitating exploring alternative technologies. As we explore remediation strategies for microplastics, particularly focusing on photocatalytic degradation, it becomes important to understand the relationships between microplastics, pollutants, and environmental processes. Photocatalytic degradation emerges as a promising remediation strategy for microplastics, leveraging the photocatalytic properties of semiconductor materials to facilitate the breakdown of plastic particles under light irradiation [38–42]. This approach provides numerous benefits, such as excellent effectiveness, precision, and the ability to expand its use in various natural and human-made settings. Using solar power and initiating oxidative processes, photocatalysts can break down microplastics into tinier pieces, decompose organic substances, and reduce their environmental effects without generating harmful substances.

Photocatalytic Degradation and Remediation 173

The escalating threat microplastics pose to ecosystems and human well-being underscores the imperative for innovative remediation approaches, and photocatalytic degradation is a promising strategy for mitigating microplastic pollution and advancing environmental sustainability [43]. The next part will explore the fundamental aspects, operational mechanisms, practical applications, and obstacles encountered in utilizing photocatalysis for addressing microplastic pollution.

9.1 PHOTOCATALYSIS: FUNDAMENTALS AND MECHANISMS

Photocatalysis stands at the top among advanced oxidation processes, leveraging semiconductor materials to propel chemical reactions when exposed to light [44]. In semiconductors, the transfer of electrons from the valance band to the conduction band triggers the chemical reaction. Although the detailed photocatalysis mechanism is complicated, the core concept remains unchanged. The electrons of the semiconductor photocatalyst are driven from the valance band to the conduction band upon exposure to light with energy equivalent to or greater than the photocatalyst's bandgap. This leaves the holes in the valance band behind [45]. In this section, we embark on a thorough journey into the core principles, mechanisms, and pivotal processes inherent in photocatalytic reactions, shedding light on the indispensable role played by catalysts in driving these transformative reactions.

9.1.1 PHOTOCATALYSIS

The essence of photocatalysis resides in the symbiotic interplay between semiconductor catalysts, light photons, and target molecules. Semiconductor photocatalysts possess a unique bandgap energy that enables them to absorb photons with energies equal to or greater than the bandgap [45–47]. Electrons are excited from the valence band to the conduction band upon light absorption, creating electron-hole pairs (Scheme 1).

$$h\vartheta + Semiconductor \rightarrow e_{CB}^- + h_{VB}^+ \tag{1}$$

Scheme 9.1: Photoexcitation process in photocatalysis: incident photons elevate electrons from the valence band (VB) to the conduction band (CB), creating electron-hole pairs.

These electron-hole pairs are reactive species capable of participating in redox reactions with adsorbed molecules on the catalyst surface or in the surrounding solution. The photogenerated electrons (e⁻) can reduce electron acceptors, while the photogenerated holes (h⁺) can oxidize electron donors (Scheme 2).

$$e_{CB}^- + O_2 \rightarrow O_2^{-\circ}$$

$$h_{VB}^+ + H_2O \rightarrow HO^\circ + H^+ \tag{2}$$

Scheme 9.2: Photocatalytic reactions: photogenerated electrons reduce oxygen molecules (O_2) to superoxide radicals ($O_2^{\cdot-}$), while photogenerated holes oxidize water molecules (H_2O) to hydroxyl radicals ($HO^{\cdot}$).

These redox reactions lead to the generation of highly reactive species such as $HO^{\cdot}$, O_2^{-}, and H_2O_2. These species exhibit strong oxidative and reductive capabilities, enabling the degradation of organic pollutants, including microplastics, into harmless byproducts such as carbon dioxide, water, and mineral salts

9.1.2 KEY PRINCIPLES AND PROCESSES INVOLVED

In photocatalytic reactions, several fundamental principles and processes shape the efficiency and specificity of the process [45–49].

9.1.2.1 Bandgap Energy

The bandgap energy of semiconductor materials dictates their ability to absorb light, directly influencing their effectiveness in photocatalysis. Catalysts with narrower bandgaps have a broader absorption spectrum, making them more adaptable for various photocatalytic applications [49].

9.1.2.2 Surface Area and Morphology

The morphology and surface area of catalysts are critical determinants of their photocatalytic performance. Materials with well-defined crystalline structures and high surface area-to-volume ratios provide adequate active sites for reactions, thus enhancing catalytic efficiency [45].

9.1.2.3 Charge Carrier Dynamics

The kinetics of reactions and the selectivity of the final product are directly influenced by the dynamics of photogenerated charge carriers, such as electrons and holes. Charge recombination is reduced by effective charge migration to the catalyst surface and separation. Consequently, this optimizes redox processes and increases the efficiency of photocatalysis. Catalytic surface interactions are complex, and the complex constructive relationship between catalyst morphology and surface properties helps to understand the underlying mechanisms that drive photocatalytic activity [50].

9.1.2.4 Reaction Environment

A wealth of research has demonstrated that the conditions surrounding a reaction directly influence how successful photocatalysis is. Variations in temperature, pH, and the inclusion of extra agents or catalysts can all significantly impact the process's efficiency. It is imperative to maintain the optimal circumstances to preserve the catalyst's stability and activity and steer the reaction down the intended reaction routes [45].

9.1.3 ROLE OF CATALYSTS IN PHOTOCATALYTIC REACTIONS

By assisting in creating, transporting, and utilizing the photogenerated charge carriers, catalysts perform an integral part in accelerating photocatalytic reactions.

Photocatalytic Degradation and Remediation 175

It has been generally observed that as catalyst loading increases, so does the rate at which pollutants and microplastics on the catalyst's surface degrade [45, 51]. Due to their appropriate energy levels and characteristics for photocatalysis, materials such as bismuth vanadate ($BiVO_4$), copper oxide (CuO), zinc oxide (ZnO), bismuth chloride (BiOCl), and titanium dioxide (TiO_2) have become popular catalysts [51–57]. Because it is stable, nontoxic, and efficient at catalyzing reactions when exposed to light, TiO_2 is one of the most examined. It has three distinct crystal forms, the most effective for photocatalysis being the anatase form. When exposed to UV and visible light, TiO_2-based catalysts exhibit remarkable organic pollutant degradation, becoming precious tools for environmental remediation [37]. Comparably, another semiconductor material with a great deal of potential for photocatalysis is ZnO. It is appealing for various photocatalytic processes due to its unique qualities, which include a large energy gap, good electron mobility, and low toxicity [53, 58]. The ability of ZnO-based catalysts to degrade pollutants, purify water, and even contribute to the production of renewable energy has demonstrated how flexible and beneficial these catalysts can be in promoting a better and microplastic-free environment.

9.2 PHOTOCATALYTIC DEGRADATION OF MICROPLASTICS: STRATEGIES AND CHALLENGES

As the problem of microplastic contamination gains momentum, creative remediation techniques are becoming more and more necessary for effectively dealing with this worldwide environmental hazard. As previously mentioned, photocatalytic degradation has shown promise among these strategies. It works by using semiconductor materials' catalytic properties to start the decomposition of microplastics under light irradiation. How this approach works, what influences its efficacy, and some practical microplastic remediation methods are discussed below.

9.2.1 OVERVIEW OF PHOTOCATALYTIC DEGRADATION METHODS FOR MICROPLASTIC REMEDIATION

In the ongoing struggle against microplastic pollution, researchers focus on the exciting area of photocatalytic degradation, which uses semiconductor materials' light-triggered properties to break down these resistant contaminants [7, 38, 59, 60]. In this discipline, experts have examined a wide range of approaches that may be roughly divided into two groups: homogeneous and heterogeneous photocatalysis. Both groups have unique benefits and challenges [61].

9.2.1.1 Heterogeneous Photocatalysis

Characterized by the immobilization of catalysts onto supporting materials to enable catalyst recovery and reuse, heterogeneous photocatalysis is a crucial strategy for tackling microplastics. To accomplish this immobilization, various techniques are used, such as physical absorption, chemical bonding, and stacking onto porous substrates, films, and nanoparticles. These systems offer potential remedies for reducing microplastic contamination by improving catalyst stability and enabling effective catalyst recovery. When exposed to enough energy, semiconductor catalysts absorb photons, which causes electron-hole pairs to form in their bulk. While positive photoholes interact with donor molecules to start a chain reaction that forms intermediates and final products, subsequent interactions with acceptor molecules result in electron transfer [62, 63].

9.2.1.2 Homogeneous Photocatalysis

Unlike heterogeneous reactions, these use molecular catalysts that are dissolved or suspended in the reaction medium. Both energy and nonenergy pathways can be a part of a homogeneous process; photo-Fenton and H_2O_2/catalyst are two examples, respectively. Homogeneous catalysts provide precise control over reaction conditions and improved catalytic activity, resulting in effective microplastic degradation [63–65]. However, it is important to note that they have unique difficulties, such as recovering the catalyst and separating the final products, which makes scaling up a little hard and challenging.

9.2.2 FACTORS INFLUENCING THE EFFICIENCY OF DEGRADATION

The optimization of photocatalytic degradation efficiency depends on several elements, including the characteristics of the catalyst, the surrounding environment, the conditions of the reaction, and the characteristics of the microplastic [66]. Achieving successful environmental remediation and improving degradation efficacy require carefully adjusting and optimizing these variables.

9.2.2.1 Catalyst Properties

Microplastic degradation's effectiveness largely depends on selecting semiconductor catalysts. In this sense, catalysts with the right band energy gaps, large surface areas, and effective charge transfer mechanisms typically work better. Also, modifying the catalyst by adding certain substances, altering its surface, or engineering junctions between different materials can make it even more effective [45, 48, 67]. These changes help separate charges better and allow the catalyst to absorb more light, enhancing its performance.

9.2.2.2 Light Source and Intensity

The type of light (wavelength), its strength (intensity), and the duration of light irradiation are super important for how fast photocatalytic reactions will happen. UV light is used most because it can excite electrons in the catalyst to higher energy levels (across the bandgap). However, researchers have also explored the potential of visible light, including sunlight, by developing catalysts that can respond to these wavelengths or by employing methods to capture and use more light energy [64, 68–70].

9.2.2.3 Environmental Conditions

Factors including but not limited to pH, temperature, and oxygen availability have notable influences on the degradation rates [68, 71]. Maintaining optimal reaction

Photocatalytic Degradation and Remediation

conditions is important for sustaining the catalyst's stability and efficiency while fostering favorable reaction pathways. Also, co-catalysts and dissolved ions can modulate reaction kinetics and influence the selectivity of reaction products as desired.

9.2.2.4 Microplastic Characteristics

The characteristics of microplastics, like their size, shape, and how their surfaces are built, i.e., surface chemistry, affect how they interact and react with catalysts and get degraded [63, 69, 72]. When microplastics are smaller, they have more surface area for catalysts to respond to, so kinetics plays a role, and they break down faster. Changes or modifications to the surface of these microplastics can also impact how well they interact with catalysts and how quickly they degrade.

9.2.3 HIGHLIGHTS OF SUCCESSFUL DEGRADATION TECHNIQUES

Many studies have showcased the effectiveness of employing photocatalytic degradation to treat microplastic pollution across various environmental settings. These successful methodologies hinge on developing innovative catalyst designs, meticulous optimization of reaction parameters, and integrating synergistic techniques to increase the efficiency of microplastic remediation efforts [65, 70, 73]. With multiple successful initiatives demonstrating its efficacy in various environmental scenarios, photocatalytic degradation is a promising strategy in the global battle against microplastic contamination. For example, using the hydrothermal approach, researchers have created zinc oxide nano-rods to degrade low-density polyethylene (LDPE) sheets, a common type of microplastic pollution [61]. This process began the degradation of LDPE under the impact of light exposure, along with the formation of new functional groups. Likewise, Li et al. [74] used polypyrrole and titanium oxide nanocomposites from the sol-gel and emulsion polymerization processes to study polyethylene plastic's degradation. Significant weight loss of polyethylene and the creation of holes from volatile organics evaporating from the polymer matrix were observed after extended exposure to sunshine. The same research group has also demonstrated using a platinum-deposited zinc oxide nanocomposite that absorbs more UV light and efficiently degrades LDP. Uheida et al. [64] investigated the possibility of immobilizing zinc oxide onto glass surfaces for the degradation of polypropylene in the presence of light, leading to a notable decrease in polypropylene size underwater. Nabi et al. [75] studied the effectiveness of titanium oxide films created with different solvents and surfactants in degradipolystyrene (PS) when exposed to UV light. Significantly, the polystyrene film containing Triton X-100 showed full mineralization, suggesting that titanium oxide plays a part in initiating UV light absorption and subsequent degradation. In the same way, Ariza-Tarazona et al. [66] produced nitrogen-doped titanium oxide from live blue mussels with promising outcomes when compared to alternative synthetic techniques for degrading high-density polyethylene. It has been demonstrated that combining photocatalysis with complementary methods can significantly speed up the degradation rate. Ariza-Tarazona et al. [66] have reported that hybrid systems, including photocatalysis when combined with ultrasonic treatment or ozonation, have shown efficacy in augmenting oxidative radical production and mass transfer, thereby increasing the rate of degradation of microplastics. Furthermore, adding carbon-based materials to photocatalytic systems has increased electron transfer rates and catalyst stability, increasing the degradation rate [76]. These reports highlight the significance of advanced catalyst designs, reaction optimization, and combining complementary methods to improve microplastic degradation. Due to their effectiveness and versatility, metal oxide-based photocatalysis, especially ZnO and TiO_2, has become popular in this field [58, 69]. Carbon-based photocatalysts are popular due to their stability and large surface areas [76, 77]. Furthermore, potential approaches to tackling the complex problems caused by microplastic contamination include creating hybrid systems that integrate photocatalysis with other methods or biological processes [66, 78]. Using photocatalytic degradation holds promise in the fight against microplastic pollution and in promoting environmental sustainability [58, 38, 59–65, 67, 69, 70–79, 80–94]. By refining catalyst designs, understanding how degradation occurs, exploring synergistic methods, and exploring these materials computationally through DFT charge transfer studies [95], experts can develop innovative solutions for tackling the complex challenges caused by microplastic contamination. It is worth mentioning that further research is necessary to overcome current limitations, enhance degradation efficiency, and enable the widespread and micro to macroscale adoption of photocatalytic remediation strategies in real-world situations.

9.3 PHOTOCATALYTIC MATERIALS FOR MICROPLASTIC REMEDIATION

The effectiveness of photocatalysis relies on the specific properties of the materials acting as catalysts, dictating their ability to initiate and accelerate degradation processes [43]. Various photocatalytic materials are commonly employed for microplastic remediation [66, 96, 97], each offering unique characteristics that contribute to their efficacy as catalysts.

9.3.1 TYPES OF PHOTOCATALYTIC MATERIAL

Photocatalytic materials encompass diverse semiconductors, each with unique properties influencing their catalytic activity and effectiveness in microplastic remediation. Some of the most used photocatalysts include metal oxides, carbon-based materials, Metal-organic frameworks, and so on [98, 99].

9.3.1.1 Metal Oxides

Metal oxides, such as TiO_2, ZnO, iron oxide (Fe_2O_3), cerium oxide (CeO_2), tungsten oxide (WO_3), and manganese dioxide (MnO_2), have been explored for their photocatalytic capabilities and are among the most widely studied photocatalytic materials. These materials offer remarkable stability and demonstrate potent photocatalytic activity, especially under UV irradiation. Their efficacy in remediation and potential to address diverse environmental challenges have made TiO_2, ZnO, and Fe_2O_3 popular (52-57). Many ongoing research endeavors also continue to uncover novel metal oxides and optimize their performance, underscoring the dynamic nature of this field and its role in sustainable development initiatives [42, 100–102].

Photocatalytic Degradation and Remediation

9.3.1.2 Carbon-Based Materials

Due to their inherent advantages and the need for sustainable remediation solutions, the exploration of carbon-based materials as photocatalysts for microplastic degradation has expanded significantly in recent years. Beyond graphene, carbon nanotubes, and activated carbon, novel carbonaceous materials have emerged as promising candidates recently [103, 104]. One example is carbon quantum dots, which, with their unique optical properties and tunable surface chemistry, have shown potential for efficient photodegradation of microplastics [105]. Carbon nitride-based materials, such as graphitic carbon nitride, have gained attention due to their robust chemical stability and visible-light-driven photocatalytic activity [106]. These advancements in carbon-based photocatalysts offer high efficiency in microplastic remediation and expand applicability across a broader range of environmental conditions, paving the way for more sustainable and effective strategies to remediate microplastics.

9.3.1.3 Metal-Organic Frameworks (MOFs)

MOFs are a family of materials consisting of organic ligands coupled with metal ions or clusters to form porous frameworks with tunable chemical properties. MOFs are a new study area in photocatalysis and have great potential [99, 107–110]. These materials are appealing for photocatalytic applications beyond their usual role due to their massive surface areas, variable frameworks, and diverse functions. Recent studies have demonstrated that MOFs can adsorb and degrade microplastics, opening the door to developing tailored catalysts with high degradation efficiency [110].

9.3.1.4 Composite or Hybrid Materials

Composite or hybrid materials for microplastic degradation are novel combinations of components that work together synergistically and strategically to improve photocatalytic activity. Utilizing the complementary effects of several remediation techniques, these new hybrid methods open the possibility of more effective and adaptable microplastic remediation methods [107, 111, 112]. Metal oxides can be combined with carbon-based materials or noble metal nanoparticles integrated into semiconductor matrices to enhance charge separation, increase light absorption range, and increase catalytic activity for microplastic degradation [113, 114]. Since hybrid materials are flexible and tunable, researchers can tailor the catalyst's properties to suit various remediation applications.

9.3.2 Properties and Characteristics of Effective Photocatalysts

Effective photocatalysts for microplastic degradation possess several key characteristics essential for their performance. Firstly, the bandgap energy of a semiconductor material determines its ability to absorb light and initiate photocatalytic reactions by generating electron-hole pairs. Materials with bandgap energies that align well with the power of incident light photons exhibit higher activity. Adjusting the bandgap energy through various material engineering techniques and doping strategies can optimize the catalyst's efficiency, specifically for breaking down microplastics [101, 114]. Secondly, the surface area determines the effectiveness of photocatalysts. A higher surface area facilitates the adsorption of microplastics onto the catalyst surface, increasing the chances of interaction between the pollutants and the reactive species generated during photocatalysis [115]. Nanostructured and porous materials offer large surface areas and improve access to active sites, thereby enhancing degradation efficiency. Moreover, the dynamics of charge carriers also determine photocatalytic performance [80, 108]. Efficient separation and transport of charges are necessary to sustain photocatalytic activity and prevent charge recombination, which can hinder degradation efficiency. Materials with well-defined crystal structures, optimized defect densities, and suitable charge carrier mobility demonstrate enhanced performance in degrading microplastics. Moreover, catalyst stability is required for long-term effectiveness in challenging environmental conditions. Materials resistant to photo-corrosion, oxidation, and leaching ensure sustained activity and minimal degradation over extended periods of operation. In addition, the redox potential of semiconductor materials determines the generation and the reactivity of reactive oxygen species (ROS), which are essential for microplastic degradation. Materials with fitting redox potential facilitate ROS generation, leading to the oxidation of microplastics and enhancing the degradation rate [52].

9.3.3 RECENT ADVANCEMENTS IN MATERIAL DEVELOPMENT FOR MICROPLASTIC DEGRADATION

It is now possible to precisely control catalyst morphology, particle size, and surface properties through template-assisted fabrication, hydrothermal treatment, and sol-gel synthesis methods. These are all advanced techniques. Higher performance is currently being attained in the decomposition of microplastics by developing nanostructured photocatalysts with improved light absorption and charge separation.

9.3.3.1 Nanostructuring and Morphology Control

Nanostructuring techniques, such as hydrothermal treatment, sol-gel synthesis, and template-assisted fabrication, allow for precise control over catalyst morphology, particle size, and surface properties. Nanostructured photocatalysts exhibit improved light absorption, charge separation, and catalytic activity, enhancing microplastic degradation performance [116].

9.3.3.2 Doping and Surface Modification

Semiconductor materials' electrical structure, bandgap energy, and surface chemistry are altered by adding hetero atoms or functional groups. The photocatalytic activity and selectivity are increased by this change. Various dopants, such as metals, nonmetals, and heteroatoms, change the dynamics of charge carriers, introduce defect states, and speed up critical redox events that lead to the disintegration of microplastics. By directing reaction pathways and enhancing pollutant adsorption, surface modification

Photocatalytic Degradation and Remediation

techniques, including co-catalyst deposition and chemical functionalization, further improve catalyst performance [114].

9.3.3.3 Hybrid and Composite Materials

The degradation of microplastics through photocatalysis is being revolutionized by combining various components to create hybrid and composite photocatalysts [111]. Through the integration of carbon-based materials, heterojunction structures, or noble metal nanoparticles into semiconductor matrices [111], researchers are enhancing charge separation, expanding the range of light absorption, and fortifying catalytic activity in the presence of both visible and ultraviolet light [114]. These adaptable hybrids mark a new chapter in photocatalyst design for microplastic remediation and promise increased degrading efficacy [117].

In conclusion, developing sophisticated photocatalytic materials greatly enhances the advancement of environmentally sustainable practices and the fight against microplastic pollution [118]. We are on the verge of developing highly efficient, selective photocatalysts capable of disassembling microplastics in various environmental situations using novel synthesis methods, tailored catalyst characteristics, and hybrid and composite methodologies. Realizing the full potential of photocatalytic materials to address the global microplastic widespread requires ongoing study into degradation mechanisms, catalyst optimization, and practical obstacles.

9.4 EXPERIMENTAL TECHNIQUES FOR ASSESSING PHOTOCATALYTIC DEGRADATION

Experimental evaluation of photocatalytic degradation is necessary to analyze the efficiency, mechanisms, and practical aspects of photocatalytic materials in tackling microplastic pollution. This section gives an overview of typical experimental setups and procedures, discusses analytical methods for determining degradation efficiency, and analyzes limitations and challenges in experimental assessment.

9.4.1 OVERVIEW OF EXPERIMENTAL SETUPS AND METHODOLOGIES

Various experimental setups and methodologies are employed to analyze and optimize photocatalytic degradation processes for microplastic remediation [45, 119]. These methodologies are key for laying down reaction processes, calculating degradation effectiveness, and analyzing photocatalyst performance. Here, we briefly present an overview of common experimental settings and analytical approaches in this field.

9.4.1.1 Batch Reactors

A batch reactor can measure photocatalytic activity in a controlled setting. Reaction vessels with stirring mechanisms are usually used in these setups to ensure that reactants and photocatalysts are mixed uniformly [101, 120]. Using batch tests, it is possible to methodically examine how different elements like temperature, pH, light intensity, and catalyst amount affect the degradation rate [121]. This makes it possible to fully comprehend the fundamental kinetics and mechanisms controlling the degradation process.

9.4.1.2 Flow Reactors

Unlike batch systems, flow reactors closely resemble real-world scenarios by simulating continuous flow conditions [122]. With these reactors, it is possible to precisely regulate reaction parameters while scaling up photocatalytic processes for real-world applications. Monitoring reactant concentrations and reaction kinetics in flow reactors allows for thoroughly examining degradation rate and reaction dynamics.

9.4.1.3 Photoreactor Design

In photocatalytic degradation processes, photoreactor design optimizes mass transport, light use, and reaction kinetics. Slurry, fixed-bed, and immobilized catalyst reactors are examples of common photoreactor designs [123]. To maximize performance and increase degrading efficiency, reactor shape, catalyst loading, and light distribution must all be optimized [124].

9.4.1.4 Sample Preparation

Adequate sample preparation is key for ensuring the accuracy and reproducibility of experimental evaluations. Important steps in the experimental setup include homogeneous coating or catalysts, homogeneous dispersion of microplastic samples, and cautious regulation of initial pollutant concentrations. Sample handling techniques can help limit experimental biases and enable confident assessments of results [125].

9.4.2 ANALYTICAL TECHNIQUES FOR ASSESSING DEGRADATION EFFICIENCY

Degradation mechanisms during photocatalytic processes are understood, and microplastic degradation is quantified through analytical methods. Various analytical techniques are used to determine degradation intermediates, track variations in pollutant concentrations, and assess photocatalyst activity [45]. Below are some prevalent analytical procedures currently being used in this context:

9.4.2.1 Chromatographic Techniques

While mass spectroscopy and liquid chromatography were typically used, high-resolution mass and gas chromatography are currently utilized as advanced techniques. Most importantly, these combined techniques have become a potent analytical tool for degradation products and intermediates. They make it possible to identify and measure organic molecules, which clarify the processes and mechanisms involved in degradation [126].

9.4.2.2 Spectroscopic Techniques

When it comes to chemical characterization, all the information about functional groups, chemical composition, excitation bands, concentrations, absorbances, and surface becomes important. Fourier-transform infrared spectroscopy (FTIR),

Photocatalytic Degradation and Remediation

fluorescence spectroscopy, and UV Vis spectroscopy are used for this purpose [127, 128]. UV Vis spectroscopy is frequently used to measure changes in absorbance or intensity of distinctive peaks and concentrations, which allows the assessment of degradation.

9.4.2.2 Electrochemical Techniques

Gaining insights into photocatalysts' electron transport characteristics, surface reactivity, and charge transfer mechanisms is important. Techniques such as electrochemical impedance spectroscopy (EIS) and cyclic voltammetry (CV) solve this problem [45, 129]. These analyses give us information on catalyst performance and reaction kinetics and supplement conventional spectroscopic and chromatographic approaches [130].

9.4.2.3 Microscopic Techniques

Transmission electron microscopy (TEM), atomic force microscopy (AFM), and scanning electron microscopy (SEM) allow visualization of catalyst morphology, particle size distribution, and microplastic interactions. Microscopic imaging provides important structural and morphological information that helps characterize catalysts and clarify degradation pathways [131–133].

In summary, integrating several experimental settings and analytical methodologies makes a thorough understanding of photocatalytic degradation processes for microplastic remediation possible. Researchers can solve the urgent environmental problem caused by microplastic pollution by utilizing these approaches to create effective and long-lasting solutions.

9.4.3 Challenges and Limitations in Experimental Assessment

Despite the advancement in assessing photocatalytic degradation as a solution to microplastic pollution, a series of challenges and limitations within experimental methodologies are still encountered. [43, 56] Even with the strides made and the adaptability of available techniques, several key hurdles persist, demanding careful consideration to ensure the accuracy and reliability of data interpretation. At the forefront of these challenges lies the complexity of environmental matrices. Sediments, natural water systems, and soil-—typical microplastic contamination environments present intricate organic and inorganic constituent mixtures. These complex matrices present a challenge as they can interfere with photocatalytic reactions, complicating the interpretation of results and undermining the credibility of assessments. To deal with these issues, standardizing experimental conditions and considering matrix effects are unavoidable measures to ensure the validity of evaluations of photocatalytic degradation efficiency. To maximize photocatalytic degradation, it becomes key to understand interfacial phenomena fully. Degradation efficiency is impacted by mass transfer restrictions, catalyst and pollutant interactions, and surface blocking. Progress has been made, but the absence of standardized techniques makes it difficult to measure microplastic deterioration precisely. A key component in evaluating photocatalytic processes is the development of reliable quantification methods and degradation standards. There are challenges in transferring lab-scale research to practical applications, such as reactor design and cost-effectiveness. Unlocking the full potential of photocatalytic remediation technologies in addressing the global challenge of microplastic contamination will require ongoing research endeavors to standardize experimental protocols, optimize catalyst performance, and overcome practical constraints.

9.5 ENVIRONMENTAL APPLICATIONS AND IMPLICATIONS

Using light and semiconductor materials, photocatalytic degradation offers a practical method of breaking down microplastics in various environmental arrays [42]. Photocatalytic remediation's real-world applications, environmental impact, and pre-considerations before widespread implementation are presented and summarized here. It is necessary to debate the possible benefits and challenges of photocatalytic degradation to clarify how it can help meet the urgent need for effective microplastic cleaning-up methods.

9.5.1 POTENTIAL APPLICATIONS AND BENEFITS

The versatility of photocatalytic degradation goes across terrestrial, freshwater, and marine environments, presenting multifaceted solutions to microplastic contamination. A few applications are discussed here:

9.5.1.1 Water Treatment

Using various light sources to activate a catalyst and start microplastic degradation processes can potentially clean water. This attempt will help reduce the harmful impact of microplastics on the ecosystem and remove them from water sources in an environmentally friendly way. The capacity of the current infrastructure can be increased by implementing those methods. These facilities' strategies will be cost-effective and appropriate for centralizing the applications[73, 134].

9.5.1.2 Soil Remediation

Photocatalytic materials can be applied to remediate microplastic soils, particularly in urban and heavy industrial areas where plastic debris accumulates. Soil amendments containing photocatalysts can facilitate the degradation of microplastics, enhancing soil quality and reducing the risk of plastic contamination to plants and organisms [135–137].

9.5.1.3 Air Purification

Photocatalytic coatings and materials can be integrated into building surfaces, road pavements, and urban infrastructure to degrade airborne microplastics and pollutants. By harnessing sunlight or artificial light sources, photocatalysts can catalyze the oxidation of microplastics, volatile organic compounds (VOCs), and other airborne contaminants, improving air quality, reducing human exposure to harmful pollutants, and preventing micropollutant-generated health issues [138, 139].

Photocatalytic Degradation and Remediation

9.5.1.4 Marine Remediation

Photocatalytic materials can be deployed in marine environments to remediate coastal areas, harbors, and aquatic ecosystems affected by microplastic pollution [55]. Floating photocatalytic devices or coatings on marine structures can facilitate the degradation of surface-bound microplastics, preventing their accumulation and dispersal in the aquatic food web [32, 38, 140–142].

9.5.2 ENVIRONMENTAL CONSIDERATIONS AND CHALLENGES

Despite its potential, photocatalytic remediation also presents several drawbacks and challenges, among which the first is limited light penetration: Photocatalytic degradation is limited by the penetration depth of light into environmental matrices, particularly in turbid waters, dense soils, or deep sediments where light attenuation is significant [143, 144]. This limitation restricts the applicability of photocatalytic processes to surface-bound microplastics and shallow environments. A further challenge is their Long-Term Stability. The prolonged stability and durability of photocatalytic materials under environmental conditions remain a concern, as catalyst deactivation, fouling, and degradation may compromise performance over time. Developing robust and durable photocatalysts capable of withstanding harsh environmental conditions is essential for ensuring the longevity of photocatalytic remediation systems. Further risks are associated with it, such as the risk of byproduct formation. Photocatalytic degradation may generate harmful side products or intermediates, particularly under certain reaction conditions or in complex pollutant mixtures. After photocatalytic remediation, the environmental risks must be evaluated, and suitable mitigation measures must be implemented. This requires realizing the route and toxicity of the degradation byproducts [144].

9.5.3 CONSIDERATIONS FOR LARGE-SCALE IMPLEMENTATION

The mounting challenge of microplastic pollution necessitates effective treatments and scalable solutions for global implementation [142, 145, 146]. Achieving widespread adoption of photocatalytic remediation demands meticulous consideration of various factors, with cost emerging as a primary concern. Assessing the cost-effectiveness of photocatalytic processes relative to conventional methods is crucial for gauging their feasibility and scalability. Rigorous cost-benefit analyses should encompass capital investments, operational expenses, maintenance costs, and long-term environmental benefits to provide accurate economic evaluations. Regulatory compliance is critical when implemented on a wide scale. To gain approval and promote acceptance of photocatalytic remediation projects, compliance with rules about environmental remediation, water quality, and public health is critical. Encouraging public trust and regulatory approval can only be gained by demonstrating effectiveness, safety, and ecological sustainability through rigorous testing and monitoring processes [147]. Community engagement is something that should be carefully considered and actively participated in. Gaining support and promoting inclusion contingent upon including stakeholders, local communities, and policymakers in developing, implementing, and supervising photocatalytic remediation programs. Clear and open communication, stakeholder engagement, and focused public education are essential for resolving issues and fostering group commitment to joint environmental goals. By prioritizing these factors, it becomes easier for photocatalytic remediation techniques to be successfully implemented on a big scale, advancing the efforts as a group to tackle the global problem of microplastic pollution.

9.6 FUTURE PERSPECTIVES AND CHALLENGES

As we look ahead, the field of photocatalytic microplastic remediation is poised for significant advancements and innovation. Emerging trends, future directions, and the key challenges that need to be addressed to realize the full potential of this technology all together need to be explored [148, 149].

9.6.1 EMERGING TRENDS AND FUTURE DIRECTIONS

The future of photocatalytic microplastic remediation is shaped by many new developments and research directions. While these developments hold enormous promise, their optimal and efficient utilization necessitates further exploration, advancement, and investigation within the field.

Advanced Photocatalytic Materials: As evident from the literature and recent publications, the primary thrust of current research and development efforts is on synthesizing advanced photocatalytic materials that are more selective, stable, targeted, tunable, and efficient in the degradation of microplastics [119]. These materials are being designed and synthesized. Current focus areas of research include nanomaterials, metal-organic frameworks (MOFs), and composite materials because of their distinctive characteristics and the potential benefits of working in tandem in photocatalysis [91–99].

9.6.1.1 Tailored Photocatalytic Systems

Customized photocatalytic systems and reactor designs are being developed to optimize the performance and applicability of photocatalytic microplastic remediation in diverse environmental settings [150]. Combining photocatalysis with other treatment technologies, such as filtration, adsorption, and biological degradation, is being explored to achieve comprehensive and synergistic pollutant removal.

9.6.1.2 Solar-Based Photocatalysis

With the increasing emphasis on renewable energy and sustainability, solar-driven photocatalytic systems are gaining traction as an eco-friendly and cost-effective approach for microplastic remediation [151]. Advances in solar photoreactors, light-harvesting materials, and energy storage technologies are enabling the harnessing of solar energy for photocatalytic degradation on a larger scale.

9.6.2 Technological Challenges and Research Gaps

Despite the promising outlook, several technological challenges and research gaps need to be addressed to overcome barriers to photocatalytic microplastic remediation: First, efficiency and selectivity are major challenges that need to be met. Improving the efficiency and selectivity of photocatalytic degradation towards different microplastic types and sizes remains a crucial challenge [152]. Achieving optimal degradation rates while minimizing the formation of undesirable byproducts/side products demands advancements in photocatalyst activity, fine-tuning reaction conditions, and a comprehensive understanding of underlying degradation mechanisms [153]. In this fast-paced era, a critical challenge is ensuring the environmental compatibility and safety of photocatalytic materials and their resultant byproducts. Safeguarding against unintended ecological impacts necessitates a rigorous assessment of the fate, transport, and toxicity degradation products. Moreover, evaluating potential risks to nontarget organisms and ecosystems is paramount for the sustainable deployment of photocatalytic remediation strategies [154]. On a practical thought, scaling up photocatalytic remediation technologies from laboratory-scale experiments to practical field (bulk level) applications presents formidable engineering and operational hurdles. Resolving challenges related to reactor design, efficient catalyst immobilization, uniform light distribution, and sustainable system maintenance is imperative for realizing practical and cost-effective solutions. While addressing these technological challenges and research gaps, interdisciplinary collaborations (a key to innovation) and innovative approaches are indispensable. By advancing our understanding of photocatalytic processes, optimizing material properties, and developing robust engineering solutions, researchers can pave the way for the effective implementation of photocatalytic microplastic remediation on a global scale [155].

9.6.3 Integration with Other Remediation Strategies

Combining photocatalysis with complementing remediation techniques creates prospects for synergistic approaches to complex microplastic pollution treatment. To illustrate, the efficiency of microplastic remediation is increased when photocatalysis is combined with biological breakdown, such as the use of enzymes or bacteria [73]. They are picturing it as a collaborative effort where biological degradation uses microbial communities and natural pathways to accelerate the breakdown of microplastics while photocatalysis breaks them down. Likewise, concentrating, or precipitating microplastics before their degradation increases the effectiveness of the degradation process when photocatalysis is combined with physical separation techniques like membrane filtration or electrostatic precipitation. It is worth mentioning that it is not just about techniques; collaboration between scientists, engineers, policymakers, and stakeholders is vital [32]. Together, it is possible to overcome the intricate problems of environmental microplastic pollution, advance research, and translate findings into workable solutions.

9.7 CONCLUSION

The rapid proliferation of microplastics in terrestrial, freshwater, and marine environments poses significant challenges to ecosystem health and human well-being [119, 156]. Microplastics, which come from many sources and have different sizes, compositions, and shapes, are now common contaminants with significant ecological effects. In addition to its adverse effect on human health, the environmental impacts of microplastics are multifaceted, encompassing direct physical harm to organisms, bioaccumulation and transfer of toxic pollutants, alteration of ecosystem functions, and disruption of biogeochemical cycles. Plus, microplastics increase the dangers to human health and the environment by acting as carriers of infections and invasive species.

The complex microplastic contamination issue necessitates multidimensional strategies incorporating legislative interventions, technical advancements, and scientific knowledge. Photocatalytic degradation has emerged as a promising remediation strategy for mitigating microplastic pollution. It leverages the catalytic properties of semiconductor materials to facilitate the breakdown of plastic particles under light irradiation.

By harnessing solar energy and catalyzing oxidative reactions, photocatalysts can effectively degrade microplastics into smaller fragments and mineralize organic components, offering a sustainable and environmentally friendly solution to microplastic pollution. Solar-driven photoreactors, customized reactor designs, and advanced photocatalytic materials fuel innovation in the field. These developments are making it possible to develop remediation solutions that are both sustainable, scalable, and affordable.

To fully achieve the potential of photocatalytic microplastic remediation, several obstacles and research gaps still need to be filled. Enhancing the efficiency and selectivity of photocatalytic degradation, ensuring the environmental compatibility and safety of photocatalytic materials, and scaling up photocatalytic technologies for field applications are critical priorities for future research and development. Integration with complementary remediation strategies, such as biological degradation and physical separation techniques, offers synergistic opportunities for enhancing the overall effectiveness of microplastic remediation efforts. Collaboration among scientists, engineers, and policymakers is essential for advancing the field and implementing comprehensive solutions to mitigate microplastic pollution on a global scale.

Photocatalytic microplastic remediation holds significant potential for defending ecosystems, biodiversity, and human health for future generations as a sustainable and practical solution to the ubiquitous issue of microplastic pollution.

REFERENCES

[1] Mofijur, M., Ahmed, S. F., Rahman, S. A., Siddiki, S. Y. A., Islam, A. S., Shahabuddin, M., & Show, P. L. (2021). Source, distribution and emerging threat of micro-and nanoplasticsto marine organism and human health: Socio-economic impact and management strategies. *Environmental Research*, *195*, 110857. https://doi.org/10.1016/j.envres.2021.110857

[2] Rafa, N., Ahmed, B., Zohora, F., Bakya, J., Ahmed, S., Ahmed, S. F. & Almomani, F. (2023). Microplastics as carriers of toxic pollutants: Source, transport, and toxicological effects. *Environmental Pollution*, 343, 123190. https://doi.org/10.1016/j.envpol.2023.123190

[3] Alimi, O. S., Farner Budarz, J., Hernandez, L. M., & Tufenkji, N. (2018). Microplastics and nanoplastics in aquatic environments: Aggregation, deposition, and enhanced contaminant transport. *Environmental Science & Technology*, 52(4), 1704–1724. https://doi.org/10.1021/acs.est.7b05559

[4] Chellasamy, G., Kiriyanthan, R. M., Maharajan, T., Radha, A., & Yun, K. (2022). Remediation of microplastics using bionanomaterials: A review. *Environmental Research*, 208, 112724. https://doi.org/10.1016/j.envres.2022.112724

[5] Jadoun, S., Fuentes, J. P., Yepsen, O., & Yáñez, J. (2023). Removal of Environmental Microplastics by Advanced Oxidation Processes. In *Microplastic Occurrence, Fate, Impact, and Remediation* (pp. 109–125). Cham: Springer Nature Switzerland. https://link.springer.com/chapter/10.1007/978-3-031-36351-1_5

[6 Cole, M., Lindeque, P., Fileman, E., Halsband, C., & Galloway, T. S. (2015). The impact of polystyrene microplastics on feeding, function and fecundity in the marine copepod Calanus helgolandicus. *Environmental Science & Technology*, 49(2), 1130–1137. https://doi.org/10.1021/es504525u

[7] Hu, K., Tian, W., Yang, Y., Nie, G., Zhou, P., Wang, Y., ... & Wang, S. (2021). Microplastics remediation in aqueous systems: Strategies and technologies. *Water Research*, 198, 117144. https://doi.org/10.1016/j.watres.2021.117144

[8] Devi, A., Hansa, A., Gupta, H., Syam, K., Upadhyay, M., Kaur, M., ... & Sharma, R. (2023). Microplasticsas an emerging menace to environment: Insights into their uptake,prevalence, fate, and sustainable solutions. *Environmental Research*, 229, 115922. https://doi.org/10.1016/j.envres.2023.115922

[9] Meng, Y., Kelly, F. J., & Wright, S. L. (2020). Advances and challenges of microplastic pollution in freshwater ecosystems: A UK perspective. *Environmental Pollution*, 256, 113445. https://doi.org/10.1016/j.envpol.2019.113445

[10] Cole, M., Lindeque, P., Halsband, C., & Galloway, T. S. (2011). Microplastics as contaminants in the marine environment: A review. *Marine Pollution Bulletin*, 62(12), 2588–2597. https://doi.org/10.1016/j.marpolbul.2011.09.025

[11] Wang, C., Zhao, J., & Xing, B. (2021). Environmental source, fate, and toxicity of microplastics. *Journal of Hazardous Materials*, 407, 124357. https://doi.org/10.1016/j.jhazmat.2020.124357

[12] Wang, F., Zhang, M., Sha, W., Wang, Y., Hao, H., Dou, Y., & Li, Y. (2020). Sorption behavior and mechanisms of organic contaminants to nano and microplastics. *Molecules*, 25(8), 1827. https://doi.org/10.3390/molecules25081827

[13] Wang, F., Wong, C. S., Chen, D., Lu, X., Wang, F., & Zeng, E. Y. (2018). Interaction of toxic chemicals with microplastics: A critical review. *Water Research*, 139, 208–219. https://doi.org/10.1016/j.watres.2018.04.003

[14] Wang, J., Liu, X., Li, Y., Powell, T., Wang, X., Wang, G., & Zhang, P. (2019). Microplastics as contaminants in the soil environment: A mini-review. *Science of the Total Environment*, 691, 848–857. https://doi.org/10.1016/j.scitotenv.2019.07.209

[15] Wang, J., Liu, X., Liu, G., Zhang, Z., Wu, H., Cui, B., ... & Zhang, W. (2019). Size effect of polystyrene microplastics on sorption of phenanthrene and nitrobenzene. *Ecotoxicologyand Environmental Safety*, 173, 331–338. https://doi.org/10.1016/j.ecoenv.2019.02.037

[16] Wang, Q., Zhang, Y., Wangjin, X., Wang, Y., Meng, G., & Chen, Y. (2020). The adsorption behavior of metals in aqueous solution by microplastics effected by UV

radiation. *Journal of Environmental Sciences, 87*, 272–280. https://doi.org/10.1016/j.jes.2019.07.006

[17] Wang, T., Hu, M., Song, L., Yu, J., Liu, R., Wang, S., ... & Wang, Y. (2020). Coastal zone use influences the spatial distribution of microplastics in Hangzhou Bay, China. *Environmental Pollution, 266*, 115137. https://doi.org/10.1016/j.envpol.2020.115137

[18] Wang, Y., Yang, Y., Liu, X., Zhao, J., Liu, R., & Xing, B. (2021). Interaction of microplastics with antibiotics in aquatic environment: Distribution, adsorption, and toxicity. *Environmental Science & Technology, 55*(23), 15579–15595. https://doi.org/10.1021/acs.est.1c04509

[19] Hartmann, N. B., Rist, S., Bodin, J., Jensen, L. H., Schmidt, S. N., Mayer, P., ... & Baun, A. (2017). Microplastics as vectors for environmental contaminants: Exploring sorption, desorption, and transfer to biota. *Integrated Environmental Assessment and Management, 13*(3), 488–493. https://doi.org/10.1002/ieam.1904

[20] Fu, L., Li, J., Wang, G., Luan, Y., & Dai, W. (2021). Adsorption behavior of organic pollutants on microplastics. *Ecotoxicology and Environmental Safety, 217*, 112207. https://doi.org/10.1016/j.ecoenv.2021.112207

[21] Tang, K. H. D. (2021). Interactions of microplastics with persistent organic pollutants and the ecotoxicological effects: A review. *Tropical Aquatic and Soil Pollution, 1*(1), 24–34. https://doi.org/10.53623/tasp.v1i1.11

[22] Müller, A., Becker, R., Dorgerloh, U., Simon, F. G., & Braun, U. (2018). The effect of polymer aging on the uptake of fuel aromatics and ethers by microplastics. *Environmental Pollution, 240*, 639–646. https://doi.org/10.1016/j.envpol.2018.04.127

[23] Hüffer, T., & Hofmann, T. (2016). Sorption of non-polar organic compounds by micro-sized plastic particles in aqueous solution. *Environmental Pollution, 214*, 194–201. https://doi.org/10.1016/j.envpol.2016.04.018

[24] Miao, L., Wang, P., Hou, J., Yao, Y., Liu, Z., Liu, S., & Li, T. (2019). Distinct community structure and microbial functions of biofilms colonizing microplastics. *Science of the Total Environment, 650*, 2395–2402. https://doi.org/10.1016/j.scitotenv.2018.09.378

[25] Li, J., Zhang, K., & Zhang, H. (2018). Adsorption of antibiotics on microplastics. *Environmental Pollution, 237*, 460–467. https://doi.org/10.1016/j.envpol.2018.02.050

[26] Dai, Y., Zhao, J., Sun, C., Li, D., Liu, X., Wang, Z., ... & Xing, B. (2022). Interaction and combined toxicity of microplastics and per-and polyfluoroalkyl substances in aquatic environment. *Frontiers of Environmental Science & Engineering, 16*(10), 136. https://doi.org/10.1007/s11783-022-1571-2

[27] Liu, W., Pan, T., Liu, H., Jiang, M., & Zhang, T. (2023). Adsorption behavior of imidacloprid pesticide on polar microplastics under environmental conditions: Critical role of photo-aging. *Frontiers of Environmental Science & Engineering, 17*(4), 41. https://doi.org/10.1007/s11783-023-1641-0

[28] Torres, F. G., Dioses-Salinas, D. C., Pizarro-Ortega, C. I., & De-la-Torre, G. E. (2021). Sorption of chemical contaminants on degradable and non-degradable microplastics: Recent progress and research trends. *Science of the Total Environment, 757*, 143875. https://doi.org/10.1016/j.scitotenv.2020.143875

[29] Gasperi, J., Wright, S. L., Dris, R., Collard, F., Mandin, C., Guerrouache, M., ... & Tassin, B. (2018). Microplastics in air: Are we breathing it in?. *Current Opinion in Environmental Science & Health, 1*, 1–5. https://doi.org/10.1016/j.coesh.2017.10.002

[30] Fournier, S. B., D'Errico, J. N., Adler, D. S., Kollontzi, S., Goedken, M. J., Fabris, L., ... & Stapleton, P. A. (2020). Nanopolystyrene translocation and fetal deposition after acute lung exposure during late-stage pregnancy. *Particle and Fibre Toxicology, 17*, 1–11. https://doi.org/10.1186/s12989-020-00385-9

Photocatalytic Degradation and Remediation 191

[31] Huang, S., Huang, X., Bi, R., Guo, Q., Yu, X., Zeng, Q., ... & Guo, P. (2022). Detection and analysis of microplastics in human sputum. *Environmental Science & Technology*, *56*(4), 2476–2486. https://doi.org/10.1021/acs.est.1c03859

[32] Jaiswal, K. K., Dutta, S., Banerjee, I., Pohrmen, C. B., Singh, R. K., Das, H. T., ... & Kumar, V. (2022). Impact of aquatic microplastics and nanoplastics pollution on ecological systems and sustainable remediation strategies of biodegradation and photodegradation. *Science of the Total Environment*, *806*, 151358. https://doi.org/10.1016/j.scitotenv.2021.151358

[33] Huang, Z., Weng, Y., Shen, Q., Zhao, Y., & Jin, Y. (2021). Microplastic: A potential threat to human and animal health by interfering with the intestinal barrier function and changing the intestinal microenvironment. *Science of the Total Environment*, *785*, 147365. https://doi.org/10.1016/j.scitotenv.2021.147365

[34] Hu, X., Biswas, A., Sharma, A., Sarkodie, H., Tran, I., Pal, I., & De, S. (2021). Mutational signatures associated with exposure to carcinogenic microplastic compounds bisphenol A and styrene oxide. *NAR Cancer*, *3*(1), zcab004. https://doi.org/10.1093/narcan/zcab004

[35] Tamargo, A., Molinero, N., Reinosa, J. J., Alcolea-Rodriguez, V., Portela, R., Bañares, M. A., ... & Moreno-Arribas, M. V. (2022). PET microplastics affect human gut microbiota communities during simulated gastrointestinal digestion, first evidence of plausible polymer biodegradation during human digestion. *Scientific Reports*, *12*(1), 528. https://doi.org/10.1038/s41598-021-04489-w

[36] Osman, A. I., Hosny, M., Eltaweil, A. S., Omar, S., Elgarahy, A. M., Farghali, M., ... & Akinyede, K. A. (2023). Microplastic sources, formation, toxicity and remediation: A review. *Environmental Chemistry Letters*, *21*(4), 2129–2169. https://doi.org/10.1007/s10311-023-01593-3

[37] Ahmed, R., Hamid, A. K., Krebsbach, S. A., He, J., & Wang, D. (2022). Critical review of microplastics removal from the environment. *Chemosphere*, *293*, 133557. https://doi.org/10.1016/j.chemosphere.2022.133557

[38] Xu, Q., Huang, Q. S., Luo, T. Y., Wu, R. L., Wei, W., & Ni, B. J. (2021). Coagulation removal and photocatalytic degradation of microplastics in urban waters. *Chemical Engineering Journal*, *416*, 129123. https://doi.org/10.1016/j.cej.2021.129123

[39] Lee, Q. Y., & Li, H. (2021). Photocatalytic degradation of plastic waste: A mini review. *Micromachines*, *12*(8), 907. https://doi.org/10.3390/mi12080907

[40] Yuwendi, Y., Ibadurrohman, M., Setiadi, S., & Slamet, S. (2022). Photocatalytic degradation of polyethylene microplastics and disinfection of E. coli in water over Fe-and Ag-modified TiO2 nanotubes. *Bulletin of Chemical Reaction Engineering & Catalysis*, *17*(2), 263–277. https://doi.org/10.9767/bcrec.17.2.13400.263-277

[41] Bratovcic, A. (2019). Degradation of micro-and nano-plastics by photocatalytic methods. *Journal of Nanoscience and Nanotechnology Applications*, *3*, 206.

[42] Parveen, K., Parvaiz, A., Subhan, M., & Khan, A. (2024). Photocatalytic Degradation of Microplastic in the Environment. In *Microplastic Pollution* (pp. 419–429). Singapore: Springer Nature Singapore.

[43] He, J., Han, L., Wang, F., Ma, C., Cai, Y., Ma, W., ... & Yang, Z. (2023). Photocatalytic strategy to mitigate microplastic pollution in aquatic environments: Promising catalysts, efficiencies, mechanisms, and ecological risks. *Critical Reviews in Environmental Science and Technology*, *53*(4), 504–526. https://doi.org/10.1080/10643389.2022.2072658

[44] Cai, M., Wei, Y., Li, Y., Li, X., Wang, S., Shao, G., & Zhang, P. (2023). 2D semiconductor nanosheets for solar photocatalysis. *EcoEnergy*, *1*(2), 248–295. https://doi.org/10.1002/ece2.16

[45] Zhang, J., Tian, B., Wang, L., Xing, M., & Lei, J. (2018). Photocatalysis. *Lecture Notes in Chemistry*, Springer, 1–17.

[46] Yang, X., & Wang, D. (2018). Photocatalysis: From fundamental principles to materials and applications. *ACS Applied Energy Materials*, *1*(12), 6657–6693. https://doi.org/10.1021/acsaem.8b01345

[47] Kisch, H. (2013). Semiconductor photocatalysis—Mechanistic and synthetic aspects. *Angewandte Chemie International Edition*, *52*(3), 812–847. https://doi.org/10.1002/anie.201201200

[48] Wenderich, K., & Mul, G. (2016). Methods, mechanism, and applications of photodeposition in photocatalysis: A review. *Chemical Reviews*, *116*(23), 14587–14619. https://doi.org/10.1021/acs.chemrev.6b00327

[49] Saravanan, R., Gracia, F., & Stephen, A. (2017). Basic principles, mechanism, and challenges of photocatalysis. *Nanocomposites for Visible Light-Induced Photocatalysis*, 19–40.

[50] Ma, J., Miao, T. J., & Tang, J. (2022). Charge carrier dynamics and reaction intermediates in heterogeneous photocatalysis by time-resolved spectroscopies. *Chemical Society Reviews*, *51*(14), 5777–5794. https://doi.org/10.1039/D1CS01164B

[51] Vinu, R., & Madras, G. (2011). Photocatalytic Degradation of Water Pollutants Using Nano-TiO2. In *Energy Efficiency and Renewable Energy through Nanotechnology* (pp. 625–677). London: Springer London.

[52] Chokejaroenrat, C., Watcharatharapong, T., T-Thienprasert, J., Angkaew, A., Poompoung, T., Chinwong, C., ... & Sakulthaew, C. (2024). Decomposition of microplastics using copper oxide/bismuth vanadate-based photocatalysts: Insight mechanisms and environmental impacts. *Marine Pollution Bulletin*, *201*, 116205. https://doi.org/10.1016/j.marpolbul.2024.116205

[53] Sun, A., & Wang, W. X. (2023). Photodegradation of microplastics by ZnO nanoparticles with resulting cellular and subcellular responses. *Environmental Science & Technology*, *57*(21), 8118–8129. https://doi.org/10.1021/acs.est.3c01307

[54] Jiang, R., Lu, G., Yan, Z., Liu, J., Wu, D., & Wang, Y. (2021). Microplastic degradation by hydroxy-rich bismuth oxychloride. *Journal of Hazardous Materials*, *405*, 124247. https://doi.org/10.1016/j.jhazmat.2020.124247

[55] Jeyaraj, J., Baskaralingam, V., Stalin, T., & Muthuvel, I. (2023). Mechanistic vision on polypropylene microplastics degradation by solar radiation using TiO2 nanoparticle as photocatalyst. *Environmental Research*, *233*, 116366. https://doi.org/10.1016/j.envres.2023.116366

[56] Zoppas, F. M., Sacco, N., Soffietti, J., Devard, A., Akhter, F., & Marchesini, F. A. (2023). Catalytic approaches for the removal of microplastics from water: Recent advances and future opportunities. *Chemical Engineering Journal Advances*, 100529. https://doi.org/10.1016/j.ceja.2023.100529

[57] Nakata, K., & Fujishima, A. (2012). TiO2 photocatalysis: Design and applications. *Journal of Photochemistry and Photobiology C: Photochemistry Reviews*, *13*(3), 169–189. https://doi.org/10.1016/j.jphotochemrev.2012.06.001

[58] Tofa, T. S., Kunjali, K. L., Paul, S., & Dutta, J. (2019). Visible light photocatalytic degradation of microplastic residues with zinc oxide nanorods. *Environmental Chemistry Letters*, *17*, 1341–1346. https://doi.org/10.1007/s10311-019-00859-z

[59] Uheida, A., Mejía, H. G., Abdel-Rehim, M., Hamd, W., & Dutta, J. (2021). Visible light photocatalytic degradation of polypropylene microplastics in a continuous water flow system. *Journal of Hazardous Materials*, *406*, 124299. https://doi.org/10.1016/j.jhazmat.2020.124299

[60] Du, H., Xie, Y., & Wang, J. (2021). Microplastic degradation methods and corresponding degradation mechanism: Research status and future perspectives. *Journal of Hazardous Materials*, *418*, 126377. https://doi.org/10.1016/j.jhazmat.2021.126377

[61] Serpone, N. (2000). Photocatalysis. In Kirk-Othmer Encyclopedia of Chemical Technology. Hoboken, New Jersey: John Wiley and Sons.

[62] Fox, M. A., & Dulay, M. T. (1993). Heterogeneous photocatalysis. *Chemical Reviews*, *93*(1), 341–357.

[63] Kim, S., Sin, A., Nam, H., Park, Y., Lee, H., & Han, C. (2022). Advanced oxidation processes for microplastics degradation: A recent trend. *Chemical Engineering Journal Advances*, *9*, 100213. https://doi.org/10.1016/j.ceja.2021.100213

[64] Li, W., Zhao, W., Zhu, H., Li, Z. J., & Wang, W. (2023). State of the art in the photochemical degradation of (micro) plastics: From fundamental principles to catalysts and applications. *Journal of Materials Chemistry A*, *11*(6), 2503–2527. https://doi.org/10.1039/D2TA09523H

[65] Prakruthi, K., Ujwal, M. P., Yashas, S. R., Mahesh, B., Kumara Swamy, N., & Shivaraju, H. P. (2022). Recent advances in photocatalytic remediation of emerging organic pollutants using semiconducting metal oxides: An overview. *Environmental Science and Pollution Research*, *29*(4), 4930–4957. https://doi.org/10.1007/s11356-021-17361-1

[66] Ariza-Tarazona, M. C., Villarreal-Chiu, J. F., Barbieri, V., Siligardi, C., & Cedillo-González, E. I. (2019). New strategy for microplastic degradation: Green photocatalysis using a protein-based porous N-TiO2 semiconductor. *Ceramics International*, *45*(7), 9618–9624. https://doi.org/10.1016/j.ceramint.2018.10.208

[67] Ge, J., Zhang, Z., Ouyang, Z., Shang, M., Liu, P., Li, H., & Guo, X. (2022). Photocatalytic degradation of (micro) plastics using TiO2-based and other catalysts: Properties, influencing factor, and mechanism. *Environmental Research*, *209*, 112729. https://doi.org/10.1016/j.envres.2022.112729

[68] Ding, L., Mao, R., Ma, S., Guo, X., & Zhu, L. (2020). High temperature depended on the ageing mechanism of microplastics under different environmental conditions and its effect on the distribution of organic pollutants. *Water Research*, *174*, 115634. https://doi.org/10.1016/j.watres.2020.115634

[69] Cao, B., Wan, S., Wang, Y., Guo, H., Ou, M., & Zhong, Q. (2022). Highly-efficient visible-light-driven photocatalytic H2 evolution integrated with microplastic degradation over MXene/ZnxCd1-xS photocatalyst. *Journal of Colloid and Interface Science*, *605*, 311–319. https://doi.org/10.1016/j.jcis.2021.07.113

[70] Chattopadhyay, P., Ariza-Tarazona, M. C., Cedillo-González, E. I., Siligardi, C., & Simmchen, J. (2023). Combining photocatalytic collection and degradation of microplastics using self-asymmetric Pac-Man TiO 2. *Nanoscale*, *15*(36), 14774–14781. https://doi.org/10.1039/D3NR01512B

[71] Ariza-Tarazona, M. C., Villarreal-Chiu, J. F., Hernández-López, J. M., De la Rosa, J. R., Barbieri, V., Siligardi, C., & Cedillo-González, E. I. (2020). Microplastic pollution reduction by a carbon and nitrogen-doped TiO2: Effect of pH and temperature in the photocatalytic degradation process. *Journal of Hazardous Materials*, *395*, 122632. https://doi.org/10.1016/j.jhazmat.2020.122632

[72] Llorente-García, B. E., Hernández-López, J. M., Zaldívar-Cadena, A. A., Siligardi, C., & Cedillo-González, E. I. (2020). First insights into photocatalytic degradation of HDPE and LDPE microplastics by a mesoporous N–TiO2 coating: Effect of size and shape of microplastics. *Coatings*, *10*(7), 658. https://doi.org/10.3390/coatings10070658

[73] Ebrahimbabaie, P., Yousefi, K., & Pichtel, J. (2022). Photocatalytic and biological technologies for elimination of microplastics in water: Current status. *Science of the Total Environment*, *806*, 150603. https://doi.org/10.1016/j.scitotenv.2021.150603

[74] Li, S., Xu, S., He, L., Xu, F., Wang, Y., & Zhang, L. (2010). Photocatalytic degradation of polyethylene plastic with polypyrrole/TiO2 nanocomposite as photocatalyst. *Polymer-Plastics Technology and Engineering*, *49*(4), 400–406. https://doi.org/10.1080/03602550903532166

[75] Nabi, I., Li, K., Cheng, H., Wang, T., Liu, Y., Ajmal, S., ... & Zhang, L. (2020). Complete photocatalytic mineralization of microplastic on TiO2 nanoparticle film. *Iscience*, *23*(7), 101326–101338. https://doi.org/10.1016/j.isci.2020.101326

[76] Fadli, M. H., Ibadurrohman, M., & Slamet, S. (2021). Microplastic Pollutant Degradation in Water Using Modified TiO2 Photocatalyst under UV-Irradiation. In *IOP Conference Series: Materials Science and Engineering* (Vol. 1011, No. 1, p. 012055). IOP Publishing, Bristol, UK.

[77] Bandara, W. L. N., de Silva, R. M., de Silva, K. N., Dahanayake, D., Gunasekara, S., & Thanabalasingam, K. (2017). Is nano ZrO 2 a better photocatalyst than nano TiO 2 for degradation of plastics?. *RSC Advances*, *7*(73), 46155–46163. https://doi.org/10.1039/C7RA08324F

[78] An, Y., Hou, J., Liu, Z., & Peng, B. (2014). Enhanced solid-phase photocatalytic degradation of polyethylene by TiO2–MWCNTs nanocomposites. *Materials Chemistry and Physics*, *148*(1-2), 387–394. https://doi.org/10.1016/j.matchemphys.2014.08.001

[79] Kinyua, E., Nyakairu, G., Tebandeke, E., & Odume, N. (2023). Photocatalytic degradation of microplastics: Parameters affecting degradation. *Advances in Environmental and Engineering Research*, *5*(3), 1–21. http://dx.doi.org/10.21926/aeer.2303039

[80] Dhull, P., Saini, N., Aamir, M., Parveen, S., & Husain, S. (2024). Function of nanomaterials in the treatment of emerging pollutants in wastewater. *Detection and Treatment of Emerging Contaminants in Wastewater*, pp. 93–112.

[81] Wang, X., Zhu, Z., Jiang, J., Li, R., & Xiong, J. (2023). Preparation of heterojunction C3N4/WO3 photocatalyst for degradation of microplastics in water. *Chemosphere*, *337*, 139206. https://doi.org/10.1016/j.chemosphere.2023.139206

[82] Chai, C., Liang, H., Yao, R., Wang, F., Song, N., Wu, J., & Li, Y. (2023). Photocatalytic degradation of polyethylene and polystyrene microplastics by α-Fe2O3/g-C3N4. *Environmental Science and Pollution Research*, *30*(58), 121702–121712. https://doi.org/10.1007/s11356-023-31000-x

[83] Ding, L., Guo, X., Du, S., Cui, F., Zhang, Y., Liu, P., ... & Zhu, L. (2022). Insight into the photodegradation of microplastics boosted by iron (hydr) oxides. *Environmental Science & Technology*, *56*(24), 17785–17794. https://doi.org/10.1021/acs.est.2c07824

[84] Kumar, R. (2023). Metal oxides-based nano/microstructures for photodegradation of microplastics. *Advanced Sustainable Systems*, *7*(6), 2300033. https://doi.org/10.1002/adsu.202300033

[85] Razali, N. A. M., Salleh, W. N. W., Aziz, F., Jye, L. W., Yusof, N., & Ismail, A. F. (2021). Review on tungsten trioxide as a photocatalysts for degradation of recalcitrant pollutants. *Journal of Cleaner Production*, *309*, 127438. https://doi.org/10.1016/j.jclepro.2021.127438

[86] Mehmood, T., Mustafa, B., Mackenzie, K., Ali, W., Sabir, R. I., Anum, W., ... & Peng, L. (2023). Recent developments in microplastic contaminated water treatment: Progress and prospects of carbon-based two-dimensional materials for membranes separation. *Chemosphere*, *316*, 137704. https://doi.org/10.1016/j.chemosphere.2022.137704

[87] Manimegalai, S., Vickram, S., Deena, S. R., Rohini, K., Thanigaivel, S., Manikandan, S., ... & Govarthanan, M. (2023). Carbon-based nanomaterial intervention and efficient

removal of various contaminants from effluents–A review. *Chemosphere*, *312*, 137319. https://doi.org/10.1016/j.chemosphere.2022.137319

[88] Kuriakose, T., Nair, P., & Das, B. (2023). Nanotechnology for Plastic Degradation. In *Modern Nanotechnology: Volume 1: Environmental Sustainability and Remediation* (pp. 361–379). Cham: Springer International Publishing.

[89] Sajid, M., Ihsanullah, I., Khan, M. T., & Baig, N. (2023). Nanomaterials-based adsorbents for remediation of microplastics and nanoplastics in aqueous media: A review. *Separation and Purification Technology*, *305*, 122453. https://doi.org/10.1016/j.seppur.2022.122453

[90] Peng, G., Xiang, M., Wang, W., Su, Z., Liu, H., Mao, Y., ... & Zhang, P. (2022). Engineering 3D graphene-like carbon-assembled layered double oxide for efficient microplastic removal in a wide pH range. *Journal of Hazardous Materials*, *433*, 128672. https://doi.org/10.1016/j.jhazmat.2022.128672

[91] Vargas, F., Lucena-Mendoza, E., Zoltan, T., Torres, Y., León, M., Angulo, B., & Tovar, G. I. (2024). Photochemical and Sonochemical Strategies in Advanced Oxidation Processes for Micropollutants' Treatments. In *Advanced Oxidation Processes for Micropollutant Remediation* (pp. 65–98). Milton Park, Abingdon, United Kingdom: CRC Press.

[92] Fazli, A., Lauciello, S., Brescia, R., Carzino, R., Athanassiou, A., & Fragouli, D. (2024). Synergistic degradation of polystyrene nanoplastics in water: Harnessing solar and water-driven energy through a Z-scheme SnO_2/g-C3N4/PVDF-HFP piezo-photocatalytic system. *Applied Catalysis B: Environment and Energy*, *353*, 124056. https://doi.org/10.1016/j.apcatb.2024.124056

[93] Zhou, T., Song, S., Min, R., Liu, X., & Zhang, G. (2024). Advances in chemical removal and degradation technologies for microplastics in the aquatic environment: A review. *Marine Pollution Bulletin*, *201*, 116202. https://doi.org/10.1016/j.marpolbul.2024.116202

[94] John, K. I., Omorogie, M. O., Bayode, A. A., Adeleye, A. T., & Helmreich, B. (2023). Environmental microplastics and their additives—A critical review on advanced oxidative techniques for their removal. *Chemical Papers*, *77*(2), 657–676. https://doi.org/10.1007/s11696-022-02505-5

[95] Sharma, N., Kour, M., Gupta, R., & Bansal, R. K. (2021). A new cross-conjugated mesomeric betaine. *RSC Advances*, *11*(41), 25296–25304. https://doi.org/10.1039/D1RA03981D

[96] Vital-Grappin, A. D., Ariza-Tarazona, M. C., Luna-Hernández, V. M., Villarreal-Chiu, J. F., Hernández-López, J. M., Siligardi, C., & Cedillo-González, E. I. (2021). The role of the reactive species involved in the photocatalytic degradation of hdpe microplastics using c, n-tio2 powders. *Polymers*, *13*(7), 999. www.mdpi.com/2073-4360/13/7/999

[97] Paiman, S. H., Noor, S. F. M., Ngadi, N., Nordin, A. H., & Abdullah, N. (2023). Insight into photocatalysis technology as a promising approach to tackle microplastics pollution through degradation and upcycling. *Chemical Engineering Journal*, *467*, *143534*. https://doi.org/10.1016/j.cej.2023.143534

[98] Qin, J., Dou, Y., Wu, F., Yao, Y., Andersen, H. R., Hélix-Nielsen, C., ... & Zhang, W. (2022). In-situ formation of Ag2O in metal-organic framework for light-driven upcycling of microplastics coupled with hydrogen production. *Applied Catalysis B: Environmental*, *319*, 121940. https://doi.org/10.1016/j.apcatb.2022.121940

[99] Mohan, B., Singh, K., Gupta, R. K., Kumar, A., Pombeiro, A. J., & Ren, P. (2024). Water purification advances with metal–organic framework-based materials for micro/nanoplastic removal. *Separation and Purification Technology*, *343*, 126987. https://doi.org/10.1016/j.seppur.2024.126987

[100] Yang, Z., Li, Y., & Zhang, G. (2024). Degradation of microplastic in water by advanced oxidation processes. *Chemosphere*, 357, 141939. https://doi.org/10.1016/j.chemosphere.2024.141939

[101] Mukherjee, A., Bose, S., & Das, P. (2024). Efficient microplastic remediation through best possible strategies: A review. *Remediation of Plastic and Microplastic Waste*, 14, 1–32.

[102] Russo, S., Muscetta, M., Amato, P., Venezia, V., Verrillo, M., Rega, R., ... & Vitiello, G. (2024). Humic substance/metal-oxide multifunctional nanoparticles as advanced antibacterial-antimycotic agents and photocatalysts for the degradation of PLA microplastics under UVA/solar radiation. *Chemosphere*, *346*, 140605. https://doi.org/10.1016/j.chemosphere.2023.140605

[103] Tewari, C., Tatrari, G., Sahoo, N. G., & Mukhopadhyay, P. (2024). Plastics Waste to Carbon-Based Nanomaterials for Water Treatment and Supercapacitor Applications. In *Applied Plastics Engineering Handbook* (pp. 219–236). William Andrew Publishing, Norwich, New York. https://doi.org/10.1016/B978-0-323-88667-3.00005-9

[104] Wang, B., Liu, W., & Zhang, M. (2024). Application of carbon-based adsorbents in the remediation of micro-and nanoplastics. *Journal of Environmental Management*, *349*, 119522. https://doi.org/10.1016/j.jenvman.2023.119522

[105] Camilli, E., Pighin, A. F., Copello, G. J., & Villanueva, M. E. (2024). Cobalt/carbon quantum dots core-shell nanoparticles as an improved catalyst for Fenton-like reaction. *Nano-Structures & Nano-Objects*, *37*, 101097. https://doi.org/10.1016/j.nanoso.2024.101097

[106] Bhanderi, D., Lakhani, P., & Modi, C. K. (2024). Graphitic carbon nitride (gC 3 N 4) as an emerging photocatalyst for sustainable environmental applications: A comprehensive review. *RSC Sustainability., 2, 265-287.*https://doi.org/10.1039/D3SU00382E

[107] Qin, J., Dou, Y., Zhou, J., Zhao, D., Orlander, T., Andersen, H. R., ... & Zhang, W. (2024). Encapsulation of carbon-nanodots into metal-organic frameworks for boosting photocatalytic upcycling of polyvinyl chloride plastic. *Applied Catalysis B: Environmental*, *341*, 123355. https://doi.org/10.1016/j.apcatb.2023.123355

[108] Zanaty, M., Zaki, A. H., El-Dek, S. I., & Abdelhamid, H. N. (2024). Zeolitic imidazolate framework@ hydrogen titanate nanotubes for efficient adsorption and catalytic oxidation of organic dyes and microplastics. *Journal of Environmental Chemical Engineering*, *12*(3), 112547. https://doi.org/10.1016/j.jece.2024.112547

[109] Jiang, R., Lu, G., Zhang, L., Chen, Y., Liu, J., Yan, Z., & Xie, H. (2024). Insight into the effect of microplastics on photocatalytic degradation tetracycline by a dissolvable semiconductor-organic framework. *Journal of Hazardous Materials*, *463*, 132887. https://doi.org/10.1016/j.jhazmat.2023.132887

[110] Pedrero, D., Edo, C., Fernández-Piñas, F., Rosal, R., & Aguado, S. (2024). Efficient removal of nanoplastics from water using mesoporous metal organic frameworks. *Separation and Purification Technology*, *333*, 125816. https://doi.org/10.1016/j.seppur.2023.125816

[111] Randhawa, K. S. (2024). Photocatalytic degradation of pollutants using advanced ceramics: Materials, mechanism, synthesis, and applications. *Journal of Inorganic and Organometallic Polymers and Materials*, 1–26. https://doi.org/10.1007/s10904-024-03068-6

[112] Sheoran, A. R., Lakra, N., Luhach, A., Saharan, B. S., Debnath, N., & Sharma, P. (2024). Nano-engineered solutions for sustainable environmental cleanup. *BioNanoScience*, 1–25. https://doi.org/10.1007/s12668-024-01370-8

[113] Khoo, V., Ng, S. F., Haw, C. Y., & Ong, W. J. (2024). Additive manufacturing: A paradigm shift in revolutionizing catalysis with 3D printed photocatalysts and electrocatalysts toward environmental sustainability. *Small*, 20, 2401278. https://doi.org/10.1002/smll.202401278

[114] Xiao, Y., TIAN, Y., Xu, W., & Zhu, J. (2024). Photodegradation of microplastics through nanomaterials: Insights into photocatalysts modification and detailed mechanisms. Materials, 17 (11), 2755 https://doi.org/10.20944/preprints202403.1239.v1

[115] Kinyua, E. M., Nyakairu, G. W. A., Tebandeke, E., & Odume, O. N. (2024). Visible light photocatalytic degradation of HDPE microplastics using vanadium-doped titania. Central Asian Journal of Water Research, 10(1), 126-141 https://doi.org/10.29258/CAJWR/2024-R1.v10-1/126-141.eng

[116] Balakrishnan, A., Jacob, M. M., Chinthala, M., Dayanandan, N., Ponnuswamy, M., & Vo, D. V. N. (2024). Photocatalytic sponges for wastewater treatment, carbon dioxide reduction, and hydrogen production: A review. *Environmental Chemistry Letters*, 22, 1–22. https://doi.org/10.1007/s10311-024-01696-5

[117] Mondal, S., Das, P., Mondal, A., Paul, S., Pandey, J. K., & Das, T. K. (Eds.). (2024). *Remediation of Plastic and Microplastic Waste.*Milton Park, Abingdon, United Kingdom: CRC Press.

[118] Namasivayam, S. K. R., & Avinash, G. P. (2024). Review of green technologies for the removal of microplastics from diverse environmental sources. *Environmental Quality Management*, *33*(3), 449–465. https://doi.org/10.1002/tqem.22131

[119] Surana, M., Pattanayak, D. S., Yadav, V., Singh, V. K., & Pal, D. (2024). An insight decipher on photocatalytic degradation of microplastics: Mechanism, limitations, and future outlook. *Environmental Research*, 247, 118268. https://doi.org/10.1016/j.envres.2024.118268

[120] Al-Yahyaey, S., Kyaw, H. H., Myint, M. T. Z., Al–Hajri, R., Al-Sabahi, J., & Al-Abri, M. (2024). Multi-channel flow reactor design for the photocatalytic degradation of harmful dye molecules. *Journal of Nanoparticle Research*, *26*(4), 72. https://doi.org/10.1007/s11051-024-05981-w

[121] Ahmed, S. F., Islam, N., Tasannum, N., Mehjabin, A., Momtahin, A., Chowdhury, A. A., ... & Mofijur, M. (2024). Microplastic removal and management strategies for wastewater treatment plants. *Chemosphere*, *347*, 140648. https://doi.org/10.1016/j.chemosphere.2023.140648

[122] Praus, P. (2024). Photoreforming for microplastics recycling: A critical review. *Journal of Environmental Chemical Engineering*, 12, 112525. https://doi.org/10.1016/j.jece.2024.112525

[123] Nikhar, S., Kumar, P., & Chakraborty, M. (2024). A review on microplastics degradation with MOF: Mechanism and action. *Next Nanotechnology*, 5, 100060. https://doi.org/10.1016/j.nxnano.2024.100060

[124] Hamd, W., Daher, E. A., Tofa, T. S., & Dutta, J. (2022). Recent advances in photocatalytic removal of microplastics: Mechanisms, kinetic degradation, and reactor design. *Frontiers in Marine Science*, *9*, 885614. https://doi.org/10.3389/fmars.2022.885614

[125] Stock, F., Kochleus, C., Bänsch-Baltruschat, B., Brennholt, N., & Reifferscheid, G. (2019). Sampling techniques and preparation methods for microplastic analyses in the aquatic environment–A review. *TrAC Trends in Analytical Chemistry*, *113*, 84–92. https://doi.org/10.1016/j.trac.2019.01.014

[126] Jiménez-Skrzypek, G., Ortega-Zamora, C., González-Sálamo, J., Hernández-Sánchez, C., & Hernández-Borges, J. (2021). The current role of chromatography in microplastic research: Plastics chemical characterization and sorption of

contaminants. *Journal of Chromatography Open*, *1*, 100001. https://doi.org/10.1016/j.jcoa.2021.100001

[127] Fakayode, S. O., Mehari, T. F., Fernand Narcisse, V. E., Grant, C., Taylor, M. E., Baker, G. A., ... & Anum, D. (2024). Microplastics: Challenges, toxicity, spectroscopic and real-time detection methods. *Applied Spectroscopy Reviews*, 59, 1–95. https://doi.org/10.1080/05704928.2024.2311130

[128] Guo, X., Lin, H., Xu, S., & He, L. (2022). Recent advances in spectroscopic techniques for the analysis of microplastics in food. *Journal of Agricultural and Food Chemistry*, *70*(5), 1410–1422. https://doi.org/10.1021/acs.jafc.1c06085

[129] Du, H., Chen, G., & Wang, J. (2023). Highly selective electrochemical impedance spectroscopy-based graphene electrode for rapid detection of microplastics. *Science of The Total Environment*, *862*, 160873. https://doi.org/10.1016/j.scitotenv.2022.160873

[130] Vignesh Kumar, T. H., & Rajendran, J. (2024). Recent progress in electrochemical methods for microplastics detection. *Microplastics and Pollutants: Interactions, Degradations and Mechanisms*, 249–263.

[131] Kalaronis, D., Ainali, N. M., Evgenidou, E., Kyzas, G. Z., Yang, X., Bikiaris, D. N., & Lambropoulou, D. A. (2022). Microscopic techniques as means for the determination of microplastics and nanoplastics in the aquatic environment: A concise review. *Green Analytical Chemistry*, *3*, 100036. https://doi.org/10.1016/j.greeac.2022.100036

[132] Gniadek, M., & Dąbrowska, A. (2019). The marine nano-and microplastics characterisation by SEM-EDX: The potential of the method in comparison with various physical and chemical approaches. *Marine Pollution Bulletin*, *148*, 210–216. https://doi.org/10.1016/j.marpolbul.2019.07.067

[133] Mariano, S., Tacconi, S., Fidaleo, M., Rossi, M., & Dini, L. (2021). Micro and nanoplastics identification: Classic methods and innovative detection techniques. *Frontiers in Toxicology*, *3*, 636640. https://doi.org/10.3389/ftox.2021.636640

[134] Sacco, N. A., Zoppas, F. M., Devard, A., González Muñoz, M. D. P., García, G., & Marchesini, F. A. (2023). Recent advances in microplastics removal from water with special attention given to photocatalytic degradation: Review of scientific research. *Microplastics*, *2*(3), 278–303. https://doi.org/10.3390/microplastics2030023

[135] Yu, Y., Luan, Y., & Dai, W. (2022). Biodegradation of microplastics in soil. , Vol. 23, p – 866-872. http://dx.doi.org/10.3233/ATDE220362

[136] Miri, S., Saini, R., Davoodi, S. M., Pulicharla, R., Brar, S. K., & Magdouli, S. (2022). Biodegradation of microplastics: Better late than never. *Chemosphere*, *286*, 131670. https://doi.org/10.1016/j.chemosphere.2021.131670

[137] Huerta Lwanga, E., & Santos-Echeandía, J. (2020). Soil remediation under microplastics pollution. *Handbook of Microplastics in the Environment*, 1–29. https://doi.org/10.1007/978-3-030-10618-8_23-1

[138] Liao, J., & Chen, Q. (2021). Biodegradable plastics in the air and soil environment: Low degradation rate and high microplastics formation. *Journal of Hazardous Materials*, *418*, 126329. https://doi.org/10.1016/j.jhazmat.2021.126329

[139] Bao, R., Pu, J., Xie, C., Mehmood, T., Chen, W., Gao, L., ... & Peng, L. (2022). Aging of biodegradable blended plastic generates microplastics and attached bacterial communities in air and aqueous environments. *Journal of Hazardous Materials*, *434*, 128891. https://doi.org/10.1016/j.jhazmat.2022.128891

Photocatalytic Degradation and Remediation 199

[140] Gao, W., Zhang, Y., Mo, A., Jiang, J., Liang, Y., Cao, X., & He, D. (2022). Removal of microplastics in water: Technology progress and green strategies. *Green Analytical Chemistry, 3*, 100042. https://doi.org/10.1016/j.greeac.2022.100042

[141] Priya, K. L., Renjith, K. R., Joseph, C. J., Indu, M. S., Srinivas, R., & Haddout, S. (2022). Fate, transport and degradation pathway of microplastics in aquatic environment—A critical review. *Regional Studies in Marine Science, 56*, 102647. https://doi.org/10.1016/j.rsma.2022.102647

[142] Ahmad, R., Ahmad, Z., Khan, A. U., Mastoi, N. R., Aslam, M., & Kim, J. (2016). Photocatalytic systems as an advanced environmental remediation: Recent developments, limitations and new avenues for applications. *Journal of Environmental Chemical Engineering, 4*(4), 4143–4164. https://doi.org/10.1016/j.jece.2016.09.009

[143] Yentekakis, I. V., & Dong, F. (2020). Grand challenges for catalytic remediation in environmental and energy applications toward a cleaner and sustainable future. *Frontiers in Environmental Chemistry, 1*, 5. https://doi.org/10.3389/fenvc.2020.00005

[144] Maleki, H., & Hüsing, N. (2018). Current status, opportunities and challenges in catalytic and photocatalytic applications of aerogels: Environmental protection aspects. *Applied Catalysis B: Environmental, 221*, 530–555. https://doi.org/10.1016/j.apcatb.2017.08.012

[145] Alalm, M. G., Djellabi, R., Meroni, D., Pirola, C., Bianchi, C. L., & Boffito, D. C. (2021). Toward scaling-up photocatalytic process for multiphase environmental applications. *Catalysts, 11*(5), 562. https://doi.org/10.3390/catal11050562

[146] Karabelas, A. J., Plakas, K. V., & Sarasidis, V. C. (2018). How Far Are We from Large-Scale PMR Applications?. In *Current Trends and Future Developments on (Bio-) Membranes* (pp. 233–295). Amsterdam, The Netherlands: Elsevier. https://doi.org/10.1016/B978-0-12-813549-5.00009-8

[147] Puri, M., Gandhi, K., & Kumar, M. S. (2023). Emerging environmental contaminants: A global perspective on policies and regulations. *Journal of Environmental Management, 332*, 117344. https://doi.org/10.1016/j.jenvman.2023.117344

[148] Sutradhar, M. (2022). A review of microplastics pollution and its remediation methods: Current scenario and future aspects. *Archives of Agriculture and Environmental Science, 7*, 288–293. https://doi.org/10.26832/24566632.2022.0702019

[149] Ali, S. S., Al-Tohamy, R., Alsharbaty, M. H. M., Elsamahy, T., El-Sapagh, S., Lim, J. W., & Sun, J. (2024). Microplastics and their ecotoxicological impacts: remediation approaches, challenges and future perspectives-A review. *Journal of Cleaner Production, 452*, 142153. https://doi.org/10.1016/j.jclepro.2024.142153

[150] Goh, P. S., Kang, H. S., Ismail, A. F., Khor, W. H., Quen, L. K., & Higgins, D. (2022). Nanomaterials for microplastic remediation from aquatic environment: Why nano matters?. *Chemosphere, 299*, 134418. https://doi.org/10.1016/j.chemosphere.2022.134418

[151] Rizwan, K., & Bilal, M. (2022). Developments in advanced oxidation processes for removal of microplastics from aqueous matrices. *Environmental Science and Pollution Research, 29*(58), 86933–86953. https://doi.org/10.1007/s11356-022-23545-0

[152] Lv, L., Yan, X., Feng, L., Jiang, S., Lu, Z., Xie, H., ... & Li, C. (2021). Challenge for the detection of microplastics in the environment. *Water Environment Research, 93*(1), 5–15. https://doi.org/10.1002/wer.1281

[153] Karim, M. E., Sanjee, S. A., Mahmud, S., Shaha, M., Moniruzzaman, M., & Das, K. C. (2020). Microplastics pollution in Bangladesh: Current scenario and future

research perspective. *Chemistry and Ecology*, *36*(1), 83–99. https://doi.org/10.1080/02757540.2019.1688309

[154] Kuspanov, Z., Baglan, B., Baimenov, A., Issadykov, A., Yeleuov, M., & Daulbayev, C. (2023). Photocatalysts for a sustainable future: Innovations in large-scale environmental and energy applications. *Science of the Total Environment*, 885, 163914. https://doi.org/10.1016/j.scitotenv.2023.163914

[155] Alimi, O. S., Claveau-Mallet, D., Kurusu, R. S., Lapointe, M., Bayen, S., & Tufenkji, N. (2022). Weathering pathways and protocols for environmentally relevant microplastics and nanoplastics: What are we missing?. *Journal of Hazardous Materials*, *423*, 126955. https://doi.org/10.1016/j.jhazmat.2021.126955

[156] Coffin, S., Wyer, H., & Leapman, J. C. (2021). Addressing the environmental and health impacts of microplastics requires open collaboration between diverse sectors. *PLoS Biology*, *19*(3), e3000932. https://doi.org/10.1371/journal.pbio.3000932

10 Challenges and Fate of Microplastics in Wastewater Treatment Processes

Chandrakanta Mall and Prem Prakash Solanki

10.1 INTRODUCTION

Research and technological development make the survival of human beings on the one hand much easier, whereas, on the other hand, the adaptation of lifestyle to available present facilities causes the generation of several hazardous materials that directly or indirectly affect the environment. Any type of change in the environment directly affects the life of whole organisms on the earth. Out of these hazardous materials, plastics are one of them and are used everywhere. Generally, the ending source of plastics is the ocean, and here, the plastics break up into very small particles, called microplastics (MPs). Other small particles are termed microbeads, which are mainly used in the manufacturing of products related to beauty and health. These microbeads enter the ocean through waterways without changing themselves. Aquatic organisms and birds mistakenly consume MPs as food.

Approximately 617.6 billion MPs are poured into the marine surroundings every 12 months from the effluent of the WWTPs. Thus, post-handling essential for the integration of MPs into the sewage slush is more than 1,202 billion MPs every year. Therefore, the supply of sewage slush must be prohibited to the land or it must be treated properly before discarding to the land.

There are several studies being conducted, but there is still not much known. Microbeads are no longer the main problem because of US legislation passed to prohibit their use, but MPs are still a massive problem today [1]. According to a United Nations Environment Programme, >430 million tons of plastic are produced yearly, and continuously the world is plunging into this plastic pollution. Nearly two-thirds of products such as facial cleansers, sunscreen, cosmetics, shower gels, hair coloring, nail polish, eye shadow, toothpaste, etc., used in everyday life are short-lived products that contain MPs that soon become waste thus affecting the environment and the human food chain [2]. It is therefore a big challenge to scientific communities to create a technique that minimizes the production of MPs.

DOI: 10.1201/9781003486947-10

This chapter discusses the classification, existence, environmental impact, and causes of MPs. The main concern of the present chapter is to summarize the challenges and fate of MPs in wastewater treatment processes (WWTPs).

10.2 MICROPLASTICS

MPs are plastics <5 millimeters (0.2 inches) in diameter. MPs do not easily break down into harmless molecules. Hundreds to thousands of years can be required for the complete disintegration of plastics. The tiny multicolored plastic beads of MPs are visible in the sand of seashores. The classification, occurrence, and effect on the environment of MPs are as follows.

10.2.1 CLASSIFICATION

A marine biologist named Prof. R. Thompson introduced the term "microplastics" in 2004 [3]. It was predicted that there are between 15 and 51 trillion individual fragments (weighing in the range from 93,000 to 236,000 metric tons) of MPs inside the oceans of the world [4]. MPs can be divided into two categories [5].

10.2.1.1 Primary Microplastics

The primary MPs are purposefully made from small pieces of plastic [6, 7]. These MPs are used in beauty treatments, makeups, and air blasting technology, and have also been used as vectors for drugs [8]. Because of their exfoliating properties, MPs are used in hand cleansers and facial scrubs. The blasting of melamine, acrylic, or polyester MP scrubbers at equipment, engines, and boat hulls to get rid of their rust and paint is included in part of air blasting generation. The scrubbers are used again and again till they reduce in size and their acerbic power is gone [9].

10.2.1.2 Secondary Microplastics

Secondary MPs result from the breakdown of larger plastic fragments into small pieces of plastic [6] due physical, biological, chem-photodegradation, and photo-oxidation (sunlight exposure) processes over many years. Thus, MPs cannot be detected by the naked eye [10]. Literature study shows that the production of MPs in sea water as well as fresh water systems might be more from the biodegradable polymers in contrast to non-biodegradable polymers [11].

10.2.2 OCCURRENCE AND SOURCES

Plastics are used in just about every facet of life and thus can be found in air, sediments, soils, freshwater, seawater, and even in organisms such as humans. MPs come into WWTPs from various home and industrial processes. The plastic-based MPs, i.e., from plastic pellets, tire fragments, street paint, etc., discharged into water streams from different industrial areas [12, 13]. The adulteration of MPs diverges geologically with the region, air-stream direction, tornados, hydrodynamic situations, ecological pressure, period, daylight, storm tidal wave, sturdy oceans, farming actions, shoreline

Challenges and Fate of Microplastics in Wastewater

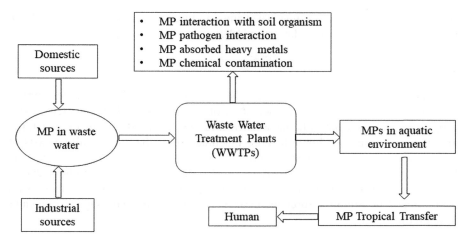

FIGURE 10.1 The sources from which the MPs enter into the WWTP.

morphology, and topography variations [14]. Figure 10.1 shows the various domestic and industrial sources from where WWTP receives MPs.

10.3 REASONS FOR THE GENERATION OF MICROPLASTICS

The use of single-us/unrecyclable plastic in Personal Protection Equipment (PPEs) expanded worldwide at some stage during COVID-19. This consequences in the blockage of manure canals, in addition to finishing up in oceans through superficial folding. According to the World Health Organization (WHO), nearly eighty-nine million routine masks are obligatory per month to avert COVID-19 increasing the manufacturing of polymeric materials [15]. The dumping of face covers fabricated with polyethylene/polypropylene/other polymeric materials consisting of nylon and polystyrene is unregulated and inappropriate which produces micro and nano-sized plastic fibers and silicon grains [16]. A study to complete a common level of cloth degression (usual 1.2± 1.3% of the preliminary weight), showed that a constantly massive wide variety of MPs/nanoplastics can be generated as of a solo mask via repeating decaying situations underneath a genuine strength level of machine-driven worsening [17].

10.3.1 AIR FLOW

The diverse shapes of rocks are a result of deterioration over time due to wind and other weather. Likewise, the abrasion process in polymeric materials may also occur, and pieces of these substances might be wafted into the air stream itself. On exposure of plastic materials of an automobile to the external atmosphere is subjected to wind movement which is amplified by the speed of the automobile in motion [18]. Since the gutter wall consists of cement and concrete, and water continuously flows through it. Thus, a trimming force like in case of polymeric surface, is also generated inside the gutter due to the continuous flow of water and this causes attrition of these cement-concrete wall over a time. The trimming force increases very much with the passage of rainwater through the gutter [18].

10.3.2 WATER FLOW

The automatic action over a polymeric surface due to overflow of water applies a trimming force and cause fragmentation of substances, such as the elements of gadgets as well as what is being washed away are exposed toward the machine-driven act of water in a home laundry machine. Since the gutter wall consists of cement and concrete, and water continuously flows through it. Thus, a trimming force like in case of polymeric surface, is also generated inside the gutter due to the continuous flow of water and this causes attrition of these cement-concrete wall over a time. The trimming force increases very much with the passage of rainwater through the gutter [18].

10.3.3 CHEMICAL OXIDATION

The polymeric materials are chemically stable in an environment parallel to ceramics, though the chemical composition of polymeric constituents is not like ceramic materials. The existence of polymeric materials is an indication of the inertness of these chemicals at some stage in and after use in the environment. Polymeric materials undergo only surface oxidation on exposure to environmental conditions and the process of oxidation is eased using the exploitation of UV radiations that are found spectrum of solar energy [18].

10.3.4 MECHANICAL FRAGMENTATION

The improper exposure to solar radiation of plastic products results in surface oxidation and the breaking of chemical bonds leads to mechanical fragmentation. The mechanical properties of plastic substances are compromised because the embrittlement occurs over years. The result of all these processes is the physical fragmentation of materials that are carried by wind as MPs. A blade tire becoming old, the fragmentation of treads occurs due to friction with sidewalks, etc. are numerous examples of mechanical fragmentation and the spread of these fragments was done by the wind, establishing additional causes of MPs [18].

10.4 MICROPLASTICS IN WASTEWATER TREATMENT PROCESSES

The sources of wastewater in WWTPs are households, industries, commercial institutions, and rainwater runoff from urban zones [19]. The number of MPs in WWTPs from household discharge is enhanced because of higher population density, modern lifestyles, and ingesting of products using MPs. According to research, since clothing is a large source of MPs there are more in the environment in the winter than in summer [20]. The abundance of MPs is linked to weather, but also with population, catchment size, combined drain systems (a system that carries wastewater and surface water into a single pipe, simultaneously), adjoining land utilization, and wastewater

Challenges and Fate of Microplastics in Wastewater 205

assets (residential, commercial, and business) [19, 21]. Michielssen et al. reported that the impact of mixed discharge small device/tools can improve the burden of MPs in WWTPs [22]. Microbeads personal care products [23], polyester, nylon, and other fibers [24], along with primary MPs from packing constituents and basic materials of the textile industry, are the key resources of MPs in WWTPs [25]. The WWTPs are widely used in filtration of dunes, UV cleansing, and organic procedures that result in photooxidation, sand scuff, and accommodation potential, resulting in the degradation of MPs into secondary MPs. However, the creation of secondary MPs as a result of WWTP methods are generally neglected.

The wastewater used in WWTP systems is complex in nature because it comprises several forms of MPs. Hence, finding precise methods for detecting MPs at the submicron scale are required. Membrane technology has shown considerable results to remove the MPs, but membrane blocking and fouling has been detected in large-scale applications. The use of strainer with opening >50 mm on MPs in WWTPs has been used to detect MPs in current research, but this type of materials are not capable in filtering the secondary MPs up to micron level [26].

10.5 CHALLENGES AND FATE OF MICROPLASTICS IN WASTEWATER TREATMENT PROCESSES

The literature regarding the separation of MPs through WWTPs shows that it may be >90%. Fragments above 300 μm are somewhat removed by WWTPs, whereas the rest of the particles in the range of the nanoscale leave the system [27]. However, the WWTPs can reduce the concentration of particles from influent in considerable amounts, yet the enormous number of MPs from the effluent of the system leak into marine milieu [28, 29]. The preliminary, primary, secondary, and tertiary methods of WWTPs can remove the MPs up to 6%, 68%, 90%, and 95%, respectively [30]. Additionally, the biological and eco-friendly methods can remove MPs up to 99% in contrast to chemical methods, although the rest are released into the environment [31].

10.5.1 CHALLENGES

There are numerous challenges associated with research on MP pollutants. For example, the lack of a stable definition of what may be considered MPs makes it hard to examine the outcomes achieved in research. Moreover, the methodologies utilized in some studies are specific and thus not always comparable. Lastly, there is a shortage of manpower needed to conduct the studies and run the equipment needed to keep MPs out of water bodies [32].

Although numerous studies have examined the occurrence of MPs in waste, the elimination of these MPs at each stage of the WWTP process has not been covered to much extent [33].

There are several methods available for the treatment of MPs, such as membrane bioreactor, electrocoagulation, membrane filtration, biofiltration, and magnetic extraction. But these methods also have some limitations, i.e., costly, non-availability to be used anywhere [34]. Though, there are several available technologies such as membrane bioreactor, electrocoagulation, membrane filtration, biofiltration, magnetic extraction, etc. to treat the MPs in WWTPs. Yet, they have their limitations including cost, ability to be used everyplace in normal available conditions such as location, space, permit, drainage, freezing zone etc., to provide to efficient services, and are also handiest used whilst excessive fine ideals are essential [34], for example, membrane bioreactors have been used in WWTPs to address MPs, but this technology is highly expensive, has high energy requirement, aeration limitation, stress on sludge, membrane material compatibility, abrasion property, lower efficiency, design and operation method. Thus, to remove all these limitations to do such research which will provide such methods which has low cost, low energy consumption, low disturbance with air, compatible with membrane used, low abrasion/erosion with high efficiency. For example, Membrane bioreactors, have been used in WWTPs to address MPs, but this technology is highly expensive, memberane fouling, very high energy requirement, aeration limitation, stress on sludge, memberane material compability, abrasion property, lower effiency, design and operation method [35]. Though, today, the results of effect of the MPs on the fitness of environment is not clear but it is vital need of time to to study, track, and keep away from additional infection (contamination) of MPs in wastewater [36].

While we know MPs can affect human health, it is important to measure the present acceptable limits and loads of MPs. The generation of MPs in water bodies is a challenge as they disturb not only the natural environment but also interrupt the food chain cycle [37]. Future research should focus on the recognition and elimination of the sources of MPs in earth, residue, etc., and the sustainable technologies in WWTPs for minimalizing the MPs [38] from the environment.

10.5.2 FATE

The WWTPs deal with the received wastewater optimally to make use of primary, secondary, and tertiary handling methods. Coagulants such as alum, ferric chloride, polymers, etc., and polymeric coagulants are applied to take away colloidal and suspended particulates in the primary treatment. The organics connected with solvated humic materials and debris can also be detached in the coagulation process.

In secondary treatment, the liquified organics are abolished aerobically via an association of microbes in suspension. The solidified slush from both primary and secondary cleansers (containing MPs in the form of polythene beads) is processed anaerobically to be dumped in the environment. In a few locations, the adsorption of activated carbon, ozonation, or purification is also carried out to finish the treatment of effluent to dispose of the concentration of organics in tertiary treatment processes. MPs in WWTPs usually consist of adsorption on suspended particulates, solvated humic materials, and primary and secondary slush [39].

The fate of MPs can be categorized as some MPs are gotten rid within the circulated currents, and some settle because of depository or gravitation processes. The fate of MPs in WWTPs principally depends on the adsorption of the MPs on suspended particulates, primary and secondary sewage, and solvated organics. The commonly used methods for the remediation of MPs are coagulation, flocculation,

TABLE 10.1
Secondary Settle-down Container with MP Units in the In/effluent and Sludge [39]

Location	Sample 1 (MPs/L)	Sample 2 (MPs/L)	Sample 3 (MPs/L)	Average (MPs/L)
Influent	220 ± 23	216± 19	184 ± 26	206 ± 31
Sludge	200 ± 41	192 ± 24	159 ± 23	183 ± 21
Effluent	106 ± 31	95 ± 17	81 ± 21	94 ± 28

and biodegradation which depend on the physical properties of the target analyte such as solubility, octanol-water partition coefficient, and Henry's constant. MPs in ineffluent and slush of secondary settle-down reservoir are summarized in Table 10.1 on the basis of results obtained in the study on the fate of MPs in a WWTP with respect to frequency and removal capacity of MPs through microfiltration membrane [40]. The influent of the system with an average of 206 MP/L was detected as the highest number of MPs as shown in Table 10.1. The results of investigations showed that the MPs were gathered in the slush at 183 MP/L.

The gathering of MPs within the slush might have been because of the adsorption of these materials with the aid of the slush matrix as per the literature [41]. Furthermore, it was observed that the number of MPs in influent/effluent of the WWTPs varies based on climate. From Table 10.1, it is clear that the quantity of MPs amplified in the influent of samples 1 and 2 was relatively higher as compared to that of sample 3, which may be due to rainwater washing the MPs into the WWTPs [42]. Temperature and operating situations including pH, dissolved oxygen, etc., also play a key part in slush flocculation [43]. Hence, the sludge floc fragments subsequently result in a low-slung settle ability of flocs at higher temperatures.

Once MPs are introduced into the soil through sewage sludge, their potential adverse impact on the soil habitat is determined by their localized distribution, which is influenced by their movement and decay. It is important to note that the size and shape of the MPs can change due to the sludge treatments, affecting their movement and environmental risk in the soil. Therefore, further research is needed to understand the impact of different sludge treatments on organic waste and their additional effects on the soil ecosystem [44].

10.6 CONCLUSION

Today lifestyle and available facilities cause the production of harmful materials that directly or indirectly disturb the life of whole organisms on the earth. Plastics are one of them, and generally, the ending of plastics is in the marine water bodies. Plastics break up into tiny particles called MPs. Air flow, water flow, chemical oxidation, and mechanical fragmentation are the main causes of MPs entering the environment.

WWTPs are the dominant cause of MP pollution in sea ecologies. Households, industries, commercial institutions, and rainwater runoff from urban zones are the sources of wastewater in WWTPs. The treatment of sewerage water using WWTPs is difficult due to the presence of numerous forms of MPs in it. A lack of a firm definition of MPs makes it hard to study them and the methodologies needed to deal with them in WWTPs. The fate of MPs depends on the adsorption of the MPs on suspended particulates, primary and secondary sewage, and dissolved organics. Thus, research should focus on developing technologies that can minimize MPs in the waste of WWTPs before the discharge of this waste into water bodies.

REFERENCES

[1] J. Q. Jiang, "Occurrence of microplastics and its pollution in the environment: A review," *Sustain Prod Consum*, vol. 13, pp. 16–23, Jan. 2018, doi:10.1016/j.spc.2017.11.003.

[2] M. Khalida, and M. Abdollahia, "Environmental distribution of personal care products and their effects on human health," *Iran J Pharm Res*, vol. 20, no. 1, pp. 216–253, Winter 2021, doi: 10.22037/ijpr.2021.114891.15088.

[3] R. C. Thompson, Olsen, Ylva, Mitchell, P. Richard, Davis, Anthony, Rowland, J. Steven, John, W. G. Anthony, McGonigle, Daniel, Russell, and E. Andrea, "Lost at sea: Where is all the plastic?" *Science*, vol. 304, no. 5672, pp. 838–838, May 2004, doi: 10.1126/science.1094559.

[4] C. Ioakeimidis, K. N. Fotopoulou, H. K. Karapanagioti, M. Geraga, C. Zeri, E. Papathanassiou, F. Galgani, and G. Papatheodorou, "The degradation potential of PET bottles in the marine environment: An ATR-FTIR based approach," *Sci Rep*, vol. 6, p. 23501, Mar. 2016, doi: 10.1038/srep23501.

[5] National Geographic Society. Last Modified October 31, 2023. https://education.nationalgeographic.org/resource/microplastics/.

[6] S. Ghosh, J. K. Sinha, S. Ghosh, K. Vashish, S. Han, and R. Bhaskar, "Microplastics as an emerging threat to the global environment and human health," *Sustainability*, vol. 15, no. 14, p. 10821, Jul. 2023, doi: 10.3390/su151410821.

[7] S. Karbalaei, P. Hanachi, T. R. Walker, and M. Cole, "Occurrence, sources, human health impacts and mitigation of microplastic pollution," *Environ Sci Pollut Res*, vol. 25, pp. 36046–36063, Oct. 2018, doi: 10.1007/s11356-018-3508-7.

[8] M. M. Patel, B. R. Goyal, S. V. Bhadada, J. S. Bhatt and A. F. Amin, "Getting into the brain approaches to enhance brain drug delivery," *CNS Drugs*, vol. 23, no. 1, pp. 35–58, Aug. 2012, doi: 10.2165/0023210-200923010-00003.

[9] M. Cole, P. Lindeque, C. Halsband, and T. S. Galloway, "Microplastics as contaminants in the marine environment: A review," *Mar Pollut Bull*, vol. 62, no. 12, pp. 2588–2597, Dec. 2011, doi: 10.1016/j.marpolbul.2011.09.025.

[10] M. Julie, B. Joel, Foster, Gregory, Arthur, Courtney, and H. Carlie, "Laboratory methods for the analysis of microplastics in the marine environment: recommendations for quantifying synthetic particles in waters and sediments," Marine Debris Program (U.S.), NOAA technical memorandum NOS-OR&R, 48, 2015. https://repository.library.noaa.gov/view/noaa/10296.

[11] X. F. Wei, M. Bohléna, C. Lindblada, M. Hedenqvist, and A. Hakonenc, "Microplastics generated from a biodegradable plastic in freshwater and seawater," *Water Res*, vol. 198, p. 117123, June 2021, doi: 10.1016/j.watres.2021.117123.

[12] F. Jabeen, M. Adrees, M. Ibrahim, A. Mahmood, S. Khalid, H. F. K. Sipra, A. Bokhari, M. Mubashir, K.S. Khoo, and P. L. Show, "Trash to energy: A measure for the energy potential of combustion content of domestic solid waste generated from industrialized city of Pakistan," *J Taiwan Inst Chem Eng,* vol. 137, p. 104223, Aug. 2022, doi: 10.1016/j.jtice.2022.104223.

[13] S. I. Zahra, M. J. Iqbal, S. Ashraf, A. Aslam, M. Ibrahim, M. Yamin and M. Vithanage, "Comparison of ambient air quality among industrial and residential areas of a typical south Asian city," *Atmosphere*, vol. 13, no. 8, p. 1168, Jul. 2022, doi: 10.3390/atmos13081168.

[14] M. Simon, N. V. Alst, and J. Vollertsen, "Quantification of microplastic mass and removal rates at wastewater treatment plants applying Focal Plane Array (FPA)-based Fourier Transform Infrared (FT-IR) imaging," *Water Res*, vol. 142, pp. 1–9, Oct. 2018, doi: 10.1016/j.watres.2018.05.019.

[15] T. A. Aragaw, "Surgical face masks as a potential source for microplastic pollution in the COVID-19 scenario," *Mar Pollut Bull*, vol. 159, p. 111517, Oct. 2020, doi: 10.1016/j.marpolbul.2020.111517.

[16] G. L. Sullivan, J. D. Gallardo, T. M. Watson, and S. Sarp, "An investigation into the leaching of micro and nano particles and chemical pollutants from disposable face masks – linked to the COVID-19 pandemic," *Water Res*, vol. 196, p. 117033, May 2021, doi: 10.1016/j.watres.2021.117033.

[17] S. Morgana, B. Casentini, and S. Amalfitano, "Uncovering the release of micro/nanoplastics from disposable face masks at times of COVID-19," *J Hazard Mater*, vol. 419, p. 126507, Oct. 2021, doi: 10.1016/j.jhazmat.2021.126507.

[18] J. M. M. Neto, and E. A. da Silva, "Sources of microplastic generation in the environment," *Int J Environ Res Public Health*, vol. 20, no. 13, p. 6202, June 2023, doi: 10.3390/ijerph20136202.

[19] E. D. Okoffo, S. O'Brien, J. W. O'Brien, B. J. Tscharke, and K. V. Thomas, "Wastewater treatment plants as a source of plastics in the environment: A review of occurrence, methods for identification, quantification and fate," *Environ Sci: Water Res Technol*, vol. 5, no. 11, pp. 1908–1931, Sept. 2019, doi: 10.1039/c9ew00428a.

[20] M. A. Browne, P. Crump, S. J. Niven, E. Teuten, A. Tonkin, T. Galloway, and R. Thompson, "Accumulation of microplastic on shorelines woldwide: Sources and sinks," *Environ Sci Technol*, vol. 45, no. 21, pp. 9175–9179, Sept. 2011, doi: 10.1021/es201811s.

[21] E. A. B. David, M. Habib, E. Haddad, M. Hasanin, D. L. Angel, A. M. Booth, and I. Sabbah, "Microplastic distributions in a domestic wastewater treatment plant: Removal efficiency, seasonal variation and influence of sampling technique," *Sci Total Environ*, vol. 752, p.141880, Jan. 2021, doi: 10.1016/j.scitotenv.2020.141880.

[22] M. R. Michielssen, E. R. Michielssen, J. Ni, and M. B. Duhaime, "Fate of microplastics and other small anthropogenic litter (SAL) in wastewater treatment plants depends on unit processes employed," *Environ Sci Water Res Technol*, vol. 2, no. 6, pp. 1064–1073, Oct. 2016, doi: 10.1039/C6EW00207B.

[23] Y. Liu, B. Wang, V. Pileggi, and S. Chang, "Methods to recover and characterize microplastics in wastewater treatment plants," *Case Studies Chem Environ Eng*, vol. 5, p. 100183, May 2022, doi: 1016/j.cscee.2022.100183.

[24] J. A. Nathanson."pollution", Encyclopedia Britannica, 17 Oct. 2024, https://britannica.com/science/pollution-environment. Accessed 18 October 2024.

[25] E. B. Jadhav, M. S. Sankhla, R. A. Bhat, and D. S. Bhagat., "Microplastics from food packing: An overview of human consumption, health threats, and alternative

solutions," *Environ Nanotechnol Monitoring Manag*, vol. 16, p. 100608, Dec. 2021, doi: 10.1016/j.enmm.2021.100608.

[26] B. Pandey, J. Pathak, P. Singh, R. Kumar, A. Kumar, S. Kaushik, and T. K. Thakur, "Microplastics in the ecosystem: An overview on detection, removal, toxicity assessment, and control release," *Water*, vol. 15, no. 1, p. 51, Dec. 2022, doi: 10.3390/w15010051

[27] S. A. Carr, J. Liu, and A. G. Tesoro, "Transport and fate of microplastic particles in wastewater treatment plants," *Water Res*, vol. 91, pp. 174–182, Mar. 2016, doi: 10.1016/j.watres.2016.01.002.

[28] S. A. Mason, D. Garneau, R. Sutton, Y. Chu, K. Ehmann, J. Barnes, P. Fink, D. Papazissimos, and D. L. Rogers, "Microplastic pollution is widely detected in US municipal wastewater treatment plant effluent," *Environ Pollut*, vol. 218 pp. 1045–1054, Nov. 2016, doi: 10.1016/j.envpol.2016.08.056.

[29] J. Talvitie, A. Mikola, O. Setälä, M. Heinonen, and A. Koistinen, "How well is microlitter purified from wastewater? – A detailed study on the stepwise removal of microlitter in a tertiary level wastewater treatment plant," *Water Res*, vol. 109, pp. 164–172, Feb. 2017, doi: 10.1016/j.watres.2016.11.046.

[30] R. M. Blair, S. Waldron, and C. G. Lindsay, "Average daily flow of microplastics through a tertiary wastewater treatment plant over a ten-month period," *Water Res*, vol. 163, p. 114909, Oct. 2019, doi:10.1016/j.watres.2019.114909.

[31] W. Liu, J. Zhang, H. Liu, X. Guo, X. Zhang, X. Yao, Z. Cao, and T. Zhang, "A review of the removal of microplastics in global wastewater treatment plants: characteristics and mechanisms," *Environ Int*, vol. 146, p. 106277, Jan. 2021, doi: 10.1016/j.envint.2020.106277.

[32] H. Westphalen and A. Abdelrasou, "Challenges and Treatment of Microplastics in Water," *Water Challenges of an Urbanizing World*, Chapter 5, Dec. 2017, Publisher: Intech Open, doi: 10.5772/intechopen.71494.

[33] F. Murphy, C. Ewins, F. Carbonnier, and B. Quinn, "Wastewater Treatment Works (WwTW) as a source of microplastics in the aquatic environment," *Environ Sci Technol*, vol. 50, no. 11, pp. 5800–5808, May 2016, doi: 10.1021/acs.est.5b05416.

[34] A. Beljanski, C. Cole, F. Fuxa, E. Setiawan, and H. Singh, "Efficiency and Effectiveness of a Low-Cost, Self-Cleaning Microplastic Filtering System for Wastewater Treatment Plants," *Proceedings of The National Conference On Undergraduate Research* (NCUR), University of North Carolina Asheville, pp. 1388–1395, Apr. 7–9, 2016, https://libjournals.unca.edu/ncur/wp-content/uploads/2021/06/2064-Beljanski-Alec.pdf.

[35] Norman Bulletin. Issue 4, March 2015. www.norman-network.net.

[36] A. Ballent, P. L. Corcoran, O. Madden, P. A. Helm, and F. J. Longstaffe, "Sources and sinks of microplastics in Canadian Lake Ontario nearshore, tributary and beach sdiments," *Mar Pollut Bull*, vol. 110, no. 1, pp. 383–395, Sept. 2016, doi:10.1016/j.marpolbul.2016.06.037.

[37] O. Setälä, V. Fl. Lehtinen, and M. Lehtiniemi, "Ingestion and transfer of microplastics in the planktonic food web," *Environ Pollut*, vol. 185, pp. 77–83, Feb. 2014, doi:10.1016/j.envpol.2013.10.013.

[38] A. S. Tagg, J. P. Harrison, Y. J. Nam, M. Sapp, E. L. Bradley, C. J. Sinclaird, and J. J. Ojeda, "Fenton's reagent for the rapid and efficient isolation of microplastics from wastewater," *Chem Commun*, vol. 53, no. 2, pp. 372–375, Dec. 2017, doi:10.1039/C6CC08798A.

[39] S. Das, N. M. Ray, J. Wan, A. Khan, T. Chakraborty, and M. B. Ray, "Micropollutants in Wastewater: Fate and Removal Processes," *Physico Chemical Wastewater Treatment and Resource Recovery*, Chapter 5, pp. 75–107, May 2017, Publisher: Intech Open, doi: 10.5772/65644.

[40] N. Yahyanezhad, M. J. Bardi, and H. Aminirad, "An evaluation of microplastics fate in the wastewater treatment plants: Frequency and removal of microplastics by microfiltration membrane," *Water Pract Technol*, vol. 16, no. 3, pp. 782–792, Apr. 2021, doi:10.2166/wpt.2021.036.

[41] X. Li, Q. Mei, L. Chen, H. Zhang, B. Dong, X. Dai, C. He, and J. Zhou, "Enhancement in adsorption potential of microplastics in sewage sludge for metal pollutants after the wastewater treatment process," *Water Res*, vol. 157, pp. 228–237, Jun. 2019, doi:10.1016/j. watres.2019.03.069.

[42] W. Xia, Q. Rao, X. Deng, J. Chen, and P. Xie, "Rainfall is a significant environment factor of microplastic pollution in inland waters," *Sci Total Environ*, vol. 732, p. 139065, Aug. 2020, doi: 10.1016/j.scitotenv.2020.139065.

[43] K. Nouha, R. S. Kumar, S. Balasubramanian, and R. D. Tyagi, "Critical review of EPS production, synthesis and composition for sludge flocculation," *J Environ Sci*, vol. 66, pp. 225–245, Apr. 2018, doi: 10.1016/j.jes.2017.05.020.

[44] A. Hooge, H. H. Nielsen, W. M. Heinze, G. Lyngsie, T. M. Ramos, M. H. Sandgaard, J. Vollertsen, and K. Syberg, "Fate of microplastics in sewage sludge and in agricultural soils," *TrAC, Trends Anal Chem*, vol. 166, p. 117184, Sept. 2023, doi: 10.1016/j.trac.2023.117184.

11 Future Perspectives of Microplastic towards Environmental Assessment

Hari Murthy

11.1 INTRODUCTION

Plastics are sourced from fossil fuels and comprise long carbon chains along with H_2, O_2, N_2, Cl, and S. High-density polyethylene (HDPE), Polypropylene chloride (PVC), Polystyrene (PS), Polypropylene (PP), and Polyethersulfone (PES) are the most widely used plastics, accounting for about 75% of plastics production [1]. Plastics can be classified into homo-chain (C-C bond in the main chain like PS, PP, PE, etc.) and hetero-chain plastics (different bonds in the main chain like PET, PA, etc.) [2]. Owing to their versatility, low manufacturing cost, waterproof, durable, and easy-to-clean characteristics, they find applications in different fields, ranging from packaging, building, electronic industries, automobile, and manufacturing to name a few [3]. Production of plastics has increased 200 times over the past 50 years with 360 million tons being produced annually, and almost 2-3% of them ending up in water bodies [4], with the numbers expected to rise to 4-5% in the next couple of decades [5]. Plastic enters the aquatic ecosystems from three zones: discharge by humans, wastewater treatment plants (WWTPs) [6], and commercial activities such as manufacturing, agriculture, and marine. The rate of plastic manufacture has overcome our ability to deal with and recycle it efficiently, thus making plastic pollution a pressing environmental issue [7]. Plastic degradation is a three-step process that involves bond breakage through light (initiation), ultimate chain scission and cross-linking (propagation), and reaction completion (termination). The photolytic degradation of plastics produces several intermediates such as acetic acid, small molecules acid, long alkanes/olefins chain, and hydrogen leading to their mineralization. This process accounts for about half of the anthropogenic waste materials released to the environment, and without appropriate waste management infrastructure, is expected to increase ten times by the next decade in the absence of. There are two approaches to reducing plastic waste generation - educating people and developing biodegradable and oxo-biodegradable plastics. Generally, microplastics (MPs) are generated from the degradation of various types of carrier bags after nine months of exposure to a natural ambiance [8].

212   DOI: 10.1201/9781003486947-11

Future Perspectives in the Field of Microplastic

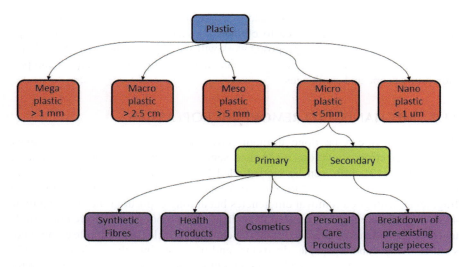

FIGURE 11.1 Classification of plastics based on particle size.

Figure 11.1 shows the various categories of plastics based on size and application [2, 3]. Once released into the environment, mechanical abrasion, weathering, biological degradation, and UV exposure can delaminate and degrade specific plastics, fragmenting them into small micro-nanoplastics (MNPs) which is a worldwide concern [1]. MNPs are further classified into primary and secondary MPs depending on their source [9].

MP because of its small size is a major environmental concern that needs immediate addressing.

MNPs have high surface area, hydrophobic properties, functional groups, and varied crystallographic junctions that allow the adsorption of organic chemicals, heavy metals, polycyclic aromatic hydrocarbons, pyrene, sulfamethoxazole [5], and antibiotics, thereby acting as potential contaminant vectors. Surface adsorption of MPs in the aquatic environment is influenced by surface chemistry and properties of the medium [4]. The nanometric size and the high surface reactivity of MPs increase their toxicity and reactivity toward the coexisting MPs.

Recent reports on the accumulation of MNPs confirm that they are a threat to terrestrial and aquatic environments, by entering the food chain affecting the internal organs of mice, and altering the biochemical processes [10]. Consequently, MPs can cause cancer, malformation in animals and humans, impaired reproductive activity, and reduced immune response [11].

Given the huge concerns around MNPs, studying the interactions of MPs is an area of great need in environmental risk assessment, to predict the lifecycle and environmental implications. The objective of this chapter is to provide an understanding of the latest trends and challenges for managing MP waste and the role that nanotechnology can play in the same. Photocatalytic nanomaterials have shown immense response to plastic degradation and can play a potentially major role in the future.

However, several areas need to be addressed before nanomaterials can be adopted wholly including but not limited to efficiency in the visible spectrum, toxicology studies, and applications. The proposed work will shed light on the existing work that has been presented on the application of nanomaterials for MP degradation and provide future scope and challenges.

11.2 MECHANISM TO REMOVE MICROPLASTICS

Plastic removal can be done through physical-chemical-biological techniques, including electrocoagulation, membrane separation, and magnetic extraction, to name a few [12]. MPs can be biodegraded by microorganisms such as *Bacillus cereus*, *Bacillus gottheilii*, *Zalerion maritimum*, and *Aspergillus flavus* to smaller fragments, with good removal efficiencies but at long exposure times ranging up to 40 days [8]. However, the underlying mechanisms of biodegradation are not entirely understood, and this process is still inefficient with the release of volatile compounds such as octadecane, eicosane, docosane, and tricosane into the environment [13]. The MP removal efficiency and development of filters are evaluated based on the MPs in the sludge runoff, effective iterations of filtration, and the effective filtration cost. The most commonly used method in most countries to remove large quantities of MPs from effluent rapidly is the sand filter, as it is cheap and efficient [5]. Sand filtration is the finishing treatment of municipal wastewater, considered as a promising wastewater treatment method for the efficient removal of MPs [14, 15] at low maintenance and energy. Sand filters fail when the source of MPs is textiles or cosmetics due to their smooth surface. WWTPs can efficiently remove the majority of MNPs but are unable to handle the huge concentration of MNPs in wastewater. It has been shown that post-treatment, about 73.8% of MPs can be accumulated in the sludge of the WWTPs [1]. Membrane bioreactor technology can completely remove MNPs from wastewater with 93% efficiency, with the efficiency reaching 79-89% if the size exceeds 100 μm. Pizzichetti et al. [16] reported a simple and low-cost system comprising three membranes for MP removal, while Poerio et al. [17] proposed a membrane technology. Bayo et al. [18] analyzed the effectiveness of membrane bioreactors and sand filtration for MP removal from urban WWTPs and concluded that the maintenance cost was higher [9], leading to abrasion and contamination [5]. These methods are costly, have high energy requirements, and complicated maintenance. Metal-organic frameworks (MOFs) with large specific surface areas have gained immense attention for MP removal because of their ease of fabrication, rapid functionalization, high association with various pollutants, and versatility to various types of MP suspensions [1]. However, mass production of MOFs is complex and costly, and their properties depend on the organic nodes and ligands, limiting their practical application. Adsorption is another process for water treatment because of its simplicity, efficiency, cost-effectiveness [9], and flexibility in selecting the adsorbents that show some affinity toward the target pollutants. Nanomaterials have been widely adopted as adsorbents because of their high surface areas, ease of synthesis and functionalization, and high association with various pollutants.

Future Perspectives in the Field of Microplastic

11.2.1 Mechanism of Photocatalytic Degradation

Photocatalysis is viable, inexpensive, environmentally friendly, and energy efficient for degradation of a wide range of polymer degradation. A surface-driven phenomenon, photocatalysis is driven by semiconductor-light interaction and has shown potential to treat effluents and MNPs. Nanostructured materials get excited with appropriate light energy creating reactive ions that effectively oxidize organic species including polymers [6]. On absorbing light, electrons get excited to the conduction band and form a hole in the valence band. These ions react with the water forming highly reactive oxygen species (ROS) such as O_2^- and OH radicals, which terminate -C-H bonds in organic compounds thereby oxidizing them [11, 17]. The addition of oxygen and removal of hydrogen results in the formation of peroxy and hydroperoxide radicals, which controls the degradation rate. There are two potential pathways for photodegradation; -one is the oxidation at the valence band, whereby holes oxidize organic compounds to CO_2 and H_2O, and generate hydroxyl radicals. In the second pathway, the electrons interact with oxygen, forming an O_2^- species, which on reacting with water generates H_2O_2 [19] that transforms into ·OH, causing degradation of MNPs. The chain cleavage occurs due to the dissociation of alkoxy radicals into carbonyl and vinyl group-containing species, leading to the complete mineralization of CO_2 and H_2O [20]. The photocatalytic degradation of MNPs is given as [8]

$$Semiconductor + h\nu \rightarrow H^+ + e^- \tag{11.1}$$

$$h^+ + H_2O \rightarrow OH^- \tag{11.2}$$

$$e^- + O_2 \rightarrow O_2^- \tag{11.3}$$

$$Pollutant + ROS \rightarrow CO_2 + H_2O \tag{11.4}$$

Figure 11.2 shows the photocatalytic process that results in the generation of ROS, which interacts with plastic materials to degrade and mineralize them to CO_2 and H_2O. The activation causes the generation of ROS, which reacts with plastics to mineralize them to CO_2 and H_2O.

Photodegradation of MNPs strongly depends on the photocatalyst used as it strongly influences the formation of ROS and degradation efficiency. Pollutant-surface ROS interaction plays a crucial role in degradation and is done in darkness until the adsorption equilibrium is reached [8], which is achievable when the pollutants dissolve in the reaction medium. It is of paramount importance to analyze the dispersion behavior and the effect of MNP size on photocatalysis, to design degradation systems that can mineralize MPs completely to CO_2 and H_2O.

The various factors that influence the generation of ROS and the photocatalytic are [21]:

1. ***Morphology***: Variations in morphologies such as surface area, porosity, and active site accessibility, impact the overall photocatalytic performance.

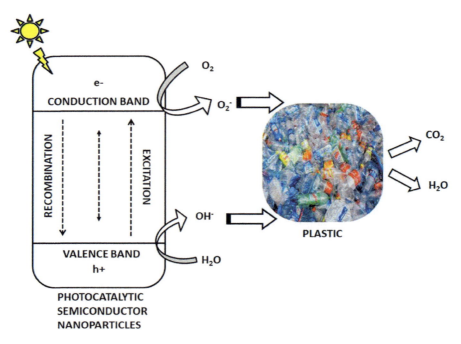

FIGURE 11.2 General mechanism of photocatalytic degradation of plastic using photocatalytic nanomaterials that activate under UV or visible light.

Optimizing morphology promotes better adsorption of MNPs, increased ROS, and efficient utilization of active sites on the catalyst surface. For example, in the absence of a catalyst, the morphology of PS shows barely noticeable change during initial irradiation but starts exhibiting changes during the first 3 h of irradiation, which increases with time [22]. Photodegradation-induced coalescence can be attributed to surface contraction and the development of cracks and cavities by the removal of intermediates, reorganization of the surface, rise in crystalline fractions, and oxygen permeation [23]. The size and density of the cavities are directly proportional to the duration of the photocatalytic degradation process, releasing volatile intermediates from the polymer particle surfaces.

2. **Surface area**: Photocatalytic performance depends on the active sites generated during excitation, with nanostructures with increased surface area reducing the distance traveled by charges on excitation, thus offering higher oxidation efficiency. Higher MP concentrations might lead to increased competition for active sites on the photocatalyst surface, influencing the overall degradation efficiency. The adsorption of MPs on a photocatalysts surface can be influenced by their size and a significant challenge in MP photocatalysis is addressing their limited or non-existent adsorption on the semiconductor due to their substantial particle size. The shape of MPs may impact their dispersion behavior and contact with the semiconductor.

3. **Temperature and pH:** Temperature and pH are important parameters for photocatalytic degradation. Semiconductor photocatalysts are influenced by temperature for charge carriers and ROS generation. Enhanced temperature increases the charge carriers, which in turn increases the reaction rate and photocatalytic activity. There is a saturation point, generally around 80 °C, beyond which charge recombination occurs reducing the efficiency.

    The catalyst and pH influence the formation of charge carriers and electrical double-layer charge at the surface and solid-electrolyte interface, respectively. Low pH levels favor the formation of hydroxyl radicals, while at higher pH levels, peroxide and superoxide are more prevalent. Improved degradation efficiency is observed at pH 3 due to increased hydroperoxide. Furthermore, pH also affects surface charge and inter-particle electrostatic attraction [24]. Mass loss at low pH (pH 3 and 0 °C) was measured to be 71.77% in HDPE MNP after 50 h of irradiation, while at 40 °C, it was 12.42%. MPs in alkaline solutions degrade faster in acidic conditions due to Coulomb repulsion [13]. Increased pH does not favor degradation - mass loss was 1.55% and 0.78% for temperatures of 0 °C and 40 °C (pH 11), respectively, with no change in morphology.

4. **Light:** UV and visible light affect the degradation of organic pollutants, with UV light having higher efficiency due to higher energy content. Lower intensity implies lower activation energy and lower degradation, and as intensity increases, more active sites are activated, increasing the degradation efficiency (up to a certain limit). It is essential to tune the MP to efficiently absorb UV or visible light.

11.3 NANOMATERIALS FOR DEGRADING MNPS

Nanomaterials are interesting photocatalytic materials because of their large specific surface area providing more active sites for oxidation and degradation of MP [21]. UV irradiation on metal oxide nanoparticles causes MP degradation to lower molecular weight biodegradable intermediates, such as carboxylic acids, ketones, and aldehydes. Transitional metal oxide nanoparticles (NPs) such as ZnO, TiO_2, CuO, Cu_2O, MnO_2, graphene, and Fe_3O_4 have significant applications in environmental areas and energy due to their high adsorption capacity, high porosity, light absorption, and unique electronic structures, making them suitable as photocatalysts [25].

11.3.1 TiO_2

Titanium dioxide (TiO_2) is among the most efficient photocatalysts due to its good photostability, high reactivity, low cost, and non-toxicity. TiO_2/plastic composite materials have been developed to produce photodegradable plastics of PP [26], PS [20], HDPE, and LDPE films [27]. Several methods have been investigated to synthesize pure, modified, doped, and nano TiO_2 to harvest a substantial portion of the light spectrum [2], such as bandgap engineering to enhance absorption, improve charge separation, and eliminate charge recombination as well as improvement in degradation efficiency [28].

Triton X-100 (TXT), due to its surface hydrophilicity and structure, demonstrated superior photocatalytic degradation irrespective of size compared to ethanol and water. TiO_2 nanofilms in Triton X-100 have exhibited a degradation efficiency of 98.4% after 12 h of UV exposure for PE and PS microspheres, with complete mineralization to carbonyl, hydroxyl, and CO_2 after 36 h [22]. Pristine TiO_2 nanorods on irradiating with a 50 W LED lamp degraded MP particles under a pH of 3 and at 0°C, with a mass loss of 60% after 50 h of irradiation [29]. The degradation under visible light was faster in an aqueous medium compared to a solid medium, though the photocatalysis was arrested in both media after irradiation for long periods, making it ineffective. Modifying pH and temperature, one can achieve HDPE microbeads weight loss percentages in the range of 6-71% [13, 29]. Protein-based N-TiO_2 with a surface area of 74.7 cm^2/g saw a total mass loss of HDPE by 4.6 % [30]. Improved PE and PS removal under visible light can be observed when TiO_2 was mixed with polypyrrole, multiwalled carbon nanotubes, copper phthalocyanine, iron phthalocyanine, and ferric stearate [31]. Titania nanotubes under UV light degraded LDPE evident by the increase in carbonyl index from 0.33 to a maximum of 2.0, though at negligible mass loss, suggesting little relevance of surface area on decomposition. Studies on C, N-TiO_2 photocatalyst with a surface area of 219.42 ± 1.82 m^2/g reacted with HDPE to introduce brittleness in its surface only at a very low temperature of 0°C, support that temperature is more critical than surface area for degradation [8]. Coating TiO_2 NPs with PE aided in degradation under visible light irradiation at 60 °C as suggested by the changes in the surface features, physical properties, and chemical compositions [32]. TiO_2-modified graphene oxide and silver NPs have been used to degrade PE [33]. Mesoporous N-TiO_2, ZnO, and Pt-ZnO photocatalysts allowed good contact between oxides and MPs [6, 29, 34]. The system was effective at driving MP degradation, although it did not reflect practical situations where it is less likely that MPs find themselves within the proximity of the semiconductor. C, N-TiO_2 semiconductor prepared from mussel proteins as a pore-forming agent [29, 35] was able to degrade PE MNPs extracted from a commercially available facial scrub [36]. Morphology of the catalyst is critical for the degradation, and significant degradation of PS MPs during 50 h under UV radiation was reported [37]. TiO_2 nanotubes with a surface area of 50 cm^2 were compared with a mixture of titanium oxide structures with a surface area of 200 cm^2, with the mixed nanograms structure exhibiting a higher elimination PS nanoplastics [37], though it is still not clear whether the surface area had a direct role to play in the degradation process or whether it was assisted by the structure as well. Degradation of PA under UV-C at 25-38 °C recorded a 97% mass loss within 48 hr [38]. TiO_2 surface becomes protonated or deprotonated depending on whether the environment is acidic or alkaline [39]. The surface will be positively charged in an acidic medium and negatively charged in an alkaline medium, with TiO_2 being a better oxidizing agent at lower pH levels. The stability of the plastic also depends on pH - HDPE is linear and crystalline and degrades at low pH [29], while PMMA and PS show good degradation at pH of 6.3 and 7, respectively [40]. TiO_2-P25/β-SiC foams effectively degraded polymers with diverse molecular structures, such as PS and PMMA, exhibiting a 50% TOC conversion after 7 hr of UV light radiation [41]. Doping platinum (Pt) into N-TiO_2 with a hydrothermal pre-treatment improved the degradation of PET MNPs [42]. Moreover, 3% Ag-doped TiO_2 demonstrated

Future Perspectives in the Field of Microplastic

enhanced performance in degrading 81% mass of MP in water samples from bottles after 4 hr of UV exposure [43].

MnO_2 is another material that can be considered for treatment due to its non-toxicity, low cost, ready availability, structural stability, and environmental friendliness, though it has very poor efficiency. Its activity is comparable to noble metals for the conversion of CO to CO_2. The photocatalytic degradation of LDPE films containing TiO_2-MnO_2 NPs suggested a significant mass loss of 21.3% when irradiated with UV for about 90 hr. The composite within the matrix makes MnO_2 NPs more active during PE oxidation and improves charge separation, while TiO_2 NPs improve the kinetics of the PE thermal decomposition between 250 and 500 °C [44].

11.3.2 ZnO

ZnO is a promising catalyst due to its excellent optical properties, high redox potential tunable MP (3.37 eV), non-toxicity, and high charge mobility. It is easy to fabricate and can be shaped into different morphologies using different synthesis methods. ZnO is better than TiO_2 due to its improved crystallinity, lower recombination, higher mobility of photogenerated electrons, and enhanced photocatalytic activity, though TiO_2 is more chemically stable. Photocatalytic performance and ROS production are more significantly determined by ZnO defects, crystallinity, morphology, and surface area. The major obstacle in the commercial application of ZnO is its reduced photocatalytic efficiency due to wide bandgap, rapid recombination, poor solar energy utilization, short charge carrier lifetime, limited active adsorption sites, and ineffective interface between ZnO-MPs in aquatic environments. ZnO under UV irradiation forms $Zn(OH)_2$ and dispersion of Zn^{2+} ions raises questions on their biotoxicity. For ZnO to be practically useful as a photocatalyst for efficient MP degradation, the research focus should revolve around efficient recovery and reuse, proper bio-assessment, bandgap engineering, doping, and the development of composite photocatalysts.

Photodegradation of PP in aqueous samples in ambiance under UV light using ZnO for 6 hours between 35 °C - 50 °C exhibited the highest degradation efficiency of 7.89% [45]. Tofa et al. [6] reported the degradation of LDPE, in aqueous medium using visible light-activated heterogeneous ZnO photocatalysts. LDPE oxidation leads to the generation of intermediates, causing surface changes and brittleness. The catalysts' surface area proved critical for enhancing LDPE degradation. LDPE surface was characterized by wrinkles, cracks, and cavities in addition to enhanced brittleness. ZnO nanorods of varying length and width were used to degrade LDPE film (residual). The total effective surface area was calculated to be 6.5, 22, 49, and 55 cm^2, respectively, which suggested that longer rods have a higher effective surface area and could be more effective for accelerated degradation.

ZnO tetrapods have exceptional optical properties (low recombination losses), with high photocatalytic activity despite their low surface area. He et al. [46] reported Ti-coated ZnO tetrapods for photocatalytic degradation of PP, PE microparticles, and PES microfibers. The rate of degradation is affected by the plastic morphology, and the presence of electron scavengers, such as H_2O_2 and $Na_2S_2O_8$, helped to maintain the photocatalytic efficiency. $Na_2S_2O_8$ exhibited better photocatalytic degradation with complete mass loss of PP, PE, and PES achieved under UV illumination after 816 hr, 480 h, and 624 h, respectively. There was improved charge separation and photocatalytic performance even at low BET surface area for TiO_2/ZnO core-shell structures (7.6 m²/g) compared to bare tetrapods (11.4 m²/g). Under simulated solar illumination, TiO_2/ZnO photocatalysts exhibit slower degradation due to reduced crystallinity. For both PE and PES, a saturation of the degradation was achieved with 100% mass loss for PE achievable. The addition of $Na_2S_2O_8$ enhances ROS generation, accelerating the degradation of PES, until it saturates. PE exhibited particle size reduction, while PES fibers showed a change in morphology, with reduced mass loss over time. Complete mass loss can be obtained for all types of MPs investigated, with varying degradation times. Uheida et al. [23] studied the visible light photocatalytic degradation for two weeks under visible light (~ four weeks of half-day sunlight) of PP MP using ZnO nanorods (NRs) coated on glass fibers, which occurred through chain scissions, causing ~65% volume reduction. Platinum (Pt) has been found to enhance visible light absorption while diminishing electron-hole recombination. ZnO-Pt exhibited 13% higher capacity and 15% higher photocatalytic performance for oxidizing LDPE MNPs under visible light due to surface modifications by the creation of oxygenated groups.

11.3.3 CuO and Cu_2O

Copper oxides (CuO, Cu_2O) are a p-type semiconductor with a narrow bandgap (~1.7-2.2 eV) and are affordable, harmless, and abundantly available, but possess a high recombination rate [25]. Due to the higher bandgap (>2.9 eV), TiO_2 and ZnO have limited utilization of visible light with reduced efficiency and energy consumption. Cu_xO with its lower MP is activated in the visible range and favors oxidation and charge generation, resulting in the degradation of various organic pollutants such as salicylic acid, orange II, and aspirin [47]. PS particles of ~ 350 nm were degraded using immobilized Cu_2O/CuO, with a 23% reduction in the PS-NPs concentration and mineralization of up to 15%. Initially, the pore generated on the surface size is smaller than the NPs, which implies no change in PS concentration due to the non-absorption of the contaminants on the photocatalyst surface. High pH (11- 12) weakens the electrostatic interactions between the oxide surface and the PS-NPs. Photolysis resulted in a 4% decrease in PS concentration, which improved to 11% for NaOH solution and 18% for NH_4F solution on using photocatalysts under visible-light irradiation and 17.1% after 48 h of UV light exposure. Cu_xO exhibited eliminations up to 23.5% under visible light irradiation. NH_4F solution showed a better response due to porous mixed Cu_2O/CuO structures promoting higher reaction surface, reducing electron recombination, and generating more oxidizing species. Photolysis exhibited a 0.6% increase in TOC, which increased significantly in photocatalytic treatments due to the breakdown of PS-NPs into intermediates. NH_4F solution showed a higher value of TOC (~12%), followed by a decay indicating mineralization of the intermediaries. Photocatalysts grown in NH_4F inhibit the formation of OH radicals and support the formation of superoxide radicals as the oxidation potential is not sufficient for OH generation. There is possible reuse of the material at least once without loss of properties due to a better I-V response, and stability under visible light. The degradation

Future Perspectives in the Field of Microplastic 221

efficiencies and mineralization were up to 18-23% and 15%, respectively, in 50 h of treatment at ambient conditions.

11.3.4 GRAPHENE OXIDE (GO)-BASED NANOCOMPOSITES

Graphene oxide (GO) is a 2D layered structure with sp2 hybridized orbitals and oxygen functional groups. It accepts photogenerated electrons and reduces recombination, thereby increasing photocatalytic activity. GO-based metal oxide NPs (GO-Cu_2O, GO-TiO_2, GO-MnO_2) and binary or ternary hybrid systems such as rGO/Fe_2O_3 or rGO/Fe_2O_3/ZnO have better degradation rates, improved photocatalytic performance, and sorption capacity [48]. The higher sorption rates are due to their small size, hydroxyl groups on the surface, and efficient electron transfer. GO directs electrons toward the absorbed MP and initiates redox degradation by generation of ROS. The mass loss of PE particles depends on the reaction time with the mass loss doubling after 4 hrs - 48.06%, 39.54%, and 50.46% for GO-Cu_2O, GO-MnO_2 and GO-TiO_2, respectively, compared with 35.66% for GO. The degradation rate was influenced by pH with optimal values observed at lower pH values and longer UV exposure time. The most efficient photocatalysts were GO-TiO_2 followed by GO-Cu_2O and GO-MnO_2.

Graphite carbon nitride (g-C_3N_4, GCN) is a polymeric semiconductor exhibiting good thermochemical, appropriate MP (~ 2.7 eV), ease of synthesis, and cheap, excellent physicochemical stability, electronic, and optical properties under visible light irradiation [12]. GCN-α-Fe_2O_3 nanocomposites improve the degradation of PE and PS MNPs significantly [49–51]. The performance for PS degradation was better than PE, due to the presence of a reactive C-H tertiary carbon bond in PS making it susceptible to photooxidative degradation and cleavage. The benzene ring on the PS surface gets activated and transfers the excitation energy transfers to the nearest C-H bonds, thereby breaking the bonds [52].

High-photocatalytic activity bismuth oxide/graphitic carbon nitride composites were synthesized [53]. CuO/GCN was shown to have increased photocatalytic activity for the degradation of salicylic acid [54] in the presence of sunlight. Mandal and Masthafa [25] proposed a novel photodegradable CuO/Bi_2O_3/GCN composite for sunlight-mediated degradation of PET MNP films. The photocatalytic activity of the nanocomposite CuO/Bi_2O_3/GCN/PET films was greater than that of pristine films, due to their increased optical absorption and efficient suppression of photo-produced charge carriers recombination. PET films with CuO/Bi_2O_3/GCN (5:5:2% wt) NPs showed a high weight loss of 41.60 ± 0.44 % after 240 h of sunlight exposure, which increased with irradiation time. A significant weight reduction of 5.73 ± 0.22% was also observed in the dark which implied that the reaction also proceeds in the absence of any photons. The reactivity is due to photo-induced electron transfer across NPs, and the weight reduction suggests bond scission with the generation of volatile substances such as H_2O, CO_2, and volatile organic molecules. The weight loss process of the non-irradiated CuO/Bi_2O_3/GCN/PET NCs film began at 110 °C and proceeded to 23.28% weight loss at 440 °C, while the photo-irradiated CuO/Bi_2O_3/GCN/PET NCs film demonstrated an initial 11.14% weight loss at 140 °C, which further yielded a 29.34% weight loss at 440 °C. Increasing the temperature to 1200 °C showed a final weight loss of 61.14% and 72.25% for the non-irradiated and radiated films. The photocatalytic mechanism behind the degradation of PET films by $CuO/Bi_2O_3/$ GCN NPs under sunlight irradiation is due to the change in the MP of $CuO/Bi_2O_3/$ GCN in addition to GCN. Following each cycle, the NPs were collected, washed multiple times with water and acetone, and dried at 80 °C for 2 hours before being reused. The reusability of $CuO/Bi_2O_3/GCN$ NPs in PET film has been studied for up to five cycles. The recovered $CuO/Bi_2O_3/GCN$ NPs were then utilized to prepare polymer NCs ($CuO/Bi_2O_3/GCN$ /PET) film, which was then exposed to solar light radiation in the open air under ambient conditions for 240 hours. In the fourth cycle, no significant loss of activity for $CuO/Bi_2O_3/GCN$ NPs was found. The first cycle had a degradation efficiency of around 86.37%, which was reduced to 78.04% in the fifth cycle. The NPs have high reusability up to the fourth cycle, confirming their sustainability. The GCN may effectively improve the separation and migration of $CuO/$ Bi_2O_3-photogenerated electrons, inhibit the conformation of photogenerated carriers, and improve the photodegradation rate of PET.

Laser-induced graphene (LIG) has a porous structure, hydrophobicity, inherent zeta potential, and low manufacturing cost making it a suitable material for dye adsorption or other organic contaminants [55]. The disadvantages of conventional LIGs for MP adsorption are overcome by preparing hybrid $Fe_3O_4/LIGPs$ on Polydimethylsiloxane (PDMS). Fe_3O_4 particles coated on the surface of the LIGPs render them hydrophobic and magnetic, making them easy to separate from the wastewater suspension. The resulting tertiary system exhibited higher MP removal efficiency with the efficiency increasing with an increase in the MP concentration. Fe_3O_4-LIGPs showed removal efficiency above 90% for three types of MPs: 99.1% for PA (50 µm), 96.8% and 92.3% for melamine (10 and 2 µm, respectively), and 98.4% for PS (10 µm) [5]. The zeta potentials of the LIG and Fe_3O_4-LIGPs were calculated over the pH range of 2.5 -12, and found to be around - 38.6 mV and - 20.1 mV, respectively. Fe_3O_4-LIGPs could adsorb the larger MPs (size ~10 µm) and trap smaller MPs (2 µm) within the structure, efficiently removing them from the suspension. The maximum adsorption capacities of the Fe_3O_4-LIGPs were 775 (melamine, 2 µm), 1050 (melamine, 10 µm), 1250 (PS, 10 µm), and 1400 (PA,50 µm) mg/g. Fe_3O_4-LIGPs showed the lowest affinity for melamine (2 µm) due to its low particle, while the H-bonding and hydrophobicity impact the MP adsorption on the surface. The highest adsorption capacity of PS and PA was at neutral pH. In acidic solutions, the adsorption capacity of the Fe_3O_4-LIGPs is affected due to the behavior of the carbon-based samples, the decrease in the hydrophobicity, and the change of zeta potential. Activation of MP surface functional groups causes the pH change, where at higher pH levels, enhanced surface charge due to the adsorption of OH^- ions results in inferior performance. The MP removal efficiency decreased gradually with heat treatment and reuse cycles- removal efficiency is around 80% PS and PA in the 4th cycle, and drops to 70% in the 7th cycle. The removal efficiency of the Fe_3O_4-LIGPs for the melamine particles of 2 µm and 10 µm size was 70% (4th cycle) and >70% (5th cycle), respectively, due to decreased magnetic force and bond breakage of the Fe_3O_4-LIGPs during the heat treatment.

Future Perspectives in the Field of Microplastic

11.4 FUTURE SCOPE

There are few studies on the dependence of pH and temperature on the photocatalytic performance of MNPs. The properties of nanomaterials need to be optimized with regard to their surface area, pore size, and chemical composition, to effectively enhance their MNP removal efficiency. Existing mechanisms such as activated carbon filters or sand filters should be able to integrate nanomaterials, to enhance their performance. Some areas for future research and development on the use of nanomaterials for MP removal from aqueous environments include:

- *Bandgap engineering*: Among the several challenges in the practical application of photocatalysts, one of the main ones is its absorption band in the UV region owing to its wide bandgap. This limited applicability reduces the utilization of entire solar energy and photocatalytic efficiency. Other challenges that need to be looked at are the rapid recombination of photogenerated charge carriers and the limited surface area and adsorption capability, which impedes ROS generation and effective interface between the photocatalyst and MNPs. Developing strategies for bandgap tuning, doping, nanocomposites, and heterostructure photocatalysts for efficient degradation is the need of the hour. Bandgap tuning increases light absorption in the visible spectrum and enhances charge and ROS generation for effectively degrading MPs.
- *Optimization of nanomaterial properties*: Pore sizes and chemical composition of the NPs or MOFs can be altered to improve the affinity and adsorption of larger MPs particles.
- *Scale-up production*: Scaling up reduces the overall cost, making the system accessible for widespread use. Scaling up of NPs is possible through process optimization, exploring new methods for large-scale production, and improving the yield of MOF synthesis.
- *Long-term stability and regeneration*: Analysis of the long-term stability of nanomaterials under varying water chemistry and temperature of aqueous environments is necessary to develop efficient regeneration methods such as heating or washing for their continued use.
- *Real-world application*: As most characterization and device performance testing of nanocatalysts is done in the laboratory, it is essential to test them in different aquatic environments, with variable water chemistry and flow rates, to assess their effectiveness and identify limitations.
- *Stability and reusability*: The development of stable and reusable nanocatalysts whose performance does not degrade over multiple use cycles is essential for large-scale applications.
- *Toxicity*: It is critical to analyze the biotoxicity of the intermediates generated during the partial or complete degradation of MPs, and develop photocatalysts that reduce their formation.
- *Selectivity*: Aquatic bodies would have multiple pollutants in them which would impact the selectivity of photocatalysts toward MNPs. One of the key areas that need to be looked at to improve the removal efficiency of the degradation system is to develop selective photocatalysts that target specific MPs.

- *Scalability*: As most research is conducted in a laboratory, scaling the size and production for practical applications without compromising their performance is a major hurdle.
- *Economic feasibility*: One of the major challenges that need immediate attention is to make the entire process economic including but not limited to reactor design and energy consumption. Addressing the cost-effectiveness of photocatalytic system would play a major role in their large-scale implementation.
- *Metal-free catalysts*: These are more environment friendly, and can remove various types of contaminants, including endocrine disruptors and antibiotics [56]. There is a lack of research on the application of metal-free catalysts, including their reusability and stability and therefore is an area that necessitates further investigation.
- *Technology Readiness level*: One important parameter to be considered before commercialization is the Technology Readiness Level (TRL) [57] of the proposed techniques. On the TRL scale, most photocatalytic degradation methods are at the lower extreme, which implies that they require significant development to be commercially applicable for MNP degradation. Additionally, the Electrical Energy per Order (EEO) examines the photocatalytic degradation methods [58]. EEO and TRL determine the practicality of photocatalytic systems in commercial settings and advanced photocatalytic reactor and system technology is the need of the hour for elevating the TRL of photocatalytic degradation methods [59].

11.5 CONCLUSION

Plastic disposal has been acknowledged as a global problem due to its ubiquitous presence and threat to the ecosystem. Worldwide manufacturing, use, and disposal of MNPs can cause serious harm to various ecosystems. Considerable progress has been made towards the sustainable decompositions of MNPs by employing the photocatalytic degradation method; however, one specific technique would not be sufficient to address all the concerns of MNPs completely from the aquatic environment and innovative technologies have to be involved to get better removal efficiency. Size and morphology influence the MNP dispersal where large, dense, and irregular-shaped particles tend to sediment while smaller, lighter, and spherical-shaped particles are retained on the surface. A knowledge base has to be created for the photocatalytic materials with respect to their size optimization, degradation efficiency, reusability, cost, and bioimpact. The development of new efficient visible-light-driven photocatalytic materials is of paramount importance. Photodegradation of MNPs is a complex process and it is essential to analyse the role of ROS in various types of plastic decomposition. Current techniques fail to degrade MPs completely and require significant time to achieve results. Nanocomposite and heterostructured photocatalysts and metal-free systems have shown improved performance with greater environmental friendliness. Several challenges need to be addressed before the photocatalytic technique can be considered as a viable technique, such as the stability, reusability, toxicity of intermediate products, scalability, and economic feasibility. To summarize, nano-photocatalytic decomposition techniques offer a method to resolve the issue of

Future Perspectives in the Field of Microplastic

MP pollution in oceans, though significant research is required before the technique can be adopted completely. Research on the development of novel materials and nanostructures for effective and sustainable removal of MP, ultimately safeguarding ecosystems and health is the need of the hour.

CONFLICT OF INTEREST

The authors confirm that there is no conflict of interest.

AUTHOR CONTRIBUTION

Hari Murthy was involved in the design, review, and writing of the manuscript.

REFERENCES

[1] Z. Honarmandrad, M. Kaykhaii, and J. Gębicki, 'Microplastics removal from aqueous environment by metal organic frameworks', *BMC Chemistry*, vol. 17, no. 1, p. 122, Sep. 2023, doi: 10.1186/s13065-023-01032-y.

[2] I. Nabi, A.-U.-R. Bacha, F. Ahmad, and L. Zhang, 'Application of titanium dioxide for the photocatalytic degradation of macro- and micro-plastics: A review', *Journal of Environmental Chemical Engineering*, vol. 9, no. 5, p. 105964, Oct. 2021, doi: 10.1016/j.jece.2021.105964.

[3] G. Astray, A. Soria-Lopez, E. Barreiro, J. C. Mejuto, and A. Cid-Samamed, 'Machine learning to predict the adsorption capacity of MPs', *Nanomaterials*, vol. 13, no. 6, p. 1061, Mar. 2023, doi: 10.3390/nano13061061.

[4] I. Uogintė, S. Pleskytė, M. Skapas, S. Stanionytė, and G. Lujanienė, 'Degradation and optimization of microplastic in aqueous solutions with graphene oxide-based nanomaterials', *International Journal of Environmental Science and Technology*, vol. 20, no. 9, pp. 9693–9706, Sep. 2023, doi: 10.1007/s13762-022-04657-z.

[5] S.-Y. Jeong, N. Sugita, and B.-S. Shin, 'Fe3O4/laser-induced graphene as an adsorbent for microplastics emitted from household wastewater', *International Journal of Precision Engineering and Manufacturing-Green Technology.*, vol. 10, no. 3, pp. 807–818, May 2023, doi: 10.1007/s40684-022-00464-6.

[6] T. S. Tofa, K. L. Kunjali, S. Paul, and J. Dutta, 'Visible light photocatalytic degradation of microplastic residues with zinc oxide nanorods', *Environmental Chemistry Letters*, vol. 17, no. 3, pp. 1341–1346, Sep. 2019, doi: 10.1007/s10311-019-00859-z.

[7] M. A. Gomez-Gonzalez, T. Da Silva-Ferreira, N. Clark, R. Clough, P. D. Quinn, and J. E. Parker, 'Toward understanding the environmental risks of combined microplastics/nanomaterials exposures: Unveiling ZnO transformations after adsorption onto polystyrene microplastics in environmental solutions', *Global Challenges*, vol. 7, no. 8, p. 2300036, Aug. 2023, doi: 10.1002/gch2.202300036.

[8] B. E. Llorente-García, J. M. Hernández-López, A. A. Zaldívar-Cadena, C. Siligardi, and E. I. Cedillo-González, 'First insights into photocatalytic degradation of HDPE and LDPE microplastics by a mesoporous N–TiO2 coating: Effect of size and shape of microplastics', *Coatings*, vol. 10, no. 7, p. 658, Jul. 2020, doi: 10.3390/coatings10070658.

[9] M. Sajid, I. Ihsanullah, M. Tariq Khan, and N. Baig, 'Nanomaterials-based adsorbents for remediation of microplastics and nanoplastics in aqueous media: A review',

Separation and Purification Technology, vol. 305, p. 122453, Jan. 2023, doi: 10.1016/j.seppur.2022.122453.

[10] J.-L. Xu, X. Lin, J. J. Wang, and A. A. Gowen, 'A review of potential human health impacts of micro- and nanoplastics exposure', *Science of The Total Environment*, vol. 851, p. 158111, Dec. 2022, doi: 10.1016/j.scitotenv.2022.158111.

[11] Campanale, Massarelli, Savino, Locaputo, and Uricchio, 'A detailed review study on potential effects of microplastics and additives of concern on human health', *International Journal of Environmental Research and Public Health*, vol. 17, no. 4, p. 1212, Feb. 2020, doi: 10.3390/ijerph17041212.

[12] C. Chai *et al.*, 'Photocatalytic degradation of polyethylene and polystyrene microplastics by α-Fe2O3/g-C3N4', *Environmental Science and Pollution Research*, vol. 30, no. 58, pp. 121702–121712, Nov. 2023, doi: 10.1007/s11356-023-31000-x.

[13] M. C. Ariza-Tarazona *et al.*, 'Microplastic pollution reduction by a carbon and nitrogen-doped TiO_2: Effect of pH and temperature in the photocatalytic degradation process', *Journal of Hazardous Materials*, vol. 395, p. 122632, Aug. 2020, doi: 10.1016/j.jhazmat.2020.122632.

[14] X.-T. Bui, T.-D.-H. Vo, P.-T. Nguyen, V.-T. Nguyen, T.-S. Dao, and P.-D. Nguyen, 'Microplastics pollution in wastewater: Characteristics, occurrence and removal technologies', *Environmental Technology & Innovation*, vol. 19, p. 101013, Aug. 2020, doi: 10.1016/j.eti.2020.101013.

[15] E. Sembiring, M. Fajar, and M. Handajani, 'Performance of rapid sand filter – Single media to remove microplastics', *Water Supply*, vol. 21, no. 5, pp. 2273–2284, Aug. 2021, doi: 10.2166/ws.2021.060.

[16] A. R. P. Pizzichetti, C. Pablos, C. Álvarez-Fernández, K. Reynolds, S. Stanley, and J. Marugán, 'Evaluation of membranes performance for microplastic removal in a simple and low-cost filtration system', *Case Studies in Chemical and Environmental Engineering*, vol. 3, p. 100075, Jun. 2021, doi: 10.1016/j.cscee.2020.100075.

[17] Poerio, Piacentini, and Mazzei, 'Membrane processes for microplastic removal', *Molecules*, vol. 24, no. 22, p. 4148, Nov. 2019, doi: 10.3390/molecules24224148.

[18] J. Bayo, S. Olmos, and J. López-Castellanos, 'Microplastics in an urban wastewater treatment plant: The influence of physicochemical parameters and environmental factors', *Chemosphere*, vol. 238, p. 124593, Jan. 2020, doi: 10.1016/j.chemosphere.2019.124593.

[19] J. M. Kesselman, O. Weres, N. S. Lewis, and M. R. Hoffmann, Electrochemical production of hydroxyl radical at polycrystalline Nb-doped TiO_2 electrodes and estimation of the partitioning between hydroxyl radical and direct hole oxidation pathways, *The Journal of Physical Chemistry B*, 101(14), pp. 2637–2643, 1997. doi: 10.1021/jp962669r.

[20] J. Shang, M. Chai, and Y. Zhu, 'Photocatalytic degradation of polystyrene plastic under fluorescent light', *Environmental Science and Technology.*, vol. 37, no. 19, pp. 4494–4499, Oct. 2003, doi: 10.1021/es0209464.

[21] E. Kinyua, G. Nyakairu, E. Tebandeke, and N. Odume, 'Photocatalytic degradation of microplastics: Parameters affecting degradation', *Advances in Environmental and Engineering Research*, vol. 04, no. 03, pp. 1–21, Jul. 2023, doi: 10.21926/aeer.2303039.

[22] I. Nabi *et al.*, 'Complete photocatalytic mineralization of microplastic on TiO_2 nanoparticle film', *iScience*, vol. 23, no. 7, p. 101326, Jul. 2020, doi: 10.1016/j.isci.2020.101326.

[23] A. Uheida, H. G. Mejía, M. Abdel-Rehim, W. Hamd, and J. Dutta, 'Visible light photocatalytic degradation of polypropylene microplastics in a continuous water flow

system', *Journal of Hazardous Materials*, vol. 406, p. 124299, Mar. 2021, doi: 10.1016/j.jhazmat.2020.124299.

[24] X. U. Zhao, Z. Li, Y. Chen, L. Shi, and Y. Zhu, 'Solid-phase photocatalytic degradation of polyethylene plastic under UV and solar light irradiation', *Journal of Molecular Catalysis A: Chemical*, vol. 268, no. 1–2, pp. 101–106, May 2007, doi: 10.1016/j.molcata.2006.12.012.

[25] J. M. Musthafa, and B. K. Mandal, $CuO/Bi_2O_3/g-C_3N_4$ nanoparticles for sunlight-mediated degradation of polyethylene terephthalate microplastic films, *Optical Materials*, vol. 154, pp.115701–115714, 2024, https://doi.org/10.1016/j.optmat.2024.115701.

[26] M. M. Kamrannejad, A. Hasanzadeh, N. Nosoudi, L. Mai, and A. A. Babaluo, 'Photocatalytic degradation of polypropylene/TiO_2 nano-composites', *Materials Research*, vol. 17, no. 4, pp. 1039–1046, Aug. 2014, doi: 10.1590/1516-1439.267214.

[27] S. Wang, J. Zhang, L. Liu, F. Yang, and Y. Zhang, 'Evaluation of cooling property of high density polyethylene (HDPE)/titanium dioxide (TiO_2) composites after accelerated ultraviolet (UV) irradiation', *Solar Energy Materials and Solar Cells*, vol. 143, pp. 120–127, Dec. 2015, doi: 10.1016/j.solmat.2015.06.032.

[28] J. Moma and J. Baloyi, 'Modified Titanium Dioxide for Photocatalytic Applications', in *Photocatalysts – Applications and Attributes*, S. Bahadar Khan and K. Akhtar, Eds., IntechOpen, 2019, doi: 10.5772/intechopen.79374.

[29] M. C. Ariza-Tarazona, J. F. Villarreal-Chiu, V. Barbieri, C. Siligardi, and E. I. Cedillo-González, New strategy for microplastic degradation: Green photocatalysis using a protein-based porous N-TiO_2 semiconductor, *Ceramics International*, 45(7), pp. 9618–9624, 2019, doi: 0.1016/j.ceramint.2018.10.208.

[30] S. S. Ali, I. A. Qazi, M. Arshad, Z. Khan, T. C. Voice, and Ch. T. Mehmood, 'Photocatalytic degradation of low density polyethylene (LDPE) films using titania nanotubes', *Environmental Nanotechnology, Monitoring & Management*, vol. 5, pp. 44–53, May 2016, doi: 10.1016/j.enmm.2016.01.001.

[31] X. Zhao, Z. Li, Y. Chen, L. Shi, and Y. Zhu, 'Enhancement of photocatalytic degradation of polyethylene plastic with CuPc modified TiO_2 photocatalyst under solar light irradiation', *Applied Surface Science*, vol. 254, no. 6, pp. 1825–1829, Jan. 2008, doi: 10.1016/j.apsusc.2007.07.154.

[32] H. Luo, Y. Xiang, T. Tian, and X. Pan, 'An AFM-IR study on surface properties of nano-TiO_2 coated polyethylene (PE) thin film as influenced by photocatalytic aging process', *Science of The Total Environment*, vol. 757, p. 143900, Feb. 2021, doi: 10.1016/j.scitotenv.2020.143900.

[33] M. H. Fadli, M. Ibadurrohman, and S. Slamet, 'Microplastic pollutant degradation in water using modified TiO_2 photocatalyst under UV-irradiation', *IOP Conference Series: Materials Science and Engineering.*, vol. 1011, no. 1, p. 012055, Jan. 2021, doi: 10.1088/1757-899X/1011/1/012055.

[34] T. S. Tofa, F. Ye, K. L. Kunjali, and J. Dutta, 'Enhanced visible light photodegradation of microplastic fragments with plasmonic platinum/zinc oxide nanorod photocatalysts', *Catalysts*, vol. 9, no. 10, p. 819, Sep. 2019, doi: 10.3390/catal9100819.

[35] H. Zeng *et al.*, 'Bioprocess-inspired synthesis of hierarchically porous nitrogen-doped TiO_2 with high visible-light photocatalytic activity', *Journal of Materials Chemistry A*, vol. 3, no. 38, pp. 19588–19596, 2015, doi: 10.1039/C5TA04649A.

[36] I. E. Napper, A. Bakir, S. J. Rowland, and R. C. Thompson, 'Characterisation, quantity and sorptive properties of microplastics extracted from cosmetics',

Marine Pollution Bulletin, vol. 99, no. 1–2, pp. 178–185, Oct. 2015, doi: 10.1016/j.marpolbul.2015.07.029.

[37] L. P. Domínguez-Jaimes, E. I. Cedillo-González, E. Luévano-Hipólito, J. D. Acuña-Bedoya, and J. M. Hernández-López, 'Degradation of primary nanoplastics by photocatalysis using different anodized TiO_2 structures', *Journal of Hazardous Materials*, vol. 413, p. 125452, Jul. 2021, doi: 10.1016/j.jhazmat.2021.125452.

[38] J.-M. Lee, R. Busquets, I.-C. Choi, S.-H. Lee, J.-K. Kim, and L. C. Campos, 'Photocatalytic degradation of polyamide 66; evaluating the feasibility of photocatalysis as a microfibre-targeting technology', *Water*, vol. 12, no. 12, p. 3551, Dec. 2020, doi: 10.3390/w12123551.

[39] U. I. Gaya and A. H. Abdullah, 'Heterogeneous photocatalytic degradation of organic contaminants over titanium dioxide: A review of fundamentals, progress and problems', *Journal of Photochemistry and Photobiology C: Photochemistry Reviews*, vol. 9, no. 1, pp. 1–12, Mar. 2008, doi: 10.1016/j.jphotochemrev.2007.12.003.

[40] P. H. Allé, P. Garcia-Muñoz, K. Adouby, N. Keller, and D. Robert, 'Efficient photocatalytic mineralization of polymethylmethacrylate and polystyrene nanoplastics by TiO_2/β-SiC alveolar foams', *Environmental Chemistry Letters*, vol. 19, no. 2, pp. 1803–1808, Apr. 2021, doi: 10.1007/s10311-020-01099-2.

[41] M. Kosmulski, 'Isoelectric points and points of zero charge of metal (hydr) oxides: 50 years after Parks' review', *Advances in Colloid and Interface Science*, vol. 238, pp. 1–61, Dec. 2016, doi: 10.1016/j.cis.2016.10.005.

[42] D. Zhou, H. Luo, F. Zhang, J. Wu, J. Yang, and H. Wang, 'Efficient photocatalytic degradation of the persistent PET fiber-based microplastics over Pt nanoparticles decorated N-doped TiO_2 nanoflowers', *Advanced Fiber Materials*, vol. 4, no. 5, pp. 1094–1107, Oct. 2022, doi: 10.1007/s42765-022-00149-4.

[43] M. F. Haris, A. M. Didit, M. Ibadurrohman, Setiadi, and Slamet, 'Silver doped TiO_2 photocatalyst for disinfection of E. coli and microplastic pollutant degradation in water', *Asian Journal of Chemistry*, vol. 33, no. 9, pp. 2038–2042, 2021, doi: 10.14233/ajchem.2021.23255.

[44] I. Kovinchuk, N. Haiuk, G. Lazzara, G. Cavallaro, and G. Sokolsky, 'Enhanced photocatalytic degradation of PE film by anatase/γ-MnO_2', *Polymer Degradation and Stability*, vol. 210, p. 110295, Apr. 2023, doi: 10.1016/j.polymdegradstab.2023.110295.

[45] Universiti Malaysia Terengganu, N. Razali, W. R. Wan Abdullah, and N. Mohd Zikir, 'Effect of thermo-photocatalytic process using zinc oxide on degradation of macro/micro-plastic in asqueous environment', *Journal of Sustainability Science and Management*, vol. 15, no. 6, pp. 1–14, Aug. 2020, doi: 10.46754/jssm.2020.08.001.

[46] Y. He, A. U. Rehman, M. Xu, C. A. Not, A. M. C. Ng, and A. B. Djurišić, 'Photocatalytic degradation of different types of microplastics by TiO_2/ZnO tetrapod photocatalysts', *Heliyon*, vol. 9, no. 11, p. e22562, Nov. 2023, doi: 10.1016/j.heliyon.2023.e22562.

[47] J. D. Acuña-Bedoya, E. Luévano-Hipólito, E. I. Cedillo-González, L. P. Domínguez-Jaimes, A. M. Hurtado, and J. M. Hernández-López, 'Boosting visible-light photocatalytic degradation of polystyrene nanoplastics with immobilized Cu_xO obtained by anodization', *Journal of Environmental Chemical Engineering*, vol. 9, no. 5, p. 106208, Oct. 2021, doi: 10.1016/j.jece.2021.106208.

[48] R. K. Upadhyay, N. Soin, and S. S. Roy, 'Role of graphene/metal oxide composites as photocatalysts, adsorbents and disinfectants in water treatment: a review', *RSC Advances*, vol. 4, no. 8, pp. 3823–3851, 2014, doi: 10.1039/C3RA45013A.

[49] Q. Xu, B. Zhu, C. Jiang, B. Cheng, and J. Yu, 'Constructing 2D/2D Fe_2O_3 /g-C_3N_4 direct Z-scheme photocatalysts with enhanced H_2 generation performance', *Solar RRL*, vol. 2, no. 3, p. 1800006, Mar. 2018, doi: 10.1002/solr.201800006.

Future Perspectives in the Field of Microplastic

[50] T. Guo, K. Wang, G. Zhang, and X. Wu, 'A novel α-Fe$_2$O$_3$@g-C$_3$N$_4$ catalyst: Synthesis derived from Fe-based MOF and its superior photo-Fenton performance', *Applied Surface Science*, vol. 469, pp. 331–339, Mar. 2019, doi: 10.1016/j.apsusc.2018.10.183.

[51] M. S. Athar, M. Danish, and M. Muneer, 'Fabrication of visible light-responsive dual Z-Scheme (α-Fe$_2$O$_3$/CdS/g-C$_3$N$_4$) ternary nanocomposites for enhanced photocatalytic performance and adsorption study in aqueous suspension', *Journal of Environmental Chemical Engineering*, vol. 9, no. 4, p. 105754, Aug. 2021, doi: 10.1016/j.jece.2021.105754.

[52] E. Yousif and R. Haddad, 'Photodegradation and photostabilization of polymers, especially polystyrene: Review', *SpringerPlus*, vol. 2, no. 1, p. 398, Dec. 2013, doi: 10.1186/2193-1801-2-398.

[53] M. Ben Abdelaziz *et al.*, 'One pot synthesis of bismuth oxide/graphitic carbon nitride composites with high photocatalytic activity', *Molecular Catalysis*, vol. 463, pp. 110–118, Feb. 2019, doi: 10.1016/j.mcat.2018.12.004.

[54] Y. Duan, 'Facile preparation of CuO/g-C$_3$N$_4$ with enhanced photocatalytic degradation of salicylic acid', *Materials Research Bulletin*, vol. 105, pp. 68–74, Sep. 2018, doi: 10.1016/j.materresbull.2018.04.038.

[55] N. H. Barbhuiya, A. Kumar, and S. P. Singh, 'A journey of laser-induced graphene in water treatment', *Transactions of the Indian National Academy of Engineering*, vol. 6, no. 2, pp. 159–171, Jun. 2021, doi: 10.1007/s41403-021-00205-2.

[56] L. Hong *et al.*, 'Recent progress of two-dimensional MXenes in photocatalytic applications: A review', *Materials Today Energy*, vol. 18, p. 100521, Dec. 2020, doi: 10.1016/j.mtener.2020.100521.

[57] N. Taghavi, I. A. Udugama, W.-Q. Zhuang, and S. Baroutian, 'Challenges in biodegradation of non-degradable thermoplastic waste: From environmental impact to operational readiness', *Biotechnology Advances*, vol. 49, p. 107731, Jul. 2021, doi: 10.1016/j.biotechadv.2021.107731.

[58] M. Solis and S. Silveira, 'Technologies for chemical recycling of household plastics – A technical review and TRL assessment', *Waste Management*, vol. 105, pp. 128–138, Mar. 2020, doi: 10.1016/j.wasman.2020.01.038.

[59] X. Pang, N. Skillen, N. Gunaratne, D. W. Rooney, and P. K. J. Robertson, 'Removal of phthalates from aqueous solution by semiconductor photocatalysis: A review', *Journal of Hazardous Materials*, vol. 402, p. 123461, Jan. 2021, doi: 10.1016/j.jhazmat.2020.123461.

Index

A

Acrylic, 4, 5
Activated carbon, 46–47, 49–50, 52–57
Additives, 22, 25, 32, 35
Adsorbents, 46
Adsorption, 42, 47–48, 50–51, 57, 86, 93, 108, 111, 117, 120, 127, 136
Aerobic, 78
Agents, 94, 97–98
Aggregation, 21–22, 47
Aquatic, 42, 45–48, 54–55
Assimilation, 74–75, 78
Atmosphere, 1, 11
Azo, 90–91

B

Bioaccumulation, 4, 7, 8
Biochar-zeolite, 47–48, 57
Biodegradation, 42, 57, 206
Biodeterioration, 74–75
Biofiltration, 205
Biofragmentation, 54, 57, 74–75
Biological, 90, 94, 99
Biomagnification, 66
Biomass, 29, 33
Biopolymers, 73–74
Bioremediation, 125, 127, 129–131, 133, 135, 137–138
Bipolar electrodes, 158–159
Blockage, 94, 97
Boron-doped diamond, 53

C

Carcinogenic, 2
Catabolic, 75, 78
Chemical, 87, 91, 92, 95, 98, 99
Chemphotodegradation, 202
Chronoamperometry measurements, 158
Cleaning, 97, 98
Coagulation, 42, 51, 53, 55, 57, 108–112, 114–116, 118, 120
Colonization, 74–75
Combustion, 72, 76–77
Conduction band, 173
Contamination, 22, 28, 35
Cyclic voltammetry, 183

D

Degrade, 29, 30
Depolymerization, 130–131
Detection, 22, 29, 32–35

E

Ecosystem, 2–5, 7–9, 11, 12, 22
Electrical energy per order (EEO), 223
Electrochemical impedance spectroscopy (ESI), 157, 183
Electrochemical oxidation, 53
Electrochemical sensors, 157
Electrocoagulation, 147–148, 150, 152, 155, 205
Electrooxidation, 53, 57
Electrostatic separation, 27
Elements, 96–97
Emission, 43, 51, 57, 67, 72, 76–78
Endogenous, 78
Energy, 86–87, 89, 92–94, 96
Environment, 21–28, 31, 34
Enzymes, 30
Extraction, 21–23, 25

F

Faradaic ion concentration polarization, 158
Fermentation, 57, 78
Filtration, 23, 24, 28, 42, 51–53, 57
Floatability, 28
Flocculation, 53, 57, 107–112, 114–116, 118, 120
Fouling, 116, 117, 119
Fragmentation, 66–67, 74–75, 114, 203–204, 207

G

Gasification, 72–73, 76, 78

H

High-density polyethylene (HDPE), 171, 211
Hydrodynamics, 7
Hydrophobic, 27–28

I

Increase, 86, 90, 93–94, 96–98
Industry, 94–95, 101
Ion depletion zones, 158
Isolate, 96

231

L

Large, 86–77, 94–97, 99
Leaching, 127–128
Low-density polyethylene (LDPE), 171, 177, 216–219

M

Magnetic extraction, 205
Magnetic nanoparticles, 47, 57
Manufacturing wastes, 160
Membrane, 24, 35, 51–53, 57, 86–94, 97–99
Membrane bioreactors, 57
Membrane filtration, 205
Mesoplastics, 4
Metal-organic frameworks, 178, 186, 213
Microbeads, 126, 130, 201, 204
Microbial electrolysis cells, 159
Microbial fuel cells, 159
Micro-nanoplastics (MNPs), 212
Microplastics, 3, 5–7, 10, 21, 32, 125, 127, 131, 134, 147, 149, 153, 157, 168, 175, 201–204, 213
Module, 96, 99, 100

N

Nanomaterials, 212–213, 215–216, 222
Nanoparticle, 6, 10
Nanoplastics, 4, 6, 10
Nephelometric turbidity units (NTU), 111
Nuclear magnetic resonance, 75
Nylon, 4, 5

O

Operating, 88, 92, 93, 97, 100
Organic, 87–89, 91–97
Ozonation, 205

P

Passage, 86, 91
Pathogen, 66
Pharmaceutical, 95, 91
Photocatalysis, 173–181, 186–187
Photocatalytic degradation, 168
Photodegradation, 1, 4, 74, 214–215, 218, 221, 223
Photooxidation, 204
Plastics, 21–22, 27–31
Pollution, 95
Polyacrylamide, 53
Polyamide, 126, 136
Polychlorinated biphenyls (PCB), 107
Polyester, 4
Polyethersulfone (PES), 126, 211

Polyethylene, 46, 126, 128, 130, 132–134, 136, 147–149, 168–169, 177
Polyethylene terephthalate (PET), 5, 21, 126, 147
Polylactic acid, 73
Polymerization, 21, 22
Polymers, 21, 24, 28, 29–31, 35
Polyolefin, 22
Polypropylene, 5, 126, 130, 132, 148–149, 153, 168–169, 177
Polypropylene chloride (PC), 211
Polystyrene (PS), 42, 47, 126, 128, 130, 132, 134–136, 168–169, 177, 211
Pore size, 23, 24
Pre-sedimentation, 115
Pristine, 79
Purification, 31–32
Pyrolysis, 66–72, 76–78

R

Radiation, 71, 75
Rapid sand filtration, 107, 109–112, 117–119
Reactive oxygen species (ROS), 214
Recycled, 94
Rejection, 86
Remediation, 67–69, 71, 73, 75–79, 147, 149–150, 153–154, 156–157, 159, 168, 172–173, 175–179, 181, 183–188

S

Sedimentation, 5, 7, 8, 10, 22, 23, 25, 27–29, 147–148, 154
Separate, 86, 88–92, 94
Sieving, 23, 24
Solar-based photocatalysis, 186
Sorption, 66
Spectrometry, 6, 10
Spiral-wound, 99, 100
Starch, 73

T

Technique, 86–96
Technology, 86–88, 90, 93–94, 96, 100–101
Technology Readiness Level (TRL), 223
Thermal degradation, 66–72, 76, 78–79
Thermoplastic, 66–73
Toxicity, 212, 216, 218, 222
Toxicology, 213

U

Uniformity coefficient, 113, 116

Index

V

Valence band, 173
Viscosity, 29
Volatile organic compounds (VOCs), 184

W

Wastewater, 4, 7, 9

Wastewater treatment plants (WWTPs), 108, 126, 211
Wastewater treatment processes, 201–202, 204
Water remediation, 107

Z

Zeolites, 47–48, 57, 70–71, 73
Zooplankton, 7, 8